6500

COMPUTATIONAL PROCESSES IN HUMAN VISION:

An Interdisciplinary Perspective

THE CANADIAN INSTITUTE FOR ADVANCED RESEARCH SERIES
IN ARTIFICIAL INTELLIGENCE AND ROBOTICS

Computational Problems in Human Vision
  Zenon W. Pylyshyn, editor

In preparation

Vision and Action: The Control of Grasping
  Melvyn Goodale, editor

COMPUTATIONAL PROCESSES IN HUMAN VISION:

An Interdisciplinary Perspective

Edited by

Zenon W. Pylyshyn
University of Western Ontario

THE CANADIAN INSTITUTE FOR
ADVANCED RESEARCH SERIES IN
ARTIFICIAL INTELLIGENCE

Zenon W. Pylyshyn, Series Editor

Ablex Publishing Corporation
Norwood, New Jersey

Library of Congress Cataloging-in-Publication Data
Computational processes in human vision.
    (The Canadian Institute for Advanced Research
series in artificial intelligence and robotics)
    Bibliography: p.
    Includes index.
    1. Vision.   I. Pylyshyn, Zenon W., 1937–
II. Series.
QP475.C63   1988   152.1'4   88-16723
ISBN 0-89391-460-6

ABLEX Publishing Corporation
355 Chestnut Street
Norwood, New Jersey 07648

This book is dedicated to the memory of Ruth MacDonald

In gratitude for her role in making the Canadian Institute for Advanced Research possible.

Table of Contents

Preface and Acknowledgments

On April 25–27, 1986 a small, congenial group of researchers, representing a variety of disciplines and specialties, met in London, Ontario, in the secluded surroundings of the University of Western Ontario's Spencer Hall. They were brought together under the auspices of the UWO Centre for Cognitive Science and the Canadian Institute for Advanced Research, to take part in informal discussions of their research and their views about current problems in understanding human vision. This volume results from a decision that the group made at the time to place some of their research results and their theoretical views on record. It is not so much a proceedings of the workshop as a product of the thinking and discussions that went on at the time and subsequently.

A major common thread that ties together more than two-thirds of the participants is their association with the Canadian Institute for Advanced Research (CIAR). CIAR is a private nonprofit institute, funded by private business—though currently with matching federal government support. Its goal is to create a critical mass of world-class researchers in Canada, within certain program areas identified by its research council. The areas which merit CIAR's attention are ones which are judged to be both intellectually exciting, and which represent important national concerns—both scientific and cultural. They are also areas in which existing strength suggested that a high standard of excellence could be sustained, as judged by international standards.

The Institute helps to build these advanced research efforts by supporting individuals through a program of Research Fellowships, Institute Scholarships, and Associateships. It also encourages the development of communities of researchers which cross narrow disciplinary and regional concerns. It does this by funding and helping to organize exchanges, visits, workshops, electronic networking, and other instruments of interaction and collaboration, both within the CIAR community (including its graduate students) and internationally.

As part of the effort to develop interactions among its members, as well as with other relevant researchers, the CIAR Program in Artificial Intelligence and Robotics (one of several programs currently being supported) brought together a group of individuals who are concerned with understanding human (as well as artificial) vision from a computational perspective. The set of papers included in this volume are (with one exception) original essays and research reports which participants of the workshop agreed to prepare especially for this book. They span a wide range of disciplinary approaches—from experimental psychophysics and psychobiology to mathematical analysis; as well as a spectrum of problems in visual processing— from how the earliest stages of processing visual signals takes place, through what the role of visual attention focusing is, to the "high level" problem of how we recognize familiar objects. They also include theoretical and empirical explorations of such general issues as the complexity of visual processing and the representation of visual knowledge.

Despite the wide range of problems addressed in these papers, they are remarkably convergent in their concern for uncovering the *computational* processes and mechanisms, as well as the general computational architecture of the visual system. The influence of work in artificial intelligence, and in particular in computer vision, is deeply reflected in the way these disparate researchers define their problems and focus their attention on relevant computational mechanisms. They all approach the understanding of human vision by asking the questions, "What computational problems are being solved? and "What biologically-implemented computational mechanisms are being used?" The result is that regardless of whether the investigator is disposed to search for psychophysical generalizations, neural pathways, or mathematical principles, the results contribute to the common enterprise.

Although a collection of papers, however topical and seminal, can hardly reflect the excitement of the personal interactions that took place during the three days of close contact among these researchers (as well as a number of visitors and graduate students), we hope that this volume will serve to indicate some of the directions that the study of human vision is taking within the computationally-oriented field known as Cognitive Science.

The participants are grateful for the support provided for this workshop and for the preparation of this volume by the University of Western Ontario, through its Centre for Cognitive Science, and by the Program in Artificial Intelligence and Robotics of the Canadian

Institute for Advanced Research. The assistance of Aileen Orthner in organizing the logistics of the workshop is gratefully acknowledged.

Zenon Pylyshyn
London, Ontario, April 1987

I
Preattentive Feature Measurement

1
Playing Twenty Questions with Nature*

Whitman Richards
Aaron Bobick

Massachusetts Institute of Technology

The Twenty Questions game played by children has an impressive reputation: in this game, participants rapidly guess an arbitrarily selected object with rather few, well-chosen questions. This same strategy can be used to drive the perceptual process, likewise beginning the search with the intent of deciding whether the object is "animal, vegetable, or mineral." For a perceptual system, however, several simple questions are required even to make this first judgment as to the Kingdom in which the object belongs. Nevertheless, the answers to these first simple questions, or their modular outputs, provide a rich data base which can serve to classify objects or events in much more detail than one might expect, thanks to constraints and laws imposed upon natural processes and things. The questions, then, suggest a useful set of primitive modules for initializing perception.

* Support for this work is provided by NSF and AFOSR under a combined grant for studies in Natural Computation, grant 79-23110-MCS, AFOSR 86-0139 and 83-12240-IST. Comments and criticisms by Shimon Ullman, Donald Hoffman, and John Rubin were greatly appreciated. Technical support was kindly provided by William Gilson. This paper is adapted from MIT A.I. Memo 660, Dec. 1982, entitled "How to Play Twenty Questions with Nature and Win." This title originated in an article by Alan Newell (1973), who pointed out the frustration of posing certain lines of questions for research on information processing, such as serial vs. parallel, peripheral vs. central, conscious vs. unconscious. I believe the Twenty Questions approach presents a worthwhile alternative for such research.

THE NAME OF THE GAME

Perceiving systems are subject to a massive bombardment of signals from the external world. Sometimes these signals are completely unexpected or unpredictable, such as when you hear a novel sound, or when I show you a postcard. Yet from this deluge of unforeseen data, the sound or scene is understood. How is this possible?

One strategy for interpreting unexpected scenes or sounds is to build a very general perceiver—a perceiver built from a hierarchy of modules of increasing complexity and scope. "Points" are aggregated into "blobs," "lines," or "edges." These elementary features then become the basis for more complex representations of shapes and regions, and their relations to one another, which are finally interpreted as "objects." Such a view is the one currently accepted by most. It seems to me a very unsatisfactory one. First the goals of the system are not really specified clearly. Presumably they include object recognition and manipulation. Yet to date no one has been able to offer a general definition of "object" that is precise enough to embody in a computer vision system. We can define special objects for which we have models, such as planes, trees, houses, or people, but not "object" in general. This difficulty currently casts doubts upon our ability to build a general purpose perceiver and raises questions about whether such a system indeed exists. Yet there is no doubt that special purpose devices can be built that match inputs to models. Here, the Twenty Questions Game can be applied profitably.

FUNDAMENTAL HYPOTHESIS (NATURAL MODES)

Before presenting one strategy for building a very general, yet special purpose perceiver, we make an important claim about the world we perceive. The claim is that the structure and events in the world are not arbitrary or random. Rather they can be seen as clustered in a multidimensional space. As a result of natural selection and environmental pressures, nature does not adopt all possible solutions to the problems it encounters (Stebbins & Ayala, 1985; Mayr, 1984). Fish and whales, although biologically quite different species, look similar because this particular body design is quite efficient for locomotion through fluids. Animals are not asymmetric and arbitrary, but are symmetric. Even chaotic systems have modal behaviors (Levi, 1986). Our fundamental hypothesis about the world is thus the following:

Principle of Natural Modes: Structure in the world is not arbitrary and object properties are clustered about modes along dimensions important to the interaction between objects and their environment.

Such modal behavior seems necessary if a perceiver is to be able to categorize correctly structures and events in the world (Marr, 1970; Bobick & Richards, 1986). Of course the scheme used by the perceiver to perform the categorization has not been specified. Nor have the modes which are appropriate (these may differ for different perceivers). Yet it is this structure of the world which will allow us to play the Twenty Questions Game profitably. The basic idea is to choose questions that are keyed to identify the natural modes of the world of interest. Correlations between the answers then permit a "natural" categorization of the event or structure.

FROM TEMPLATES TO QUESTIONS

In simple worlds such as many industrial settings and laboratories, the modalness is quite apparent because the range of objects and views is quite limited and well-defined. An open-end wrench appears on the conveyor belt, or a red "cube" is placed on a table. Because the "object" of interest is known in advance and controlled, simple template matching usually suffices to solve these tasks. This is a very primitive form of the game we propose, where the perceiver's questions are tailored to the world. Examples can also be found in natural environments: the blowfly feeds when its receptors identify the ring structure of a sugar, and rejects the hydrocarbon chains of alcohols (except Inositol, which is an unnatural ring alcohol [Hodgson, 1961]!). Other examples are the hungry fledgling gull that responds immediately to the looming red spot on its parents' beak; the mating call of the cricket (or bee), which is so precisely engineered that a simple pattern of pulses can be tailored to reflect even subtle species differences. Such examples are countless (Tinbergen, 1951; Wilson, 1971). In each case, an important primitive goal, such as feeding or the reproduction of the species, is achieved successfully in a very direct and reflexive manner only because the environment is highly structured.

Our strategy is to capitalize on this structure of the world, to build perceptual modules that as directly as possible identify whether a mode is reflected in the sensory data. To begin, we consider the three most obvious natural modes in our environment: structures that are either animal, vegetable, or mineral. That these are indeed separate kingdoms receives much biological support (see Figure 1). Although there are at least two other independent modes (fungi and slime), we consider these not relevant to our categorization goals.

Consider now the classical children's game of Twenty Questions, where the goal is to identify an object. By "object" we mean an entity constructed from properties that exhibit natural mode behavior. (Objects composed of random structures and properties thus lie outside our domain.) Our first questions attempt to identify whether the object is animal, vegetable, or mineral. Subsequent questions attempt to determine the size, shape, or mass, or the sounds "it" might make, how "it" moves, or perhaps its function. The final questions then become very specific and detailed. If we are clever and shrewd in our choices, we rapidly converge to the object. A perceptual system we propose is designed along similar lines. Imagine that for our first set of questions we identify a dozen or two—let's say twenty—very general but independent attributes of "things," where an "attribute" is a particular modal property present in the world. We simply ascertain whether each attribute is present or not. Then 2^{20} or roughly a million different types of events could be crudely categorized (Web-

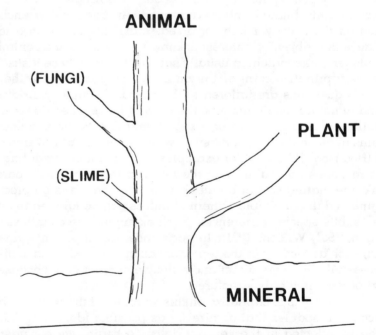

Figure 1.   Tree of Life, showing the animal and plant kingdoms. Recently, biologists have added Fungi, Protozoa, and Slime as separate branches. (Woese, 1981).

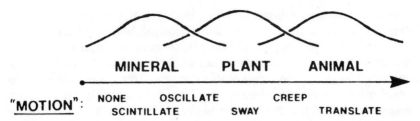

"MOTION" :

NONE	OSCILLATE	CREEP
SCINTILLATE	SWAY	TRANSLATE

Figure 2.   Qualitative types of motion or mobility associated with different kinds of living or inanimate objects, crudely ordered along the "Tree of Life" dimension shown in Figure 1.

ster's Dictionary only lists 60,000 words total.) Certainly, such assertions all computed in parallel would form a useful way of initializing the perceptual process, providing an initial description of the events or contents of a scene. Can we indeed find such questions that are powerful and general, yet are simple enough to be computed from the sense data? Let's play a slightly modified version of the Twenty Questions Game to explore its power.

PLAYING THE GAME

Imagine that an "object" has just entered our field of view, emitting some distinctive sounds. Our task is to identify as quickly as possible the general nature of the object. Loosely speaking, we would like to distinguish a man from a cat or a bird, but monkeys and men or clouds and smoke may be confused.[1] The principal rule of the game is that all our "questions" must be ones for which the answers can plausibly be computed from the sense data.

In the classical Twenty Questions game, our first question was, "Is it animal (or vegetable or mineral)?" How can we answer this question from the sense data? In fact, there are many ways to determine whether the "event" arose from an animal, vegetable, or mineral. For example, animals translate, rocks or plants do not (Figure 2). Animal sounds are different from the sounds of minerals (running water or falling rocks) or of the wind through the trees. Plants and animals have different shapes or colors; they "feel" different. Many of these attributes can be computed from the sense data using foreseeable technology.

---

[1] To specify rigorously the precision required of the Twenty Questions game is an important issue, but one which requires a clearer statement of the objectives and goals of the inquirer.

Surprisingly, the answers to the first set of questions posed to determine whether the event is animal, vegetable, or mineral tell us much more than just which of these three categories the event falls into. Consider Game 1 (shown in Appendix I). Our first question—"Is it moving?"—gave the answer *translation*, implying *animal*. The second question yielded the answer "legs"—confirming the *animal* interpretation. Yet the answer to the third question—that the emitted acoustic frequencies are broadband, rather than narrow-band as expected for an animal—causes us to question whether the "event" indeed arises from an animal. In this particular game, which is a transcript of one actually played, eight more questions are required to pinpoint the object. By playing such games, we see the power of an appropriate set of questions. Although the answers are restricted to a choice of triples[2] the collection of such answers is sufficient to narrow down an object or event much more precisely than just whether it is animal, vegetable, or mineral. The animal-vegetable-mineral distinction merely serves as a useful dimension along which values of various properties or attributes can be represented. In some sense, it is a dimension of "stuff" or "behavior." Mineral "stuff," plant "stuff," and animal "stuff" each represent different branches of the Tree of Life (Figure 1). We will see later that these fundamentally different properties will be useful descriptors of attributes outside their kingdom of origin. The utility of the animal-vegetable-mineral dimension for "stuff" thus goes far beyond what is implied by our first game.

CRITERIA FOR TWENTY QUESTIONS

Table 1 summarizes some useful preliminary questions that address various properties of natural things.[3] The first column is the attribute measured or extracted from the raw sense data. The next three columns indicate the initial three output states of the question box or module. The animal-vegetable-mineral categories serve to guide the choice of the type of output assertion to be computed. The final

---

[2] In practice, a default response may be necessary on occasion. Thus each question requires 2-bits for the answers. More answer categories may be counter-productive if one wishes to create an indexable representation for memory that can be efficiently accessed (Dirlam, 1972).

[3] The list makes no distinction between "shape," "stuff," and "structure," although the strategies for computing these properties are clearly quite different. For example, see Rubin & Richards, 1982; Hoffman & Richards, 1982. Chemical attributes are not included because localization for scene segmentation is usually difficult.

Table 1. Example Questions and the Three General Categories of Their Answers.

ATTRIBUTE (Question)	AUDIO-VISUAL			(REF)
	MINERAL	PLANT	ANIMAL	
acoustic frequency	none or broadband (lo)	broadband (hi)	narrowband	1
acoustic modulation	none	pseudo-sine	interrupted	2
frequency change	no	no	yes	3
motion	none	sway	lateral	4
support	no "leg"	one "leg"	several "legs"	5
symmetry	irregular	3-D (one axis)	mirror (bilateral)	6
axis	none	vertical	horizontal?	7
"texture"	irregular (2-D wideband)	fractal	1-D parallel (hair)	8
"color"	yellow, brown, blue	green, red	agouti	9

ATTRIBUTE	TACTILE			(REF)
	MINERAL	PLANT	ANIMAL	
heat emmission/ absorption	cold	neutral	warm	10
texture	rough	rough and smooth	soft, smooth	11
hardness	rigid	crunchy, crisp	soft, elastic	
			hairy, feathers	12
movement	none	passive (bend)	active (wriggles)	13
adhesion/ viscosity	none (dry or wet)	sticky	oily	14

9

(fourth) column gives a reference in Appendix II as to how feasible it is to compute these outputs, using current or foreseeable technology.

Our preliminary choice of questions has been guided by several considerations. The first, already mentioned, is the computational feasibility. A second is the degree to which an attribute can encode a useful modal property of a "thing." Here we rely upon our intuitions about which properties are likely to exhibit highly modal behavior. For example, the sound an object makes reflects something about the structure of the source. If the sound is narrowband, then the source must have a tuned resonant cavity, which neither plants nor minerals have. All candidate objects from these two kingdoms can then be rejected—a rather strong assertion (see Rubin & Richards, 1982). Furthermore, because the size of the cavity determines the fundamental frequency of the sound, some indication of the source size can be inferred from the pitch. An elephant "roars" because it has a large resonant cavity whereas the mouse "squeaks" because of necessity it must have a small cavity. The sounds an animal can emit thus depend critically upon its size and therefore encodes size, if one wishes to examine this attribute in more detail. We see immediately that the simple question "What is the *pitch* of the source?" not only may tell us whether the object is animal, plant, or mineral, but also provides some information about its size. Translatory visual motion information can be similarly utilized to indicate animal size, as shown in Figure 3. Such questions about the pitch of a sound or the rate of motion are ideal questions because their refinement provides still more useful and quantitative information about the object. This ability to refine a question to extract more information was a third factor which influenced our selection of questions.

Other selection criteria for our Twenty Questions relate to the perceiver-object relation. Obviously one would like an object representation to be independent of the viewer's position or the particular disposition of the object. Yet most of our sense data seem to depend critically upon our particular view. For example, image intensities on the retina are seriously confounded with the orientation and reflectivity of the surfaces that reflect the light; or auditory intensities will depend upon the source distance and the intermediate absorbing and reflecting media. Is it at all reasonable, then, to hope to find descriptive attributes of objects and "things" that are insensitive to our particular viewpoint or position? Of the five basic physical variables—charge, mass, length, time, and temperature—only time is independent of the observer's position and the medium in which he exists. The best examples of viewer-independent attributes of an event or "thing" will thus be those where the temporal pattern encodes the property. The sparkle of water, the scintillating pattern of

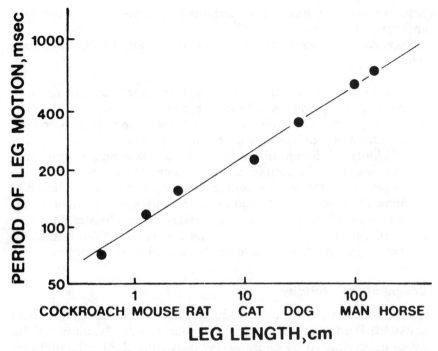

Figure 3. The rate at which the legs move encodes leg length and hence animal size, as shown by the high correlation between size and gait. (Adapted from McMahon, 1975). This desirable property of the motion Question we call "conveyance".

fluttering leaves on a tree, the gait of an animal, the chirp of a cricket—all are important characteristics of the "object" whose pattern remains the same regardless of where the perceiver is located. The dynamical environment is thus a critical ingredient of the Twenty Questions game.

However, it should be stressed that all Questions, including those based on the temporal dimension, are designed to deliver answers about the external world, and do not just report the state of the sensorium. *Has "legs"* does not just mean that there are two or more roughly parallel elongated blobs in the image. The latter are features. (We view feature as an image-based data structure, whereas attributes or properties are assertions about an event or structure in the world external to the observer.) The inference process posed by the Twenty Questions is thus exceedingly difficult, and one which we believe must rely on the modal character of the natural world. (See Bennett, Hoffman, & Prakash, 1988.) However, the positive and very optimistic point we are making is that, given this modal character, only a few well-chosen modules can serve as a basis for a successful perceiver.

To summarize, we have five major criteria for our choice of questions:

1. *Computational Validity*—The representation of the attribute must be easy and reliable to compute.
2. *Conveyance*—One should be able to refine the attribute to yield a more metrical measure of an object property.
3. *Modality*—Different attributes or questions should be capturing different modal qualities of the "events" or "things."
4. *Viewer Independence*—Representations of attributes should be insensitive to the particular relations between the perceiver and the "object", i.e., to object distance, scale, or disposition.
5. *Configuration Independence*—The attributes should be independent of the particular state or configuration of the object.

Computational Validity

Given the above criteria, how do we know when they have been satisfied? Particularly difficult in this regard is the "modality" of the set of questions, to be addressed shortly, and their computational validity. The best evidence for the ability to answer one of the Twenty Questions is an example of a machine system that will deliver the correct answer. The references in Appendix II document the feasibility of designing sensors or information processors that can answer the question posed.

In the audio-visual realm, narrow-band sensors that measure the frequency of the acoustic spectrum have been available for many years (Flanagan, 1972). The measurement of acoustic frequency and intensity changes is thus readily accomplished for isolated sound sources. Not so easily achieved, however, is the isolation of a sound source, although this is a task performed reliably by the most simple natural binaural system (Howard & Templeton, 1966; Knudsen & Konishi, 1979). As long as the environment does not have more than one or two competing sources, the source direction or isolation can be found fairly reliably using either signal onset times or intensity differences, or both (Altes, 1978; Searle, Davis, & Colburn, 1980). Additional work needs to be done in this area, however, for source isolation (and direction) is a critical computation that must precede many of the acoustic questions, especially if one wishes to determine details about the physical properties of the source (i.e., is it metallic, wood, or rustling leaves?), or the nature of animal sounds (Klatt, 1977).

Similarly, for vision, a rather powerful input representation is also required before the Twenty Questions game can proceed with reasonable success. Although lateral motion or scintillation or sway can be computed crudely for a region using only primitive intensity information (Thompson & Barnard, 1981; Ullman, 1981), the exact shape of the region cannot yet be found reliably (Horn & Schunck, 1981; Hildreth, 1982). "Edge"-finding algorithms are still quite primitive, and confuse many types of intensity changes such as surface markings, shadows, or occluding edges. For vision, the most useful data base for the Twenty Questions game would be Marr's primal sketch (Marr, 1976; Marr & Hildreth, 1980), which is still unavailable and poses many quite difficult computational problems. Furthermore, it is still a feature-based representation and does not make assertions about properties in the world. Thus, although questions such as "number of supports" or "symmetry type" seem feasible in the long term (Brady & Asada, 1984; Richards, Dawson, & Whittington, 1986), as yet we do not have a sufficiently powerful "primal sketch" to permit these questions to be answered reliably.

More tractable are questions about the surface properties such as its roughness or composition, although obstacles also occur here. Many sensors are available to measure the spectral composition of reflected light, but we must remember that a reliable determination of the spectral reflectance of a surface also requires knowledge of the source illumination. Fortunately, this is rather constant in natural environments, and our crude color question is computationally feasible (Judd & Wysecki, 1975; Myrabo, Lillesaeter, & Hoimyr, 1982). Remote measures for surface roughness or quality, on the other hand, are still rather primitive and far from robust, although several recent studies, particularly in the remote sensing area, show promise of providing practical applications (Moon & Spencer, 1980; Milana, 1981). Tactile sensing, on the other hand, appears quite tractable, with several impressive recent advances in detecting surface properties (Hillis, 1982; Raibert & Tanner, 1982).

In sum, it is still uncertain the extent to which the technology of the near future can give reliable answers to all the posed questions. Those that concern "shape" appear particularly difficult, whereas those that address the "stuff," composition, or size of the object seem more tractable. The challenge is obvious.[4]

[4] In many cases the property-based questions can not be entirely decoupled from the shape descriptors, at least for vision. For example, many grouping tasks for connecting isolated contour segments may require that a property tag be attached to the contour descriptor (such as its codon type). This requirement complicates the integrated structure of the set of Twenty Questions, but does not obviate the need for them.

Modality

We have criteria and constraints on the types of questions we should ask, but we still have not found a rule or procedure that tests whether our questions are independent in the sense that they capture different modal properties in the world. At best, we have suggested that the behaviors or properties of objects within each of the three kingdoms will differ, yet this is clearly not the case in practice. Very often a property, such as a hard "shell" (rock), or soft "feathers" (grass) may appear in more than one kingdom.

The problem of orthogonality and modalness is further complicated by the wide scale of sizes over which objects and events may exist—from the amoeba to the dinosaur; from the blade of grass to the giant Sequoia, or from the tiny grain of sand or speck of dust to Mount Everest. This enormous range of scales has led to the application of different natural laws to solve similar problems. The amoeba locomotes one way, the elephant another; the speck of dust behaves differently from a massive stone when subject to the wind or forces of nature. At any one scale, however, where size and mass are comparable, the behaviors are similar, at least to the degree that the "stuff" is the same. As the "stuff" differs, then the behavior will differ. Hence, the nature of the "stuff" becomes a dimension along which different behaviors or attributes may be categorized at any one scale. The log placed on water acts differently from stone because its "stuff" differs. The animal-plant-mineral distinction is thus basically a crude dimension to a property list. To the extent that the properties are independent, the questions will be independent. We appeal to the process of natural selection to converge upon an optimal set of questions that captures these different properties.

SUCCESSES AND FAILURES

The strategies and remaining problems encountered with the Twenty Question approach become more apparent as the game is played. Ideally, one would like to have available a massive dictionary against which the game could be played on a computer. In this way, the "top-down" and "bottom-up" inferences might be made more explicit, while at the same time, the evolution of the best questions (and their priorities) could be examined. In lieu of this, Appendix I presents two sample games to show what inferences may be (or are) drawn from successive questions when the game is played serially.

(Of course, any biological implementation may elect to ask many of the questions in parallel.)[5]

Several problems become immediately apparent when one plays the game. For example, often one can be badly misled by the first or second question. If the answer to "motion" is "none," obviously one cannot immediately infer that the thing is not an "animal," for it may be a stationary animal, lying down. Similarly an animal in such a state will seem to have "no legs" and will emit no sounds. Clearly our deductions will be way off in this case. Have we therefore missed the mark?

Once again, we must consider the rather primitive goals of the Twenty Questions game: namely, to provide a crude classification of "things," often as they bear upon our survival. Certainly if the animal is not moving, then its immediate threat as a predator is less than if it is looming toward us. Given the alternatives, ones attention is focused upon the most active events in the environment.

Finally, the dimensions and attributes of our Twenty Questions have been driven by the natural, biological environment. The man-made world is quite different. In some sense, its qualities, although largely made of mineral "stuff," may extend the mineral-plant-animal dimension further to the right. Automobiles or planes translate more swiftly; their bodies are more resilient and "metallic." Yet what natural animals possess these same qualities? If there are none, then our original Twenty Questions strategy can still be applied successfully even in the world of man-made objects.

APPENDIX I: EXAMPLE GAMES*

"I'm thinking of an object. It is in its natural habitat (which is the same as yours), and is behaving in its most natural way. What is the object?

"The only questions you are allowed to ask and receive answers to are those which could be used by a rather simple sensory device, that is, one which is feasible to build today. For simplicity, the de-

---

[5] We must be careful about comparing the performance of a serial Twenty Questions Game (Siegler, 1977) with that obtained with parallel questioning. In the former, the earlier questions influence the context applied to succeeding questions whereas answers obtained in parallel share the same context. Here we have totally neglected the control problems in playing the Twenty Questions game—but for some discussion of this problem, see Bobick & Richards, 1986.

* NOTE: The original A.I. Memo also included a set of questions that determined the habitat.

vice will have only three outputs (plus a default if no firm answer is possible).

"Each of the output states indicates a different quality of the "thing" or dimension relevant to your question. For example, if you ask "Is it moving?," the relevant dimension is whether it behaves like an animal, plant or mineral, in which case it will either translate, sway, or not move at all.

"There are three main dimensions that you may use to frame your questions. One characterizes the basic biological structure from mineral to plant to animal. The second dimension pertains to the habitat or environment, ranging from arctic to temperate to tropic. A third dimension captures a different aspect of the location of the "thing" in the environment, namely, is it in the air, or the ground, or subterranean—below ground or under water."

Game 1

Habitat: (previously determined to be temperate environment, green rolling hills. Elevation of "thing" is on the ground.)

	ANIMAL	PLANT	MINERAL
Q1: Is it moving?	*translates*	sway	no
Q2: How many supports?	2, 4 or > 4	1	0
Implication: Confirms animal— has four "legs".			
Q3: What acoustic frequencies are emitted?	narrowband	*broad*	*broad*

Implication: Disconfirms animal. "Thing" makes low frequency, broad-band sounds, moves and has 4 legs. Must be big. Elephant or cow?

Q4: Acoustic source			
	point	extended	extended

Implication: Confirms "animal" or isolated object.

Q5: Visually symmetric?

	mirror	3D	irregular

Q6: What major axis?

	horizontal	vertical	none

Implication: Still seems to be some kind of large animal with horizontal major axis.

Q7: Modulation of acoustic intensity?

	interrupted	pseudo-sine	*none*

Implication: A large, moving animal with horizontal major axis that continually emits a steady broadband sound.

Q8: Color?

	agouti	green, red	yellow, brown, *blue*

Implication: "Animal" is blue. This is unlikely.

Q9: Texture?

	1-D parallel	fractal	irregular

Answer: None of the above. (Note that with two bits for answers, we have room for the default category.)

Q10: Hardness?

| | soft, elastic | crunchy, crisp | *rigid* |

Implication: Large animal with horizontal axis that moves on ground and emits a steady sound; surface is blue and hard like a "mineral," but the texture is not hairy or irregular. Car?

Q11: (Scale of dimension) What is rate of leg motion?

Answer: Zero

Implication: Object has no legs, but moves (on wheels?). Confirms a car.

Game 2

	ANIMAL	PLANT	MINERAL
Q1: Is it moving?	*translates*	sway	*no*
Q2: What acoustic frequencies are emitted?	narrow-band	broad	broad
		(None of the above.)	
Q3: How many supports?	2, 4 or > 4	1	0

Implication: Animal at rest or a mineral.

Q4: Visually sym-
  metric?

mirror	3-D	irregular

Q5: Texture?

fine	*smooth*	rough

Implication: Nei-
  ther an animal
  nor mineral.

Q6: Hardness?

soft	crunchy	*rigid*

Q7: Color?

brown	green, red	*yellow-white-blue*

Implication: Hard,
  whitish-blue,
  mirror symmetric
  object with a
  smooth surface
  that lies flat on
  ground without
  support and
  makes (is mak-
  ing) no sound. (A
  round, white,
  smooth rock?)

Q8: What is its ele-
  vation?

Answer: Subterranean

Q9: What is its im-
  mediate environ-
  ment?

solid	*soft*	*liquid*

Implication: Object
  is in moist soil
  and partially
  submerged un-
  der water. (As if
  in a pond or lake
  or *ocean?*) Oys-
  ter, clam or
  snail?

Q10: What is its
approximate
size?
Answer: Slightly
smaller than a
man's hand.

              Confirms oyster or clam.

APPENDIX II

1. *Acoustic Frequency.* Comb filtering has been used for several years to separate sound sources (Shields, 1970; Flanagan, 1972; Zwicker & Terhardt, 1979). Unless many broadband sources are active simultaneously at S.P.L.'s comparable to the narrow-band sources, this question can be answered with available technology (Klatt, 1977). As initially formulated (Richards, 1980), the question simply addresses whether the source is broadband or not (such as wind through the trees, rushing water, or an animal cry). Much more useful, but also much more difficult, would be to extract the physical properties of the source—i.e., its acoustic "color": Is it metallic, wood striking wood, or a footfall?

2. *Acoustic Modulation.* Tracking a sound source to determine its modulation characteristics (Atal, 1972) also requires localization (as may Question #1). For narrow-band, harmonic sources with different spectral signatures, such localization is possible provided there are only a few competing sources (Altes, 1978). Again, as in Question #1, work should be undertaken to understand how the "textural" properties of the source can be extracted from the modulations. For example, is the source "harsh" or grating, or like clacking sticks, or "suave" and "smooth," or "roaring" like a brook or lion.

3. *Frequency Change.* Here again, as in Questions #1 and #2, localization is helpful but not as necessary because only Animals are generally capable of producing sounds of variable frequency. Simple ⅓ octave filtering should allow the detection of frequency change (Flanagan, 1972; Klatt, 1977.)

4. *Motion.* The motion of an "object" can be both visual and auditory. Clearly the detection of auditory movement requires localization (Altes, 1978; Searle *et al.*, 1980), and may be difficult. Visual motion detection has progressed enormously over the

past 10 years, and can be detected with simple systems provided the background is stationary (Horn & Schunck, 1981; Thompson & Barnard 1981; Ullman, 1981; Hildreth, 1982). More work is still required, however, to use motion to segregate a visual scene, especially if sway or scintillation is to be disambiguated from translation or rotation.

5. *Support.* Although a powerful question, to estimate the numbers of "legs" supporting a region is quite complicated. First, the ground plane must be determined (see Question #20); secondly the candidate "support" must be recognized (e.g., leg or trunk); and finally a region should be identified as being supported although it may have a different color or texture. In the case of stationary supports, the local parallelism of the vertical occluding edges of the support may serve as a basis for determining the supporting member (Stevens, 1980). What to do in the case of animal motion, however? Also, shrubs clearly may have many "supports." The computational validity of this attribute is questionable, therefore, although a strong assertion would be quite useful.

6. *Symmetry.* Given that the occluding contour can be determined from an image, then mirror symmetry can be answered from available technology (Kanade, 1981; Hoffman & Richards, 1982). To determine 3D symmetry also requires a depth map, which is computable if binocular vision is available (Grimson, 1981). The difficult part of this question, therefore, is extracting the occluding contours, which at present can be done only on restricted classes of images (Davis & Rosenfeld, 1981; Binford, 1981; Richards, Nishihara & Dawson, 1982).

7. *Axis.* Again, as in Question #6, the orientation of a region can be answered rather easily (Ballard & Brown, 1982, Witkin, 1981) provided either that the occluding contour can be found, or the approximate areal extent of the region can be determined, such as by its spectral or textural qualities.

8. *"Texture".* The intent of this question is to determine whether the surface property of the region is typical of rocks or metals, grass or shrubs, or animal skin, hair, or feathers. Schemes for disambiguating such surface properties have only recently been considered (Horn, 1977; Milana, 1981; Moon & Spencer, 1980; Cook & Torrance, 1982; Rubin & Richards, 1982) This is an area ripe for research.

9. *"Color".* The value of spectral information in assessing food quality (Francis & Clydedale, 1975), printing inks or pho-

tographic reproductions (Judd & Wysecki, 1975) and in remote sensing (Chance & Lemaster, 1977; Lintz & Simonett, 1976; Myrabo et al., 1982) have provided a variety of practical tools.

10. *Heat Emission/Absorption.* The determination of surface temperature relative to one's own body temperature is a simple sensory ability if contact is used (Hertzfeld, 1962). Of course remote sensing is also possible here, as performed in surveillance or Landsat imagery (Lintz & Simonett, 1976; Barbe, 1979; Trivedi, Wyatt, & Anderson, 1982).

11. *Texture.* Passive touch sensing is coming close to obtaining the resolution required to determine surface roughness, as well as the texture pattern of the surface. At present, grid resolutions of 16×16 per cm^2 have been obtained (Hillis, 1982; Raibert & Tanner, 1982).

12. *Hardness.* The measurement of hardness of a point on a surface is a routine metallurgical technique and is trivial (Cox & Baron, 1955; O'Neill, 1962). The difficult task is to devise a skin-like sensor for the rigidity using force-feedback and the pattern of deformation. Recent progress in touch-sensing suggests that such sensors may be forthcoming in a few years, with possible applications for testing food ripeness (Kato, Kudo, & Ichimaru, 1977; Harmon, 1982).

13. *Movement.* The Hillis (1982) touch sensor could, in principal, be redesigned to measure whether a grasped object is wriggling or breathing. Whitney (1979) and Harmon (1982) also provide reviews describing the spectrum of compliant sensors now available.

14. *Adhesion/Viscosity.* Although a variety of rheometers are available to measure the viscocity and flow of fluids and gases (Van Wager, 1963), I do not know of a skin-like sensor that measures "stickiness" or "oiliness." Again, compliant sensors in this area, although perhaps relatively straightforward compared to remote sensing, will probably await commercial needs.

REFERENCES

Agin, G. (1972). Representational description of curved objects. Stanford AI Project Memo AIM-173, Stanford University.

Altes, R. A. (1978). Angle estimation and binaural processing in animal echolocation. *Journal of Acoustical Society America, 63*, 155–183.

Atal, B. S. (1972). Automatic speaker recognition based on pitch contours. *Journal of Accoustical Society of America, 52,* 1687–1697.

Ballard, D. H., & Brown, C. M. (1982). *Computer vision.* New Jersey: Prentice-Hall.

Barbe, D. F. (1979). Smart sensors. *Proc. Soc. Photo-Opt. Instrum. Eng., 178.*

Bennett, B., Hoffman, D., & Prakash, K. (1988). *Observer Theory.* Cambridge, MA: MIT Press.

Binford, T. (1981). Inferring surfaces from images. *Artificial Intelligence, 17,* 205–244.

Bobick, A., & Richards, W. (1986). Classifying objects from visual information. MIT AI Memo 879. Cambridge, MA: Massachusetts Institute of Technology.

Brady, M., & Asada, H. (1984). Smooth local symmetries and their implementation. *International Journal of Robotics, 3,* 36–61.

Chance, J. E., & LeMaster, E. W. (1977). Suits reflectance models for wheat and cotton: theoretical and experimental tests. *Applied Optics, 16,* 407–412.

Cook, R. L., & Torrance, K. E. (1982). A reflectance model for computer graphics. *ACM Transactions on Graphics, 1,* 7–24.

Cox, C. P., & Baron, M. (1955). A variability study of firmness in cheese using the ball-compressor test. *Journal of Diary Research, 22,* 386–390.

Davis, L., & Rosenfeld, A. (1981). Computing processes for low level usage: A survey. *Artificial Intelligence, 17,* 245–263.

Dirlam, D. K. (1972). Most efficient chunk sizes. *Cognitive Psychology, 3,* 355–359.

Flanagan, J. L. (1972). *Speech analysis: Synthesis and perception.* Berlin: Springer-Verlag.

Francis, F. J., & Clydedale, F. M. (1975). *Food colorimetry: Theory and applications.* Westport, CN: AVI Publishing.

Grimson, W. E. L. (1981). *From images to surfaces.* Cambridge, MA: MIT Press.

Harmon, L. (1982). Automated tactile sensing. *International Journal of Robotics Research, 1*(2):3–32.

Hertzfeld, C. M. (1962). *Temperature, its measurement and control in Science and Technology. Vol. 3.* Reinhold, N.Y.

Hildreth, E. (1982). The integration of motion information along contours. IEEE Proceedings of a Conference of Computer Vision Representation and Control, September, pp. 83–91.

Hillis, W. D. (1982). A high resolution image touch sensor. *International Journal of Robotics Research, 1*(2), 33–44.

Hodgson, E. S. (1961). Taste receptors. *Sci. Amer., 204,* 135–144.

Hoffman, D. D., & Richards, W. A. (1982). Representing smooth plane curves for recognition: Implications for figure-ground reversal. Proceedings of the National Conference on Artificial Intelligence, August 18–20, and MIT AI Memo 630, Cambridge, MA: Massachusetts Institute of Technology.

Horn, B. K. P. (1977). Image intensity understanding. *Artificial Intelligence, 8,* 201–231, and MIT AI Memo 335, Cambridge, MA: Massachusetts Institute of Technology.

Horn, B. K. P., & Schunck, B. G. (1981). Determining optical flow. *Artificial Intelligence, 17,* 185–203.

Howard, I. P., & Templeton, W. B. (1966). *Human spatial orientation.* London: Wiley.

Judd, D. B., & Wysecki, G. (1975). *Color in business and industry.* New York: Wiley.

Kanade, T. (1981). Recovery of the three-dimensional shape of an object from a simple view. *Artificial Intelligence, 17,* 409–460.

Kato, I., Kudo, Y., & Ichimaru, I. (1977). Artificial softness sensing—An automatic apparatus for measuring viscoelasticity. *Mechanism and Machine Theory, 12,* 11–26.

Klatt, D. (1977). Review of the ARPA speech understanding project. *Journal of Acoustical Society of America, 62,* 1345–1367.

Knudsen, E. I., & Konishi, M. (1979). Mechanisms of sound localization in the barn owl. *Journal of Comparative Physiology, 133,* 13–21.

Levi, B. (1986). New global formalism describes paths to turbulence. *Physics Today, 39*(4), 17–18.

Lintz, J., & Simonett, D. S. (1976). *Remote Sensing of the Environment.* Reading, MA: Addison-Wesley.

Marr, D. (1970). A theory of cerebral neocortex. *Proceedings of the Royal Society of London B, 176,* 161–234.

Marr, D. (1976). Early processing of visual information. *Phil. Trans. R. Soc. Lond. B., 275,* 483–524.

Marr, D. (1982). VISION: *A computational investigation into the human representation and processing of visual information.* San Francisco: Freeman.

Marr, D., & Hildreth, E. (1980). A theory of edge detection. *Proceedings of the Royal Society of London, B, 207,* 187–217.

Mayr, E. (1984). Species concepts and their applications. In E. Sober (Ed.), *Conceptual issues in evolutionary biology: An anthology* pp. 531–541. Cambridge, MA: MIT Press.

McMahon, T. A. (1975). Using body size to understand the structural design of animals: Quadrupedal locomotion. *Journal of Applied Physiology,* 1975, 619–627.

Milana, E. (1981). Apparatus for testing surface roughness. *Patent Nos. 4, 290, 698.*

Moon, P., & Spencer, D. E. (1980). An empirical representation of reflection from rough surfaces. *Journal of IES, 9,* 88–91.

Myrabo, H. K., Lillesaeter, O., & Hoimyr, T. (1982). Portable field spectrometer for reflectance measurements. *Applied Optics, 21,* 2855–2858.

Newell, A. (1973). You can't play 20 Questions with nature and win: Projective comments on the papers of this symposium. In William Chase (Ed.), *Visual information processing.* New York: Academic Press.

O'Neill, H. (1962). *Hardness measurement of metals and alloys*. London: Chapman and Hall.

Raibert, M., & Sutherland, I. (1983). Machines that walk. *Scientific American, 248*, 44–53.

Raibert, M. H., & Tanner, J. E. (1982). Design and implementation of a VLSI tactile sensing computer. *International Journal of Robotics Research, 1*, 3–18.

Richards, W., Dawson, B., & Whittington, D. (1986). Encoding contour shape by curvature extrema. *Jrl. Opt. Soc. Am. A*, Series 2, Vol. 3.

Richards, W., Nishihara, H. K., & Dawson, B. (1982). Cartoon: a biologically motivated edge detection algorithm. AI Memo 668, Mass. Inst. of Tech. Artificial Intelligence Laboratory, Cambridge, MA.

Richards, W. (1980). Natural computation: Filling a perceptual void. Presented at the 10th Annual Conference on Modelling and Simulation. April 25–27, 1979, University of Pittsburgh. Proceedings, N. G. Vogt & M. H. Mickle (eds.), 10:193–200.

Rubin, J. M., & Richards, W. (1982). Color vision and image intensities: When are changes material? *Biological Cybernetics, 45*, 215–226.

Schunck, B. G., & Horn, B. K. P. (1981). Constraints on optical flow computation. *Proceedings of IEEE Conference on Pattern Recognition and Image Processing*, August, 1981, 205–210.

Searle, C. L., Davis, M. F., & Colburn, H. S. (1980). Model for auditory localization. *Journal of Acoustical Society of America, 60*, 1164–1175.

Shields, V. C. (1970). *Separation of Added Signals by Digital Comb Filtering*. Master's thesis, Massachusetts Institute of Technology, Cambridge, MA.

Siegler, R. S. (1977). The twenty-question game as a form of problem solving. *Child Development, 48*, 395–403.

Stebbins, G. L., & Ayala, F. J. (1985). Evolution of Darwinsim. *Scientific American, 253*(1), 74.

Stevens, K. (1978). Computation of locally parallel structure. *Biological Cybernetics, 29*, 19–28.

Stevens, K. (1980). Surface perception by local analysis of texture and contour. MIT AI Technical Report 512, Cambridge, MA: Massachusetts Institute of Technology.

Thompson, W. B., & Barnard, S. T. (1981). Lower-level estimation and interpretation of visual motion. *IEEE Computer*, 14:20–28.

Tinbergen, N. (1951). *The study of instinct*. Oxford: Clarendon Press.

Trivedi, M. M., Wyatt, C. L., & Anderson, D. R. (1982). A multispectral approach to remote detection of deer. *Photogrammetric Engineering and Remote Sensing, 48*, 1879–1889.

Ullman, S. (1981). Analysis of visual motion by biological and computer systems. *IEEE Computer Magazine, 14*(8), 57–69.

Van Wager, J. R. (1963). *Viscosity and flow measurement; A laboratory handbook of rheology*. New York: Interscience.

Whitney, D. E. (1979). Discrete parts assembly automation—An overview. *Trans. ASAT, 101*, 8–15.

Wilson, E. O. (1971). *The insect societies.* Cambridge, MA: Belknap Press.
Witkin, A. P. (1981). Recovering surface shape and orientation from texture. *Artificial Intelligence, 17,* 17–45.
Woese, C. R. (1981). Archaebacteria. *Scientific American, 244,* 98–122.
Zwicker, E., & Terhardt, E. (1979). Automatic speech recognition using psychoacoustic methods. *Journal of Acoustical Society of America, 65,* 487–498.

2
Aspects of Visual Texture Discrimination*

Rick Gurnsey

Department of Psychology
&
Department of
Computer Science
University of Western
Ontario

Roger A. Browse

Department of Psychology
&
Department of Computer
and Information Science
Queen's University

On the simplest view, textural segmentation may be characterized as involving (a) the measurement of certain image properties and (b) detection of differences in these properties between neighbouring regions. Apparently following this simple view, questions have been asked about which image properties permit effortless discrimination of textures by humans. We show results which indicate that several other factors must be taken into account in a satisfactory theory of textural segmentation. As might be expected, the probability of discriminating two textures depends on how long subjects have to examine the textural display. Contrary to what might be expected, however, texture discrimination is not symmetric; the probability of discriminating two textures, in many cases, depends on which forms the foreground (disparate) region and which forms the background. It would seem then, that discrimination cannot be based entirely on local differences between textured regions; a difference signal should be indifferent to the "sense" of the boundary. Furthermore, the ability to discriminate two textures depends on the amount of practice that subjects have had with the materials and procedure. In general, with naive subjects, the probability of discriminating two textures and the probability of a discrimi-

---

* This research was supported by a grant from the Natural Sciences and Engineering Research Council of Canada (A2427). We are grateful to Dr. Brian Butler and Marion Rodrigues for their comments.

nation asymmetry arising, is related to differences in rather simple image properties such as scale, contrast, and orientation.

Texture may be considered an image property which, like colour, contrast, and disparity, provides a basis upon which images may be parsed into coherent regions prior to interpretation. This view is explicit in the computational vision literature and most textbooks on the subject (e.g., Ballard & Brown, 1982; Levine, 1985; Navatia, 1982) include chapters on the use of texture analysis in image segmentation. One motivation for the view of textural segmentation as a precursor to interpretation is the apparent ability of humans to spontaneously segment regions composed of different textures (e.g., Julesz, 1984).

Perhaps because of the work on image segmentation in the computational vision literature, the study of human textural segmentation has undergone somewhat of a renaissance (e.g., Beck, 1982; Beck, Pradzny, & Rosenfeld, 1983; Caelli, 1982, 1985; Caelli & Moraglia, 1985; Enns, 1986; Grossberg & Mingolla, 1985; Gurnsey & Browse, 1987a; Julesz, 1981, 1984; Julesz & Bergen, 1983; Nothdurft, 1985a, 1985b, 1985c; Richards, 1979). Most texture experiments involve artificial textures, usually constructed from micropatterns chosen because they have certain theoretically relevant properties. Presumably, the textural properties found to permit discrimination in the laboratory are also important in more usual environments. In general, however, we shall consider artificial textures comprising two spatially disjoint regions, such as those in Figure 1.

Central to many investigations of texture processing is the rather simple question: What limits our ability to segment two textures? Answers to this question typically concentrate on the properties of the textures involved; however, we shall show that there are several other theoretically relevant aspects of human visual texture discrimination[1] that are not addressed by current theories.

A SIMPLE MODEL OF TEXTURE SEGMENTATION

The simplest theory of textural segmentation might suggest a two stage process involving (a) the measurement of some image proper-

---

[1] We shall try to maintain a distinction between the terms *segmentation* and *discrimination*. *Segmentation* refers to the process of segmenting image regions from each other on the basis of textural properties of the regions. We will use the term *discrimination* in the context of experimental tasks in which subjects are ostensibly required to segment two textures in order to perform the task.

Figure 1. Examples of textural displays typical of those used in texture discrimination experiments.

ties and then (b) the assertion of boundaries between image regions differing in terms of these measured properties. The simple model leads to the reasonable assumption that the less similar two regions are, the greater the probability that they will be discriminated. Conversely, if two regions do not differ significantly in terms of textural properties they will be treated as one. This simple model of textural segmentation appears to be implicit in studies of textural segmentation in both human and machine vision. Accordingly, there have been many proposals, both psychological and computational, regarding the kinds of image properties or features that could form the basis of a "feature" space in which textures may be defined. A short list of such proposals would include first and second order statistics (Julesz, Gilbert, Shepp, & Frisch, 1973), textons (Julesz, 1984), fractal dimension (Pentland, 1984), families of orthogonal filters (Caelli, 1985; Laws, 1980), and simple, nonrelational properties that are discriminable with peripheral vision (Beck, 1972). Although it would seem a straightforward matter to evaluate any such proposal by manipulating the suggested image properties to see how discriminability is affected, this is not always the case.

THE PSYCHOPHYSICS OF TEXTURE DISCRIMINATION

The discriminability of two textures could be investigated with a two-alternative forced choice procedure where subjects are required to decide whether a textural display consists of one texture or two. However, given unlimited inspection time, a subject could foveate several regions of the display and compare textural properties across regions in a very deliberate manner, that is, visual search. Typically, investigators are interested in textures that can be discriminated "when attention is distributed over the whole image," "in parallel," "spontaneously," "effortlessly," or "preattentively." Therefore, inspection time is generally restricted to a single eye fixation (usually less than 200 msec) in an effort to prevent item-by-item search (e.g., Caelli & Julesz, 1978). However, this precaution may not eliminate the deliberate scrutiny of the elements composing textural displays.

Recently, there has been interest in the ability of humans to move visual attention around the visual field *within a single fixation*. Such a capability figures significantly in Treisman's *feature integration theory* and Julesz's *texton theory*. Both theories adhere to the attentive/preattentive distinction popularized by Neisser (1967). Although the theories differ in detail, many of the same properties are ascribed to the preattentive and attentive visual systems in both. The preattentive system distributes processing over a wide area of the visual display but is sensitive only to differences in simple image characteristics such as line orientation. The attentive visual system is capable of determining more abstract relationships between simple items; for example, whether two line segments are arranged so as to form a T or an L. Preattentive vision is implicated in textural segmentation whereas attentive vision is implicated in visual search. Converging evidence that attention may be repositioned within a fixation comes from the work of Posner and others (Posner, 1980; Shulman, Remington, & McLean, 1979; Tsal, 1983). Jolicoeur and Ullman (Jolicoeur, 1986; Ullman, 1984; Jolicoeur, Ullman & Mac-Kay, 1986) also argue for the existence of fast serial processes whose spatial locus may be repositioned within a fixation.

The relevant point for the present discussion is that the heavy machinery of attentive vision can be repositioned several times within a fixation. Therefore, if the presentation of a textural display is sufficiently long or the location of the boundary between two regions is predictable, the possibility exists that textures will not be discriminated on the basis of distributed, parallel processing but by scrutinizing individual elements of a textural display. That is, we cannot rule out the possible involvement of attentive vision in the texture discrimination task.

A second difficulty associated with textural psychophysics is determining the effective stimulus when textures are described in terms of informally defined features. To illustrate the problem, consider the hypothesis that micropattern *size* is an important factor influencing discrimination. Such a hypothesis might be investigated by using textures composed of various sizes of white circles on a dark background. Presumably, as size differences between the circles in the two regions increase, the two regions would become more discriminable. The difficulty however, is in attributing discriminability to size itself. As the circles get bigger then the brightness of the region also increases. If brightness were controlled by making the bigger circles correspondingly dimmer, then a contrast difference would be introduced (the small circles would have greater contrast with the background). Furthermore, the larger circles would be closer together, introducing another potentially important "feature,"—proximity. It would seem to be a general problem for feature-based approaches to determine exactly what "feature" is controlling discrimination.

EXPERIMENTS WITH TEXTURES

The potential involvement of a search strategy and difficulties in defining the controlling features of discrimination lead us (Gurnsey & Browse, 1987a) to question certain assumptions of Julesz's (1984) texton theory. Texton theory defines *textons* as visual features such as (a) elongated blobs having particular orientations, lengths, widths, and intensities (b) the terminators of these blobs and (c) crossings of line segments. The theory holds that only textures differing in texton composition will be *preattentively* discriminable.

The first difficulty with the theory is that textons, as defined by the theory, have not been shown to be either necessary or sufficient to account for "preattentive" texture discrimination. Reportedly discriminable texture pairs differing in texton composition (such as terminators and line crossings) also differ in other characteristics such as size and shape (configurational differences). Texton theory rejects the suggestion that configurational differences between micropatterns could have a significant effect on discrimination; however, in our view, the evidence does not support such a rejection. Secondly, although it is claimed that only texton-differing micropatterns will be preattentively discriminable, no behavioural criteria have been offered to differentiate those textures that are preattentively discriminable and those that are not. Furthermore, there are very few published data to support the theory; most demonstrations of discriminable textures have been of the look-and-see-for-yourself variety.

The Methodology

To assess the theory we ran several experiments (Gurnsey & Browse, 1987a) in which three factors were manipulated: (a) the characteristics of the micropattern pairs, (b) the time available to inspect the image, and (c) which member of a particular pair formed the foreground (disparate region) and which formed the background (see Figure 1). To investigate the importance of texton differences (in particular, terminator and line-crossing textons) we chose two main types of micropattern pairs: (a) those having different texton composition but similar configurational characteristics and (b) those having different configurational characteristics but no texton differences. Also, several pairs involved both configurational and texton differences.

The displays used in our experiments were identical in structure to those shown in Figure 1. The displays were presented for 67-, 100-, 133-, or 167-msec and followed immediately by the presentation of a random-dot mask. On each trial the disparate region appeared in an unpredictable quadrant of the display. The subjects' task was to identify the disparate quadrant and our dependent measure was the probability of correctly making this discrimination. Similar methodologies have been used in the past by Beck and Ambler (1972), Enns (1986), Julesz (1980) and Olson and Attneave (1970). The displays were presented in a random order so that it was unpredictable which texture pair would be shown on a particular trial, and it was also unpredictable which member of the pair would form the disparate region. Therefore, subjects were operating under a great deal of uncertainty from trial to trial. Each experiment involved six micropattern pairs, two foreground-background conditions for each pair, and four stimulus durations for a total of 48 experimental conditions each of which was presented eight times (twice in each quadrant) to each of 12 naive subjects.

Because the disparate region appeared in an unpredictable quadrant on each trial, and because there was a great deal of uncertainty about which textural display would appear on any given trial, we expected our procedure to diminish the effectiveness of rapid visual search. However, without explicit constraints on the capabilities of such search processes (e.g., upper limits on the speed with which attention can be repositioned and the complexity of the operations that may be carried out within an attentional fixation) we can never rule out their possible involvement with absolute certainty.

Results and Discussion

Several results emerged from the experiments described above which have implications for texton theory in particular and theories of textural segmentation in general. First, as might be expected, there was great variability in the overall discriminability of the different texture pairs. No support was found for the notion that terminator- and line crossing-textons play an important role in discrimination when configurational differences between micropatterns composing the textures are controlled. To the contrary, textures composed of different arrangements of the same line segments (isotexton textures) proved to be the most easily discriminated. Second, the discriminability of two textures is asymmetrical in many cases. That is, the probability of correctly identifying the disparate quadrant is often dependent on 'which texture forms the disparate region. It would appear that the simple model based on measuring local differences in image properties between neighbouring regions is inadequate to account for the asymmetry; a difference signal should be indifferent to the "sense" of the boundary. Third, performance improves with exposure duration. Fourth, the experience that subjects have had with the materials and procedure has a great effect on performance. Textures that are nondiscriminable for naive and moderately practiced subjects are easily discriminated after massive practice. We start the discussion by considering discriminability of the various pairs that were studied.

Micropattern pairs. Table 1 shows the 18 micropattern pairs[2] used in the three experiments. Pairs 1–6, 7–12 and 13–18 were presented to different groups of subjects. The third column of the table shows the probability of a correct detection averaged over all stimulus durations and foreground/background conditions. The fourth column (labelled UT) indicates the number of unshared line-segments, -crossings, and -terminators between the members of each pair. The correlation between columns three (probability of a correct detection) and four (number of unshared textons) is −.51. If the number of unshared textons played an important role in discrimination then we would have expected a high positive correlation. Micropatterns that differed in terms of gross shape (pairs 1, 3, 7, 8, 9, 15) were, in general, relatively easy to discriminate. Therefore, in this stimulus set, it would seem that textons, as usually defined, are less important than configurational differences.

Pairs 1, 8, and 15 were particularly discriminable. The members of these pairs differed significantly in terms of the sizes of their minimum enclosing circles. It might be suggested, then, that a size statistic of this kind is computed for each element in the display and

[2] It should be noted that all textural elements were presented in random orientations as in Figure 1. Therefore, orientational differences between regions should not contribute to discrimination. However, it is obvious that orientation provides a very strong basis for segmentation (Beck, 1982).

Table 1.   The eighteen micro-pattern pairs used by Gurnsey & Browse (1987a). Pairs 1–6, 7–12, and 13–18 were presented to separate groups of subjects. The third column, P(corr), shows the probability of a correct detection averaged over all exposure durations and the foreground/background condition. Column 4, UT, shows the number of unshared textons between the members of each pair. Column 5, D, shows a difference measure computed from the grey-level histograms of Gaussian blurred versions of textures composed from the members of each pair (see text).

Pair	Patterns	P(corr)	UT	D
1	┌ +	.736	3	274
2	┌ ┬	.421	1	164
3	+ ┬	.496	2	173
4	⊓ ⊔	.410	2	143
5	⊓ ⊔	.275	1	100
6	⊔ ⊔	.297	1	82
7	☰ ☲	.527	0	196
8	☰ ☲	.797	0	432
9	☲ ☲	.548	0	298
10	☰ E	.382	4	141
11	☰ ∓	.547	2	140
12	E ∓	.419	4	130
13	╰╯	.392	3	235
14	╲╱	.340	3	165
15	⊐ ⊐	.806	0	235
16	☰⊏	.264	6	139
17	⊏⊒	.337	4	243
18	⊜⊐	.246	3	82

discrimination is a function of how different the two textures are on such a measure. This suggestion is unsatisfactory however, because it is peculiar to the situation where textures are composed of micro-patterns and therefore provides no insight into segmentation in general. Rather, we note that, by definition, the smaller member of each pair is more compact than the larger member; the same line segments are concentrated in a more confined area. This difference in compactness suggests that the members of these pairs would stimu-

---

[3] Note the correspondence here with Julesz's (1965) notion of first order statistics. When Julesz used the term "first order statistics" he was referring to grey-level distributions. Julesz considered the grey-level distributions of textures composed of black dots on white backgrounds which amounts to stating the probability of a black or white dot occurring. We have considered the grey-level distributions of Gaussian blurred images which involve many more grey-levels.

late different simple detectors; for example, simple center surround operators of the difference-of-Gaussians (DOG) variety (Rodieck & Stone, 1965; Wilson & Bergen, 1979). To illustrate this point we Gaussian-blurred textures composed of the micropatterns used in our experiments. Each point in the blurred image may be thought of as the output of a receptive field that integrates intensity values in its neighbourhood according to a Gaussian weighting function. The grey-level histograms corresponding to each blurred texture were constructed[3] and from these we computed the simple difference

$$\text{measure D} = \sum_{i=1}^{255} | H_1(i) - H_2(i) | \text{ (where there were 256 grey levels,}$$

and $H_1(i)$ is the number of occurrences of grey-level i in the histogram of one of the blurred textures) to measure the difference between two histograms (H_1 and H_2). This statistic is shown in column 5 of Table 1 in arbitrary units. The correlation between column 5 (D) and column 3 (the probability of a correct detection) is .77, accounting for almost 60 percent of the variability in the overall probability of a correct detection.

Because the extent to which two textures differently stimulate an ensemble of circulary symmetric, Gaussian-shaped receptive fields is related to their discriminability, it seems reasonable to pursue (in future) the possibility that an explanation of the discriminability of these sorts of textures might be given in terms of the outputs of DOG-type center-surround operators. In fact, Beck (1986) and Bergen (1986) have already demonstrated that the responses of center-surround operators can form the basis of discrimination in several interesting cases. It should be noted that, although we have concentrated on the nature of textural properties that may relate to discriminability, a specification of the processes that utilize such information is also required (Caelli, 1985).

Discrimination asymmetry and stimulus duration. Figure 2 shows the discriminability of pairs 1, 2 and 3 from Table 1 as a function of exposure duration, and which pattern of each pair forms the disparate region. It is clear from Figure 2 that the probability of a correct discrimination depends on which member of a pair forms the foreground region and which forms the background. Significant discrimination asymmetries were found for pairs 1, 2, 3, 4, 7, 8, 11, 13, 14, and 17 (Gurnsey & Browse, 1987a). If textural segmentation only requires comparing textural properties in neighbouring image regions then there would no reason to expect the asymmetry. A difference

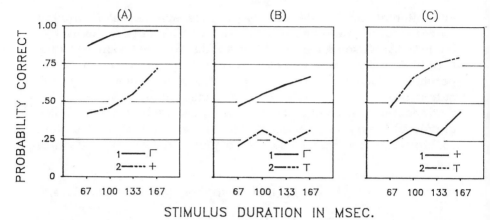

STIMULUS DURATION IN MSEC.

Figure 2. The probability of a correct discrimination for pairs 1, 2, and 3 (from Table 1) as a function of exposure duration and which member of each pair forms the disparate region. (Adapted from Gurnsey & Browse, 1987a).

signal should be insensitive to which side of the texture boundary the two textures involved happen to fall. Additional assumptions would be needed if an explanation based on computed differences is to be maintained (as we shall discuss later).

The only report of an asymmetry in a texture discrimination task that we know of was provided by Beck (1982) who used displays comprising two types of textural elements dispersed throughout the same region; that is, the two textures do not form spatially disjoint regions such as shown in Figure 1. Several examples of asymmetries have been reported in visual search tasks (e.g., Frith, 1974; Julesz, 1981; Richards & Reicher, 1978) the most recent of which is by Treisman and Souther (1985). We will consider their explanation of search asymmetry below since it is potentially relevant to our results. One important aspect of their theory is that it does *not* rely on the computation of local differences.

In one experiment, Treisman and Souther found that reaction time (RT) to detect a circle with an intersecting line in a background of circles was independent of the number of distracting circles (a classic diagnostic in Treisman's model for preattentive vision) whereas RT to detect a circle in a background of circles with intersecting lines was linearly dependent on the number of background distractors (a classic diagnostic of attentive vision). Treisman and Souther's explanation of the search asymmetry follows *feature integration theory* (Treisman & Gelade, 1980). The theory posits that visual displays are first analyzed into their constituent features (such as color, orientation, and size) which are represented in retinotopic feature maps. Even though these maps are retinotopic, the positional information they contain is not directly available to "higher" processes.

> The information which can be retrieved without focussed attention is simply the presence and the amount of activity in any given *prespecified* map (Treisman & Souther, 1985, p. 307; emphasis added.)

A unique, prespecified feature may be quickly reported as present by determining if there is activity in that map during the presentation of the stimulus; if there is no activity the target may just as quickly be reported as absent. If the features of the to-be-detected item are a proper subset of the features of the distractors then that item must be searched for by scrutinizing individual elements or small groups of elements with focused attention. Because the circle with intersecting line "popped out" of a background of circles, it would be assumed that the circle with intersecting line contained some feature not shared by the background circles.

It is not clear, however, how the theory would account for the asymmetry in our experiments, which required correctly localizing the disparate quadrant, since all the subject has access to is the amount of activity in a prespecified map.[4] More importantly, the theory says nothing about how the asymmetry arises when subjects are uncertain (as they were in our experiments) which texture would be forming the disparate region on any given trial; viz, they would be unable to prespecify the relevant feature map.

If we are to assume the availability of a measure representing the degree to which some feature is present in the image (Treisman & Souther, 1985), then local differences could be expressed in terms of this more global measure. Thus, some form of Weber's law ($\Delta I/I = k$) could be invoked to characterize the paradoxical asymmetry in the texture discrimination task. The seeming paradox is that, locally, all that can be measured at a boundary is contrast and because of this no asymmetry should arise. However, if local contrast is proportional to some *space average* value then we would expect asymmetries. For example, given two sorts of textural elements differing only in brightness, then the space average luminance is greater

---

[4] This is not exactly correct. The subject also has access to a "master map of locations, in which the positions of any discontinuities in stimulation are coded without specific information on the nature of the discontinuity." (Treisman & Souther, 1985, p. 307). A textural boundary might be located by consulting such a map; however, since all that is available is the presence and location of the boundary, this says nothing about the asymmetry.

when the brighter elements form the background and the dimmer elements form the foreground, than in the obverse case (assuming the foreground region is smaller than the background). Therefore, the contrast at an edge is greater relative to the space average luminance when the bright elements form the foreground region. Such a conceptualization may be easily generalized to other, physical differences between textured regions, or targets and distractors. Thus, rather than assuming that the asymmetry is due to subjects being aware of the presence or absence of activity in some feature map (Treisman & Souther, 1985), asymmetries in both visual search and texture discrimination could be explained in terms of the relationship between local contrast and this more global measure.

It would seem wise to look to differences in simple, measurable image properties between textures—or between targets and distractors—rather than appealing to differences in highly specific feature (e.g., terminators, line-crossings and closure) as an account of discrimination in general and the asymmetry in particular. For example, in the case of circles and circles with intersecting lines, the asymmetry may be due to simple luminance factors in exactly the manner outlined above.

Treisman (1986) reports that elements differing along a *quantitative dimension* yield an asymmetry when the stimulus having greater magnitude on the dimension is the target. For example, Treisman reports that long lines pop out of short ones but not vice versa. Beck (1982) showed similar results with line segments. In one case it was found that one long line among three short lines was more easily detected than one short line among three long lines when viewed with peripheral vision. In another demonstration it was shown that long lines distributed through a background of short lines were more easily segmented from the background than were short lines distributed through a background of long lines. Treisman also reports that darker greys pop out of lighter greys and two line targets pop out of single line distractors. In our results a similar pattern emerged. When an asymmetry existed, it was the texture composed of "larger" micropatterns (measured by the radius of the minimum enclosing circle) that was more easily discriminated. We hesitate, of course, to attribute the asymmetry to size differences described in this way, but as mentioned earlier, we can imagine larger micropatterns stimulating different simple operators which encode scale information. For example, DOG-type organizations might encode the required scale information.

Another example from our own work (Gurnsey & Browse, 1987b) may be interpreted in a similar fashion. We found that randomly arranged circles were more easily discriminated among regularly placed circles than were regularly placed circles in a background of randomly placed circles (see Figure 3). In a field of randomly placed circles, large empty spaces tend to occur between elements and, in other cases, the circles form clusters, creating large "compound" micropatterns. These two cases (large gaps and large compound micropatterns) do not occur in the same way in the regularly placed circles and this difference may underly the asymmetry. Figure 3 should be compared with Figure 1. We noted earlier (see Figure 2, panel A) that Ls were more easily discriminated in a background of +s than were +s among Ls. The +s, by virtue of being more compact, form a more regular pattern when composing a texture than do the Ls. The same explanation given for the randomly- and regularly-placed circles could be applied to the case of the Ls and +s.

In general, we suggest that the most generality may be captured by characterizing differences between micropatterns in terms of image properties that may be measured by computationally simple mechanisms. Exhausting simple possibilities such as these seems logically necessary before positing the contribution of "higher order properties" such as closure, terminators, and line crossings, for which no computational definitions are forthcoming.

The account of asymmetrical texture discrimination outlined above involved both local and global factors. That is, local measurements were expressed in terms of average, global properties. An alternative suggestion is that interactions between local coding

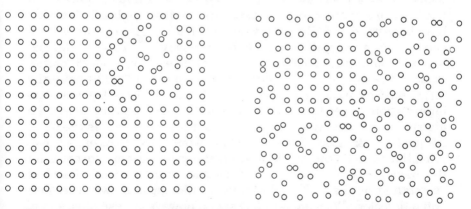

Figure 3.   Two textures composed of identical textural elements differing only in the regularity of their placement. When presented tachistoscopically, the disparate region composed of randomly placed elements shown on the left is more easily discriminated than the region composed of regularly placed elements shown on the right.

mechanisms result in one region *masking* another. Such an account might involve the notion of activation spreading from strongly stimulated detectors to more weakly stimulated detectors. In Caelli's (1985) analysis, textural segmentation involves (a) *convolution* of the input image with a family of orthogonal masks (e.g., Frei & Chen, 1977), (b) an *impletion*—or relaxation—process (e.g., Hummel & Zucker, 1983) whereby excitation from highly stimulated detectors spreads to nearby, weakly stimulated detectors of similar type (activity due to weakly stimulated detectors dies out if no excitation is received from those that are highly stimulated), and (c) *grouping* areas which have reached stable states with similar patterns of activity. Some process analogous to the impletion operation could be responsible for one region masking another. Consider just one detector type that is strongly stimulated in texture A and weakly stimulated in texture B. When A is the foreground, activity from the strongly stimulated detector will spread to the weakly stimulated detectors in region B. The extent of this spread will depend on the nature of the impletion process; however, the effect will be to "enlarge" the disparate region somewhat and perhaps increase its probability of being detected. When A is the background and B is the foreground, the high activity from the strongly stimulated detectors will spread from A into B, thereby "swamping" (Treisman, 1986) the disparate region and decreasing the probability of its being detected by some process looking for textural differences between regions.

It should also be noted that performance generally improves with increased exposure duration (see Figure 2 for example); however, a wide range of curves relating exposure duration to discriminability have been observed (Gurnsey & Browse, 1987a). It is clear from these curves and the overall discriminabilities shown in Table 1 that texture discrimination is not an all-or-none process. The general performance increase with increased exposure duration could be naturally interpreted as indicating that processing *differences* between textured regions takes time and the amount of time required to make a discrimination depends on how different the two textures are (e.g., Bergen & Julesz, 1983). Such an account would have to assume something like the masking process outlined above in order to account for the asymmetry. At short durations the asymmetrical effects of "spreading activation" might prevent differences between regions being detected. With longer processing, however, these effects may be overcome and performance would improve.

Practice. Figure 4 shows a comparison of the performance of a highly practiced subject, a moderately practiced subject, and the average of 12 naive subjects on three micropattern pairs (pairs 13, 14,

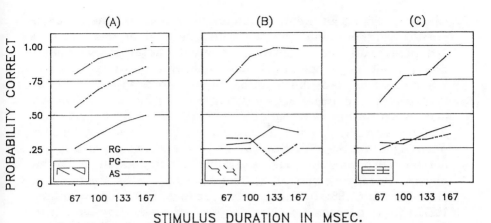

Figure 4. The probability of a correct discrimination for three texture pairs as a function of exposure duration and the practice level of the subjects involved. RG had had extensive practice with many different textures, PG was moderately practiced and AS represents the average of twelve naive subjects.

and 17 from Table 1). The highly practiced subject (RG) had been through tens of thousands of experimental trials with many different textures before the data presented here were recorded. The moderately practiced subject (PG) had been through 2400 trials involving the textures he was eventually tested on and 448 trials involving other textures. The naive subjects (AS) participated in one experimental session (as described earlier).

Each panel of Figure 4 shows the discriminability of micropattern pairs as a function of exposure duration, averaged over the foreground/background condition for RG, PG, and AS. Pair 13, shown in panel (A), is an example of what Julesz has reported as a discriminable, isodipole texture pair. It is clear that in our study, the naive subjects had a great deal of difficulty with this pair. After moderate practice however, PG found this pair to be rather easy to discriminate. RG's performance, on the other hand is essentially on the ceiling for all but the shortest exposure durations. In this case at least, performance appears to improve in proportion to practice. Panel (B) shows another isodipole pair that the naive subjects have great difficulty with. Moderate practice does not improve PG's performance over that of the average subject. However, massive practice again puts RG on the ceiling. Essentially the same pattern holds for pair 17, shown in panel (C), except that RG found this pair more difficult than the other two.

Figure 4 shows that practice is a factor to be taken into account when we talk about *the discriminability of textures.* The effects of practice further highlight the difficulty with terms such as "effortless" and "preattentive" when applied to texture discrimination. The important theoretical question that has not arisen before is: What improves? Is it the case that highly practiced subjects (e.g., RG) employ a parallel segmentation process but simply become more sensitive to the output of simple operators? Or, do such subjects develop highly efficient, rapid serial processes such as compiled visual routines (Ullman, 1984)? With extreme practice it may not be reasonable to consider texture discrimination a psychophysical task, any more than tachistoscopic letter identification (e.g., Mewhort, 1985) or automatic search (e.g., Schneider & Shiffrin, 1977) should be considered psychophysical tasks. However, it remains to be seen if the same textural properties ultimately limit the performance of both practiced and naive subjects.

DISCUSSION

In the past, theories of textural segmentation have focused on the aspects of textures that limit our abilities to discriminate them. It is clear from the results presented here that several other factors also limit performance. One point worth noting is that the term "preattentive" adds nothing to our understanding of performance in the texture discrimination task. We cannot set some criterion that separates those textures that are preattentively discriminable from those that are not. Rather than classifying texture pairs as discriminable or nondiscriminable, we should instead attempt to relate quantitative descriptions of two textures to the probability of a correct discrimination when one is embedded within the other. Of course, given the effects of stimulus duration and practice we must further relate the probability of discrimination to some configuration of these conditions as well.

We are in agreement with Beck (1972, 1982) who has consistently argued that differences in simple, nonrelational properties determine the ease with which two textures will be discriminated. These same characteristics appear to play some role in the asymmetry as well. For image properties such as contrast, scale, and orientation, there exist computational proposals as to how they might be extracted from an image. Using such methods, computational models of the segmentation process may be constructed and we would expect to be able to make quantitative statements about how manipulations of these simple variables interact and contribute to performance. Such models would actually have to produce a segmentation of the image (or report a discrimination) and, therefore, some account must be given of the processes that use the "simple image characteristics" that we have been discussing (Caelli, 1985). A number of design choices would be available, ranging from completely parallel to completely serial or some combination of both (e.g., Bergen & Julesz, 1983). An open theoretical issue is to constrain the properties attributable to search processes that may be operative in the discrimination task.

Finally, a recent development in the study of textural segmentation is Caelli's (1986) suggestion that, the visual system *adaptively* constructs filters to segment two textures. The nature of the constructed filters is dependent on the properties of the particular textures involved. An apparent consequence of the approach is that it will not be empirically determinable whether or not some "feature" generally controls the discrimination of two textures. We showed in Table 1, for example, that when configurational differences between micropatterns are eliminated, textures differing in terms of the line-crossing "feature" are poorly discriminated. Caelli would argue that we have shown *one case* in which the existence of line-crossings does not control performance *but* there may be other situations in which the adaptively derived filter *is* essentially detecting differences in line-crossings. Such a view would lead us to be very pessimistic about the possibility of capturing *general* textural properties that determine the discriminability of two textures. We are not sure, however, what exactly the behavioral consequences of adaptive filtering should be. If adaptability implies that we can't infer general textural properties that influence discrimination then there would appear little point in studying texture discrimination since no principled explanation would be possible. However, if we can make general, predictive statements about the ease with which textures will be discriminated, then adaptive filtering would be one of several possible implementation strategies.

REFERENCES

Ballard, D. H., & Brown, C. M. (1982). *Computer vision.* Englewood Cliffs, NJ: Prentice-Hall.

Beck, J. (1972). Similarity grouping and peripheral discriminability under uncertainty. *American Journal of Psychology, 85,* 1–19.

Beck, J. (1982). Textural segmentation. In J. Beck (Ed.), *Organization and representation in perception.* Hillsdale, NJ: Erlbaum.

Beck, J. (1986). Unresolved issues in texture segregation. Paper presented at the Center for Visual Science Symposium, Rochester, N.Y.

Beck, J., & Ambler, J. (1972). Discriminability of differences in line slope and in line arrangement as a function of mask delay. *Perception and Psychophysics, 12*, 33–38.

Beck, J., Prazdny, K., & Rosenfeld, A. (1983). A theory of textural segmentation. In J. Beck, B. Hope, & A. Rosenfeld (Eds.), *Human and machine vision* (pp. 1–38). New York: Academic Press.

Bergen, J. (1986). Texture segregation and the psychophysics of texture. Paper presented at the Center for Visual Science Symposium, Rochester, NY.

Bergen, J., & Julesz, B. (1983). Rapid discrimination of visual patterns. *IEEE Transactions on Systems, Man and Cybernetics, SMC-13*, 857–863.

Caelli, T. (1982). On discriminating visual textures and images. *Perception and Psychophysics, 31*, 149–159.

Caelli, T. (1985). Three processing characteristics of visual texture segmentation. *Spatial Vision, 1*, 19–30.

Caelli, T. (1986). Correlational mechanisms for spatial vision. Paper presented at the CIAR workshop on vision, April, 1986.

Caelli, T., & Julesz, B. (1978). On perceptual analyzers underlying visual texture discrimination: Part 1. *Biological Cybernetics, 28*, 167–175.

Caelli, T., & Moraglia, G. (1985). On the discrimination of Gabor signals and discrimination of Gabor textures. *Vision Research, 25*, 671–684.

Enns, J. (1986). Seeing textons in context. *Perception and Psychophysics, 39*, 143–147.

Frei, W., & Chen, C. (1977). Fast boundary detection: A generalization and a new algorithm. *IEEE Transactions on Computers, C-26*, 988–998.

Frith, U. (1974). A curious effect with reversed letters explained by a theory of schemata. *Perception and Psychophysics, 16*, 113–116.

Gurnsey, R., & Browse, R. A. (1987a). Micropattern properties and presentation conditions influencing visual texture discrimination. *Perception and Psychophysics, 41*, 239–252.

Gurnsey, R., & Browse, R. A. (1987b). Micropatterns, the elements of texture experiments, and their interactions; Paper presented at the first annual Queen's Vision Conference, Kingston Ontario.

Grossberg, S., & Mingolla, E. (1985). Neural dynamics of perceptual grouping: textures, boundaries and emergent segmentations. *Perception and Psychophysics, 38*, 141–171.

Hummel, R., & Zucker, S. W. (1983). On the foundations of relaxation labelling processes. *IEEE: PAMI, 3*, 267–287.

Jolicoeur, P. (1986). Routine computations in the perception of spatial relations. Paper presented at the CIAR workshop on vision, London Ontario.

Jolicoeur, P., Ullman, S., & MacKay, M. (1986). Curve tracing: A possible basic operation in the perception of spatial relations. *Memory and Cognition, 14*, 129–140.

Julesz, B. (1980). Spatial nonlinearities in the visual perception of textures with identical power spectra. *Philosophical Transactions of the Royal Society of London, B 290*, 83–94.

Julesz, B. (1981). Textons, the elements of texture discrimination, and their interactions. *Nature, 290*, 91–97.

Julesz, B. (1984). Toward an axiomatic theory of preattentive vision. In G. Edelman, W. Einer, & W. Cowan (Eds.), *Dynamic aspects of neocortical function*. New York: John Wiley & Sons.

Julesz, B., & Bergen, J. (1983). Textons, the fundamental elements in preattentive vision and perception of textures." *Bell Systems Technical Journal, 62(6)*, 1619–1645.

Julesz, B., Gilbert, E. N., Shepp, L. A., & Frisch, H. L. (1973). Inability of humans to discriminate between visual textures that agree in second order statistics-revisited. *Perception, 2*, 391–405.

Laws, K. (1980). Textured image segmentation. USC Image Processing Institute, Los Angeles, CA., Report 940.

Levine, M. D. (1985). *Vision in man and machine*. New York: McGraw-Hill.

Mewhort, D. J. K. (1985). Information stores and mechanisms: Early stages in visual processing. In H. Heuer & A. F. Sanders (Eds.), *Issues in perception and action* (pp. 335–357). Erlbaum. Hillsdale, NJ.

Navatia, R. (1982). *Machine perception*. Englewood Cliffs, NJ: Prentice-Hall.

Neisser, U. (1967). *Cognitive psychology*. New York: Appleton-Century-Crofts.

Nothdurft, H. C. (1985a). Orientation sensitivity and texture segmentation in patterns with different line orientation. *Vision Research, 25*, 551–560.

Nothdurft, H. C. (1985b). Sensitivity for structure gradient in texture discrimination tasks. *Vision Research, 25*, 1957–1968.

Nothdurft, H. C. (1985c). Discrimination of higher-order textures. *Perception, 14*, 539–543.

Olson, R. K., & Attneave, F. (1970). What variables produce similarity-grouping? *American Journal of Psychology, 83*, 1–21.

Pentland, A. P. (1984). Fractal-based descriptions of natural scenes. *IEEE Transactions on Pattern Analysis and Machine Intelligence, PAMI-6*, 661–674.

Posner, M. I. (1980). Orienting of Attention. *Quarterly Journal of Experimental Psychology, 32*, 3–25.

Richards, J. T., & Reicher, G. M. (1978). The effect of background familiarity in visual search: An analysis of underlying factors. *Perception and Psychophysics, 23*, 499–505.

Richards, W. (1979). Quantifying sensory channels: Generalizing colorimetry to orientation and texture, touch and tones. *Sensory Processes, 3*, 207–229.

Rodieck, R. W., & Stone, J. (1965). Analysis of receptive fields of cat retinal ganglion cells. *Journal of Neurophysiology, 28*, 833–849.

Schneider, W., & Shiffrin, R. M. (1977). Controlled and automatic human information processing: I. Detection, search and attention. *Psychological Review, 84*, 1–66.

Shulman, G. L., Remington, R. W., & McLean, J. P. (1979). *Journal of Experimental Psychology: Human Performance and Perception, 5,* 522–526.

Treisman, A. (1986). Features and objects in visual processing. *Scientific American, 225,* 114b–125.

Treisman, A. M., & Gelade, G. (1980). A feature integration theory of attention. *Cognitive Psychology, 12,* 97–136.

Treisman, A. M., & Souther, J. (1985). Search asymmetry: A diagnostic for preattentive processing of separable features. *Journal of Experimental Psychology: General, 114,* 285–310.

Tsal, Y. (1983). Movements of attention across the visual field. *Journal of Experimental Psychology: Human Performance and Perception, 9,* 525–530.

Ullman, S. (1984). Visual routines. *Cognition, 18,* 97–159.

Wilson, H. R., & Bergen, J. R. (1979). A four-mechanism model for threshold spatial vision. *Vision Research, 19,* 19–32.

3
Correlational Mechanisms for Spatial Vision

Terry Caelli

Department of Psychology
and
Centre for Machine Intelligence and Robotics,
The University of Alberta

INTRODUCTION

The aim of this paper is to review the ways in which correlational-type computations are used to represent the processing of spatial information in biological and computer visual systems, specifically in the areas of texture segmentation and pattern recognition. The author hopes that from this review the many ambiguous and unclear representations proposed to model these processes will be given firmer computational bases and that it will relieve some of the mysticism that often surrounds them. The author hopes too that spatial vision will be seen to have a series of unifying principles of which correlational processes play an important role.

In its most general sense, correlation is concerned with the relationships between at least pairs of events, symbols, or variables. In the case of two-dimensional images, physical correlations exist within the image, visual processing mechanisms, and between the image and what is computed within the machine or vertebrate brain. Indeed, the notion that there are cells along the visual pathways which are tuned to the processing of specific luminance profiles is a prime example of how correlational processes are used to represent what is coded in images. That is, the notion of "receptive field," in its definition and measurement, assumes that the activity of cells (within various vertebrate visual neural networks) can be approximated by the degree to which its receptive field profile *matches* the input signal. In fact, the assumption is usually used in its inverse form by defining the receptive field in terms of the luminance profile which elicits the maximum response (Rodieck, 1973). This notion of detectors has also permeated psychophysics where the equivalent "channels" or "perceptive fields" are determined from the degree to which signals are known to facilitate (summation), or inhibit (mask) the detection of others (Breitmeyer, 1984).

Two questions follow naturally from this use of correlational measures of activity. What types of spatial correlations are processed in a given task? This is equivalent to the question of how many detectors are functioning in a given visual task. Two, how is the information from such detectors used in texture segmentation and pattern recognition?

Much of vision research, focused upon the first question, has been concerned with the enumeration of spatial detectors or channels by means of psychophysical paradigms involving gratings, point, or bar sources of light and, in general, highly artificial stimuli. From these studies we have learned that the visual system (a) processes relatively uncorrelated information via different "mechanisms," and (b) that the number of such mechanisms is not definitive, finite in number, or fixed in nature. That is, the set of frequency bandwidths or spatial filters are clearly adaptive to the signal and so, by definition, cannot be ultimately determined from such basic psychophysical tasks with one singular type of signal. This is particularly true with gratings which are, by no means, the "canonical" signal for the visual system. This is because receptive fields, by definition, are band-limited in their response characteristics and are *not* characterized optimally by their responses to singular grating patterns (See Caelli & Moraglia, 1985).

Even if the question of enumerating such detectors was definitively answered in such artificial environments, just how such responses would be utilized in the conscious brain when processing textures, or enabling pattern recognition, is of equal importance. Indeed, it is impossible to construct an unequivocal spatial encoding model without explicitly defining what information is used from the outputs of such detectors, and this question is still unresolved in both computer and human pattern recognition research; an issue of focal interest to this paper.

Table 1 summarizes the candidate decomposition processes and response measures extant in the literature for spatial vision and for the encoding stages of typical computational visual systems. The decomposition processes shown across the top columns refer to the possible ways in which the input signal is decomposed by percep-

Table 1

Shows the classification of decomposition processes and registration techniques currently used in biological and computational vision to depict the encoding processes relevant to texture segmentation and pattern recognition.

Spatial Encoding Models

		Decomposition Processes			
		None: Complete image	Isotropic filters (size specific)	Orientation size specific	Adaptive filters
R E G I S T R A T I O N	T E C H N I Q U E	Complete	Matched filters	1-D "channels" as outputs of retinal ganglion cells	2-D "channels" simple cells orientation, size specific	Feature filters are signal dependent
		Rectified Activities		Outputs registered as deviations from baseline response only		
		Zero-Crossings		Outputs precisely localized in space: generalized "edges"		
		Thresholding		Only responses above specified level are registered		

tual encoders to enable texture segmentation or pattern recognition to occur. By "Complete" is meant that the full image at a given perceptual level of resolution (usually defined by sampling frequency) is represented internally, per se. This encoding procedure does not necessarily imply that a form of decomposition does not occur (for example, via receptive fields). Rather, at the *explicit* stage of computation, this decomposition approach contends that the *explicit code* is already the "internally resynthesized (neural) image."

The "Isotropic filter" decomposition (column 2, Table 1) refers to the idea that multiple views of the image are made through circular-surround receptive-type filters known to exist in the vertebrate retinal ganglion cells. It is these filters which are two-dimensional generalizations of the primitive "channels" developed by, for example, Wilson and Bergen (1979), to explain aspects of spatial vision (see

Daugman, 1983, for a review). Secondly, such filters underly the primative $\nabla^2 G$ encoders utilized by Marr and Hildreth (1980) in their edge extraction schemes. These isotropic filters do not explicitly encode orientation; rather orientation selectivity comes about by the response of the filters to the anisotropic nature of the signal to be processed. This distinction between explicit code and implicit response is of crucial importance to many aspects of texture segmentation and pattern recognition.

The third decomposition process, the "Anisotropic filters," (Table 1, column 3) refers to the involvement of filters (receptive fields) which are selectively sensitive to edge and bar orientations and sizes contained within the incoming signal. Such filters are reminiscent of the simple cell receptive fields in the vertebrate visual cortex as recorded by Hubel and Wiesel and many others (see Hubel & Wiesel, 1977), and make *explicit* an orientation respose to the input signal. Figure 1 shows typical candidate detectors for these two decomposition processes.

The problem with such anisotropic decomposition rules is that the dimensionality of the feature space increases remarkably—relative to the isotropic case. Indeed, in a large number of experiments that we have completed (Caelli & Hübner, 1983; Caelli, Brettel, Rentschler & Hilz, 1983; Caelli, 1983), at least 500 energy-detecting filters are apparently required in order to resynthesize the input signal—if they merely acted as energy detectors. Even if 500 filters were adequate to represent the processes of texture segmentation and pattern recognition, then it seems most unlikely that optimization schemes and analytic treatment of such encoding procedures could be enacted in such high dimensional (explicit) feature spaces. For such reasons a third type of adaptive decomposition is envisaged (Table 1, Column 4). That is, the explicit filters (or features) utilized by visual systems, in solving perceptual problems, are few in number and adaptive to the class of signals and problem to be solved. Such filters may come about by the combination of lower-order fixed detectors, as illustrated in the previous types of decomposition processes. One algorithm for generating such filters will be discussed in the following section on texture segmentation.

Issues concerned with the types of decomposition processes employed in specific problems cannot be divorced from the types of registration techniques used in encoding the outputs of such filtering mechanisms. The rows of Table 1 illustrate four of the more commonly used registration methods. The "Complete" registration technique corresponds to the popular notion of "neural image" where the presence of certain features (as defined by the detector profiles) is directly related to the strength of the filter output, per se (see Ade-

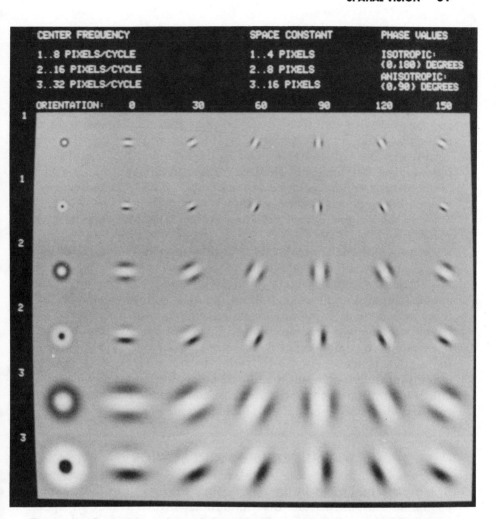

CENTER FREQUENCY   SPACE CONSTANT   PHASE VALUES

1..8 PIXELS/CYCLE   1..4 PIXELS   ISOTROPIC:
2..16 PIXELS/CYCLE   2..8 PIXELS   (0,180) DEGREES
3..32 PIXELS/CYCLE   3..16 PIXELS   ANISOTROPIC:
                                                               (0,90) DEGREES

ORIENTATION: 0   30   60   90   120   150

Figure 1.   Sets of hypothetical detectors corresponding to feasible retinal ganglion cell (column 1) receptive fields and simple cells (columns 2 to 7). Apertures (space constants), center frequencies, and phase values are relative to an underlying 64×64 pixel format. The correlation (zero shift) between even and odd detectors was zero and neighboring orientations was no better than 0.5. The orientation specific detectors are spaced at 30° apart and have a fixed orientation turning width of ±15° over all three different center *radial* frequencies: the bandwidths increasing with increases in centre frequencies.

lson & Bergen, 1985, for an example). The "Rectified activities" registration technique (row 2, Table 1) refers to the notion that evidence for activity, or the existence of certain features, comes about by the deviation of the output image from a specified baseline level. Of recent interest is the third registration technique, localization through "zero-crossings" (row 3, Table 1). Here the location of specific features is determined by the positions of zero-crossings of the detector responses to the image. This, of course, would only be viable for band-pass filters, that is detectors which have center-surround receptive field profiles (see Yuille & Poggio, 1986; Toree & Poggio, 1986, for details). This type of registration measure is of particular interest to the registration of image edge information at a variety of scale values determined by the bandwidth of the filtering process (see Witkin, 1983). The final registration technique is a simple nonlinear version of the complete output model: "thresholding," where the evidence for the existence of a certain feature is only taken to be salient if that activity exceeds a certain thresholding value.

From a biological perspective, to this stage we have little evidence that any *singular* registration technique is exclusively used by the vertebrate visual system in encoding images or responding to spatio-temporal stimulations. Since neural electrophysiological activity occurs as a temporal modulation, the spatial mapping of the receptive field is a process inferred by specific assumptions about the statistics of such temporal response characteristics to the incoming signal. However, in the following sections we hope to demonstrate that from a computational and perceptual point of view, specific registration techniques are apparently more useful for the solution of specific perceptual problems.

VISUAL TEXTURE SEGMENTATION

It is generally agreed in the psychophysical literature that texture segmentation, or texture discrimination, involves that of spontaneously ("pre-attentively") observing disjoint textured regions with no prior information about local image structures or micro-pattern shapes. Computational models for this process therefore involve the quantitative association of texture (or image) positions to only one of n texture classes. The aim of this section is to demonstrate how correlational mechanisms may be used in biological texture segmentation to attain this goal.

For a biological visual system endowed with a multitude of detectors (as defined above), it seems reasonable to presume that visual texture segmentation may come about by the comparison or correlation of detector outputs over such textured regions. This proposal has received experimental and mathematical attention over the past decade with one-dimensional grey-scaled textures (Richards, 1979; Harvey & Gervais, 1978) and two-dimensional textures (Caelli & Julesz, 1978; Caelli, 1982, 1983, 1985). However, only until recently has a *full computational model* been proposed which actually produces segmentation as a function of such "texton" (Julesz, 1981) outputs, and this paper reviews the above analyses in a number of ways.

Texture segmentation is viewed as having three component processes: (a) spatial decomposition, (b) dynamical associative processing, and (c) classification of textured regions. The specific aims of this model are to enable segmentation when the textures consist of sparse micro-patterns; to create networks which will extract, or adapt to, the predominant features of the texture, and to use a classification procedure which is adaptive to the outputs of such detectors.

TIE MODEL

The initial process of texture segmentation is envisaged to involve the registration of the input (foveal) textures through the parallel outputs of many detectors whose responses are determined by some nonlinear transformation of their cross-correlation with the input. Assuming a relatively fixed "retinal pre-processor," having opponent center-surround receptive fields, the primary information to be processed must have differential, or band-pass, components emphasized. Further to this, we assume the existence of a relatively fixed primary projection area where such image information is further classified (encoded) by cortical edge and bar detectors (anisotropic detectors, Table 1, Figure 1) whose outputs are a nonlinear function of the cross-correlation of the detector's profile with the input image. That is, we assume that the response (R_i (x,y)) of a detector d_i at retinotopic position (x,y) is determined by:

$$\sum_{\alpha,\beta} d_i(\alpha, \beta) = \text{const}, \tag{1}$$

and

$$R_i(x,y) = \text{const} + \gamma\psi\{d_i o I\}, \ \gamma \equiv \text{scalar}, \tag{2}$$

o denoting cross-correlation between the detector and the image (I).

$$d_i o I(x,y) = \sum_{\alpha,\beta} d_i(\alpha,\beta)I(x + \alpha, y + \beta), \tag{3}$$

and

$$-1 \leq \psi_\delta\{z\} = \frac{1 - e^{-\delta z}}{1 + e^{-\delta z}} \leq +1, \ \delta \equiv \text{constant}. \tag{4}$$

In our simulations we have used

$$R_i(x,y) = 128 + 127\psi_\Delta\{d_i o I\}, \ \delta = 0.03, \ \Sigma d_i = 0, \tag{5}$$

to fit in with an 8-bit response range. The nonlinear transducer enables one to move smoothly from square wave ("ideal" edge and bar) to Gaussian modulated sinusoids (Daugman, 1983) representations for edge and bar, or orientation detectors, via δ in (4). Orientations and sizes of the detectors were chosen to fit with a large number of experimental results on human texture discrimination showing the inability of observers to resolve image orientations to better than $\pm 5°$ (Caelli, 1982; Beck, 1983). With evidence that such receptive or "perceptive" fields are limited in size to $\pm \frac{1}{8}$ octave, or $1\frac{1}{2}$ cycles to $1/e$ decay of a Gaussian aperture, we have generated 24 fundamental orientation detectors over 7×7 pixel kernels (relative to 128×128 pixel textures) satisfying these profile constraints for both edge (odd) and bar (even) detectors (see Figure 1).

We secondly assume that the response profile for each detector is 'rectified' into an "activity" profile (see Table 1, row 2):

$$A_i(x,y) = |R_i(x,y) - \text{const}|, \text{ where const} = 128, \text{ in this case.} \tag{6}$$

In contrast to representing texture codes by detectors defined at different size scales, in Gaussian pyramids (etc., see Adelson & Bergen, 1985) which would be capable of responding to texture regions in areas greater than the actual micro-pattern size—the approach adopted here remains at the resolution of the basic texture—though this is not a necessary restriction. Further, the process of texture region "filling-in" (impletion or region growing) is seen as a dynamical process involving the iterative activity of activated detectors in terms of how their responses may spread over contiguous regions in an accumulative fashion. This is a relaxation process (Hummel & Zucker, 1983) where the strength of a given spatial response is reinforced or inhibited as a function of neighbouring collaborative or competitive evidence. In particular, it is assumed that the activity of a given detector d_i determined by (6) is updated dynamically by the following (associative) "texture processing equation":

$$A_i^{t+1}(x,y) = \frac{1}{rs} \sum_\alpha^r \sum_\beta^s A^t \left(x + \alpha - \frac{\gamma}{2}, y + \beta - \frac{s}{2} \right) + \sum_{\substack{j=1 \\ i \neq j}}^n W_{ij}^t A_j^t(x,y) \tag{7}$$

where (α, β) index the "region of influence or size rxs" at each iteration. W_{ij}^t is the coupling, or associativity, between two activity profiles at time t which can be either fixed or adaptive. For the fixed case we have used the well-known "mass-action" formulation for detector coupling (Grossberg, 1982), where we have set:

$$\forall t, \ W_{ij}^t = \begin{cases} k \ldots (i, j) \text{ being (edge, bar) pairs of the same orientation} \\ -\left(\dfrac{1}{n-2}\right) k \ldots \text{ otherwise,} \end{cases} \tag{8}$$

for n detectors and k being the coupling strength such that

$$\forall i, \ \sum_{j=1}^{n} W_{ij}^t = 0. \tag{9}$$

It seems, in general, unlikely that the (neural) connectivity between such detector planes can be defined by a single stationary matrix W_{ij}^t over space and time. Like Hebb (1949) and Fukushima (1984), we assume that the process of perceptual learning and adaptation involves the dynamic updating of W_{ij}^t as a function of the detector's response strengths and correlations. For these reasons, our interests are also focused upon investigating formulations for W_{ij}^t such as:

$$W_{ij}^t = {}^t r_{ij}^P \{ \alpha_{ij} A_j^t(x,y) + b_{ij} \} \tag{10}$$

where α_{ij} and b_{ij} correspond to slope and intercept regression coefficients of A_j on A_i, p the degree to which this correlated information is associated with the response of A to result in "new" detector profiles. This dynamical system converges to strong "attractor" detectors which (as will be shown) have receptive fields related to the first few eigenvalues of the coupling matrix. Using (10) in (7) requires normalization as:

$$A_i^{t+1}(x,y) = \frac{1}{\left[1 + \displaystyle\sum_{\substack{j=1 \\ i \neq j}}^{n} {}^t r_{ij}^P \right]} \cdot$$

$$\left[\frac{1}{rs} \sum_{\alpha,\beta}^{r,s} A_i^t(x + \alpha, y + \beta) + \sum_{\substack{j=1 \\ i \neq j}}^{n} {}^t r_{ij}^P \left[\alpha_{ij}^t A_j^t(x,y) + b_{ij}^t \right] \right] \tag{11}$$

for $0 < p$ and even. The first component:

$$\frac{1}{rs} \sum_{\alpha,\beta}^{r,s} A_i^t(x + \alpha, y + \beta),$$

being an averaging process (moving spatial window), is clearly "local" and restricted to a given detector plane. That is, activity within a given plane spreads as a function of the spatially neighbouring activity of the same detector type and converges to a mean level of activity. Secondly, this activity is reinforced as a function of the degree to which *other detectors exhibit similar graded responses* over the full texture regions—a *global cooperative component*—represented by the last component in (11): "synergesis."

This has the effect of combining *correlated* detector responses and converging to common ("attractor") profiles, so reducing the number of different detectors. Again, it should be noted that the solutions are critically dependent on the input signal. In the case of texture segmentation, where broad spatial regions have to be "labeled," inhibitory forms of W_{ij} seem inappropriate as they differentiate the detector responses in further ways. This, in turn, would not produce the percept of *contiguous* spatial regions, and would be more useful in pattern recognition where it is precisely these differentiated dimensions which are needed (see below).

Finally, we consider the network to "complete" its activity when it reaches near equilibrium state as measured by:

$$\frac{1}{m^2 n} \sum_{i=1}^{n} \sum_{x,y}^{m,m} \left[\frac{A_i^{t+1}(x,y) - A_i^t(x,y)}{A_i^t(x,y)} \right] \leq \delta, \tag{12}$$

δ being near zero (in our case $\delta=0.02$). Here n corresponds to the number of detectors and (m,m) to the image size.

It should be noted that formulation (11) is an example of an associative network whose coupling undergoes adaptation, and if we consider the problem of texture segmentation as primarily involving the extraction of the main dimensions for segmentation, then it is the eigenvalues of W_{ij} which are critical. In the simulations to be reported we shall observe the behavior of the eigenvalues of W_{ij} to investigate how these adaptive processes are changing the *dimensionality* of the problem.

Since textured regions are proposed to appear as a function of position response differences in detector outputs ("feature space"), the appropriate classification process seems to be the minimum distance classifier (MDC, Ahmed & Rao, 1975). This method determines whether a pixel falls into one of two textured regions as a function of whether it is closer to the centroid of the texture. The MDC determines the discriminant (function) hyperplane which constitutes the locus of points equidistant between both centroids, and of the form:

$$g(a_1, \ldots, a_n) = \underbrace{\sum_{j=1}^{n} [\bar{a}_{1j} - \bar{a}_{2j}]a_j - \tfrac{1}{2} \sum_{j=1}^{n} [\bar{a}_{1j}^2 - \bar{a}_{2j}^2]} \qquad (13)$$

where $(a_1. . a_n)$ correspond to the feature dimensions—in this case, detector outputs. $\bar{a}_{ij}$ corresponds to the mean value for group (texture) i on feature j, while a_j corresponds to given input texture pixel feature weights. The pixel is classified as a function of the sign of $g(a_1. . a_n)$.

To introduce a degree of "fuzziness" or "segmentation strength" into this procedure, it would seem appropriate to use the distance between the means over the standard deviation of both sets (or average student's t statistic):

$$\bar{t} = \frac{1}{n} \sum_{j=1}^{n} t_j, \text{ where } t_j = \frac{\bar{a}_{1j} - \bar{a}_{2j}}{\sqrt{s_{1j}^2/n_1 + s_{2j}^2/n_2}} \qquad (14)$$

and n_1, n_2 correspond to the number of pixels in each textured region, $s._i^2$ to the appropriate variance statistic.

This function not only indicates that adding *common* features to the textures would decrease *perceptual* segmentation, but would also decrease if more variability in detector outputs was observed over either, or both, regions.

SUMMARY AND SIMULATIONS

We first summarize the main properties of the model:

(T_1) Detector activity is determined by the rectified response profiles as a result of detector cross-correlation with the incoming texture, according to (1)–(7).

(T_2) The activity of a given detector at time t and position (x,y) is determined by the degree to which neighboring regions are also active with respect to this detector and the activity of others.

(T_3) The associativity between detector arrays (i,j) is adaptive to their responses.

(T_4) A perceptual classification is made after this dynamical system reaches equilibrium.

(T_5) Classification of positional information into textured regions is accomplished by a form of the minimum distance classifier, weighted by the total texture "entropy."

All three processes (convolution, impletion/co-operativity, and classification) have been implemented to quantitatively observe the behavior of the system with four critical texture pairs consisting of grey-scaled textures differing in granularity, simple textures differing in orientation, and those differing in micro-pattern spatial characteristics: *T,L,* and so on. We have chosen the latter two pairs since it has been (a) shown that they differ in discriminability, and (b) it has been *proposed* that they *require* "end-of-line" and intersection detectors to result in discrimination (Julesz, 1984)—the latter we can disprove. These are shown in Figure 2 together with the outputs of

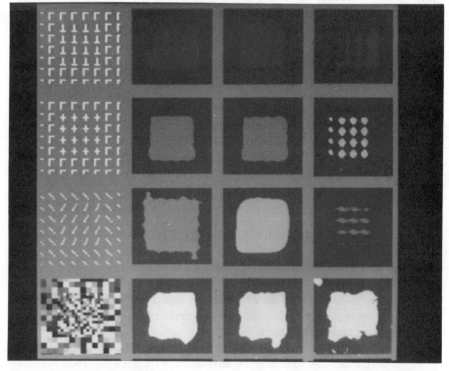

Figure 2.   Summarizes the classification of four textures (left hand column) as a function of no coupling ($w_{ij} = 0$, in (7): column 2), with adaptive coupling ((11), column 3) and inhibitory coupling ((8): column 4). Here the "strength" of segmentation is depicted by the contrast between center and surround texture regions (from Caelli, 1988).

the classification procedure, using (14) to represent the relative "strength" of discrimination (Convergence usually occurred within 5–7 iterations).

To illustrate the effects of associativities on decreasing the dimensionality of the classification process, Figure 3 shows the eigenvalues for the solutions shown in columns two and three of Figure 2. Such reductions in "dimensionality" are clearly related to the iterative process converging to common strongly active detector profiles and inhibiting less active and isolated ones.

What connects the texture processing equation (7) and the "texton" approach is that such cooperative networks *decrease* the dimensionality of the problem to the more strongly active—though "adaptively generated"—detectors or dimensions. That is, the profile of each detector in the process described by (8) is not stationary but, rather, is adapted by the energy it is designed to process and the activity of other units. Indeed, *the actual profile at any time is recoverable by inverse filtering* (Rosenfeld & Kak, 1976).

This model for texture segmentation is algebraically similar to a *class* of models for pattern recognition based on the associative (coupled) activities of large numbers of computational units whose activity profiles adapt to the signal and network states (see Kohonen, 1977; Fukushima, 1984). The main difference lies in how each computational component is interpreted, and the involvement of a classification scheme at the end, which actually produces the textured regions. In this sense the model is not formally dependent upon the initial edge and bar detectors chosen, but rather on the ways in which their outputs are correlated over space and time according to (7).

In the present texture-processing model the nature of the decomposition, and so dimensions, of a given texture segmentation task is dependent on the signal and the type of coupling operating between the computational units. If the visual system (or, indeed, the scientist) were to choose detectors which satisfied *absolute* orthogonality ($d_i * d_j \equiv 0$, $*$ being convolution) then, from a mathematical perspective the ideal detector conditions would be present and the associative processes defined by (11) and (12) would not be required. Some form of Impletion, or region-growing, process would be required, along with a classification algorithm and this condition has already been examined in a previous paper (Caelli, 1985). However, we can assume that the visual system is not that precise in creating detectors which are, a priori, orthogonal. Rather, the idea is that the visual system converges on, or "attends to", the important differentiating features by adaptation processes like those described here—being signal-dependent and network specific (see Caelli, 1988 for more details).

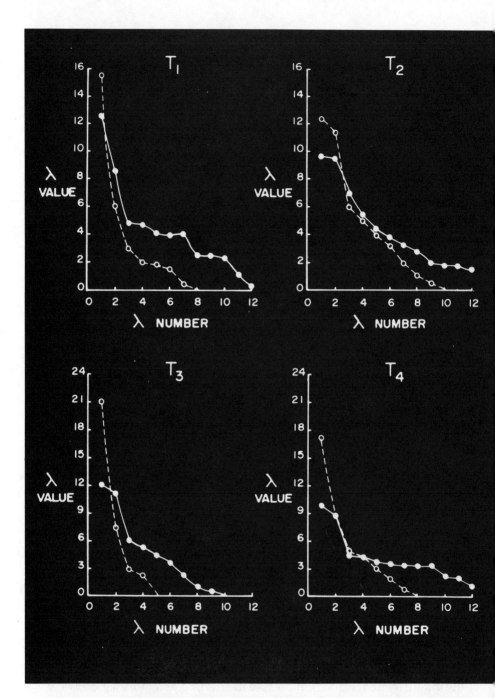

Figure 3. Eigenvalues for the nonassociative (solid lines: column 2, Figure 1) and associative (dashed lines: column 3, Figure 1) segmentation processes. T_1 to T_4 correspond to the four textures shown up column 1 of Figure 2 (from Caelli, 1988).

VISUAL PATTERN RECOGNITION

One important distinction between texture segmentation and pattern recognition is the degree to which the observer has prior knowledge of the image information to be processed. In texture segmentation no prior knowledge is assumed since the segmentation process is presumed to be purely signal-driven. Pattern recognition, on the other hand, involves the detection and localization of signals specified a priori. Of course, the degree to which prior knowledge is given determines, to a large extent, the signal matching or detection algorithms involved. However, recent findings, both in electrophysiology and human psychophysics suggest that specific types of encoding procedures may well be used in biological systems and have significant relevance to the development of computer pattern recognition systems which emulate the human observer.

Since prior knowledge is an essential ingredient of pattern recognition systems, most computer and human experiments involve two stages: (a) a learning stage, and (b) a classification stage. The learning stage involves giving the observer samples from various classes of patterns to be recognized. In the classification stage, the observer is required to classify new input signals, or indeed the old samples, in terms of the classes presented in the learning phase. Figure 4 illustrates the essential ingredients of such pattern recognition systems in the light of the decomposition and registration techniques defined in Table 1. Here the observer is presumed to construct internal prototypes for each category as a function of (a) the samples employed in the learning phase, and (b) the decomposition processes involved. These latter processes are defined by the filters d_1 to d_n, and nonlinear decision process shown in Table 1. The consequences of this multiple view are that, for a given class, the observer is presumed to have (within the encoding processes of the visual system) multiple "neural images" for each prototype. Since positional information is retained in this view by the assumption that such detectors are mapped over the full image, similarity between a given input and a given prototype would come about by the cross-correlation of the various neural image of samples relative to the various prototypes. Class membership would then be determined by the maximum magnitude of the cross-correlation vectors as illustrated in Figure 4.

Clearly, there are three assumptions of critical importance to this approach to pattern recognition: (a) that decomposition occurs, (b) that some specific registration technique is employed of a nonlinear kind, and (c) that cross-correlation between such encoded features represents the *internal comparative process*.

◉

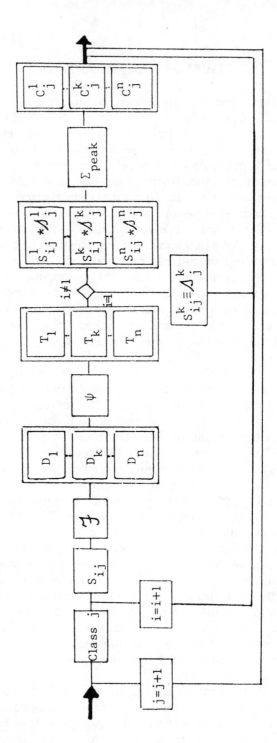

(A) *Learning Phase.*

62

Figure 4
LEGEND

s_{ij} — sample i, class j
D_k — filtered sample via detector k
T_k — transformed version of D_k
— filtering operation
S_{ij}^k — kth detector version of S_{ij}
ψ — transformation process

$*$ — cross-correlation
Σ_{peak} — sum wrt peak shift
C_j^k — prototype for class j, detector k
$\frac{k}{j}$ — reference (for alignment) signal for class j,
P_j^k — disjunctive version of C_j^k.

Prototype Formation Scheme

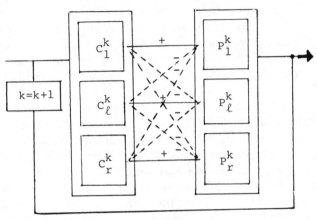

(B) *Disjunctive (Selective Attention) Stage.*

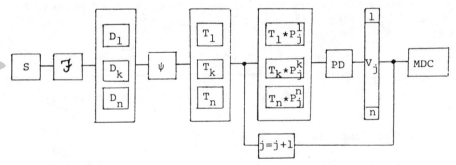

(C) *Classification Stage.*
— PD: peak detector
— MDC: minimum distance classifier
— other terms as before.

Figure 4. **(a) Shows learning stage for pattern recognition process. Here prototypes are constructed from the sample class and the decomposition/registration processes (Table 1) employed.**
(b) Shows classification stage where a new image is classified as belonging to the class whose prototypes correlates (x) best with the image (from Caelli, Bischof and Liu, 1988).

In relation to the decomposition issue, recent evidence in the behavior of observers in detecting signals in white and nonwhite noise suggests that the visual system is clearly bandpass in its response characteristics. This suggests a predominance of involvement of center/center-surround receptive fields which, when distributed over the visual field, can be emulated by bandpass filters (Caelli & Moraglia, 1986). Further to this, we (*ibid.*) have been able to show that in highly nonstationary images as, for example, more naturally occurring images, the use of edge-only information extracted from such bandpass filters seems to be more representative of what codes are particularly attended to in the matching of signals to their locations in images. This process is congruent with recent analytic results (Liu & Caelli, 1986) which show objectively that the form of the "optimal filter" for such situations is bandpass in nature (Caelli, Bischof & Liu, 1988).

These results strongly suggest that the observer compares a signal prototype with candidate signals in an image via cross-correlation. However, the form of the cross-correlation is of crucial importance. As recently shown by Caelli and Nawrot (1988), the involvement of a pre-whitener is essential for adequate modeling of human matching performance. As shown in Figure 5, the cross-correlation process must take into account the background energy of the signal in order to avoid spurious matches. This process is termed "pre-whitening" and involves a division of the direct cross-correlation function by an equivalent estimate of the background noise as defined by equations 15, 16 and 17.

$$O \leqslant N(x,y) = \frac{\displaystyle\int_\beta \int_\alpha s(\alpha,\beta)\, I(x + \alpha, y + \beta)\,d\alpha d\beta}{\left[\displaystyle\int_\beta \int_\alpha s^2(\alpha,\beta)\cdot d\alpha d\beta. \int_\beta \int_\alpha I^2(x + \alpha,\, y + \beta)\,d\alpha d\beta\right]^{1/2}} \tag{15}$$

Here the signal energy E_s is defined by:

$$E_s = \int_\beta \int_\alpha s^2(\alpha,\beta)\,d\alpha d\beta, \tag{16}$$

and the "pre-whitener" is defined by:

$$\int_\beta \int_\alpha I^2(x + \alpha, y + \beta)\,d\alpha d\beta. \tag{17}$$

From these simple examples we see that the main differences between texture segmentation and pattern recognition involves the

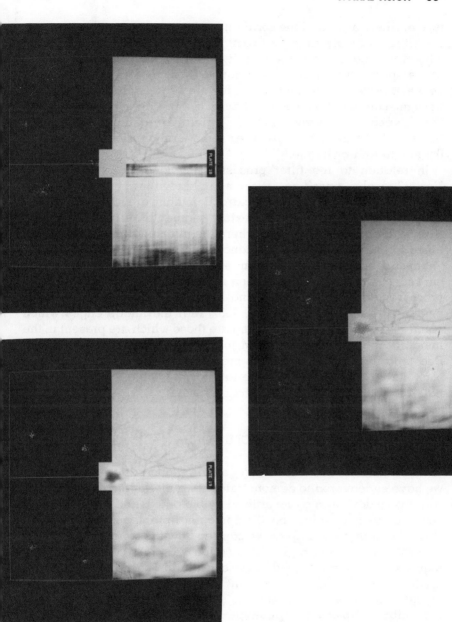

Figure 5.   Shows four tumor signals (center of plates) embedded in mammogram images (top left). Top right images correspond to the normalized cross-correlations, while the bottom images are a density plot of 160 responses for each observer as to the location of the signal in the image (from Caelli & Nawrot, 1988).

type of filters used and the correlations involved. In pattern recognition, it is relatively clear that some form of matched filtering occurs whereby evidence as to the identity ("what") and location ("where") of the specified pattern is determined by some form of correlation between patterns and images. In texture segmentation the correlations are computed between outputs of filters—directly. In this sense the latter can be regarded as more "primitive" than pattern recognition which, by definition, requires some form of prior knowledge of the signal and/or image.

In relation to the filter processes involved (correlations at a "lower" level of encoding), it seems unlikely that edge processing, edge correlations, or boundary extraction alone are sufficient to simulate the processes of human texture segmentation. This is not the case with pattern recognition where the matching process is highly tuned to image edges, or luminance gradients, in general.

These reflections upon pattern recognition are not exhaustive and our recent work has focused upon more detailed evaluations of the processes defined in Table 1 and Figure 4. For example, we have evidence (Caelli & Nagendran, 1987) that the image edges which result in reliable pattern matching are those which are present in the image at many scales (common to a wide variety of filter bands). Decomposition and registration processes (Table 1) which optimize such stable edges seem appropriate for the "attentive features" of pattern prototypes for the matching process.

CONCLUSIONS

In these examples of texture segmentation and pattern recognition, we have endeavored to demonstrate how a hierarchical set of cross-correlation devices may be utilized by biological and computational visual systems to solve specific perceptual problems in two dimensions. From these examples we can see how the concept of multiple views, associativity, and classification via cross-correlation techniques can be implemented; these convolution-based operations are readily programmable in current pipeline pixel processor technology (see Caelli & Nagendran, 1987, for an example in pattern recognition). However, it remains to be seen whether such direct hierarchical sets of correlational devices can be enumerated within the vertebrate visual system as having clearly defined physiological substrata. We have seen that without the concept of correlation the very notion of receptive field would not exist, nor would the concept of classification or recognition via feasible comparative devices. Rather than being an insidious process, however, correlation is an explicit algorithm which may or may not prove to be successful in the future—at least in this algebraic form. However, for the moment, spatial correlation—in this highly cascaded and associative form—seems a focal candidate for the encoding and processing of spatial information within the vertebrate visual system and within the context of computational vision.

REFERENCES

Adelson, E., & Bergen, J. (1985). Spatial temporal energy models for the perception of motion. *Journal of the Optical Society of America (A)*, 2(2), 284–299.

Ahmed, N., & Rao, K. (1975). *Orthogonal transforms for digital signal processing*. Berlin: Springer.

Beck, J. (1983). A theory of textural segmention. In Jack Beck (Ed.), *Human and machine vision* (1–38). New York: Academic Press.

Breitmeyer, B. (1984). *Visual masking: An integrative approach*. New York: Oxford University Press.

Caelli, T. M. (1982). On discriminating visual textures and images. *Perception and Psychophysics, 31,* 149–159.

Caelli, T. M. (1983). Energy processing and coding factors in texture discrimination and image processing. *Perception and Psychophysics, 34*(4), 349–355.

Caelli, T. M. (1985). Three processing characteristics of visual texture segmentation. *Spatial Vision, 1*(1), 19–30.

Caelli, T. M. (1988). An adaptive computational model for texture segmentation. *IEEE Transactions on Systems, Man and Cybernetics,* (in press).

Caelli, T. M., Bischof, W. F., & Liu, Z.-Q. (1988). Filter-based models for pattern classification. *Pattern Recognition,* (in press).

Caelli, T. M., Brettel, H., Rentschler, I., & Hilz, R. (1983). Discrimination thresholds in the two-dimensional spatial frequency domain. *Vision Research, 23*(2), 129–133.

Caelli, T. M., & Hübner, M. (1983). On the efficient two-dimensional energy coding characteristics of spatial vision. *Vision Research, 23*(10), 1053–1055.

Caelli, T. M., & Julesz, B. (1978). On perceptual analyzers underlying visual texture discrimination: Part 1. *Biological Cybernetics, 28,* 167–175.

Caelli, T. M., & Moraglia, G. (1985). On the detection of Gabor signals and discrimination of Gabor textures. *Vision Research, 25*(5), 671–684.

Caelli, T. M., & Moraglia, G. (1986). On the detection of signals embedded in natural scenes. *Perception and Psychophysics, 39*(2), 87–95.

Caelli, T. M., & Nagendran, S. (1987). Fast edge-only matching techniques for robot pattern recognition. *Computer Vision, Graphics, and Image Processing, 39,* 131–143.

Caelli, T. M., & Nawrot, M. (1986). *On the localization of signals in images.* Manuscript submitted for publication.

Caelli, T. M., & Nawrot, M. (1988). On the localization of signals in images.

Journal of the Optical Society of America A: Optics and Image Science, 4(12), 2274–2280.

Daugman, J. (1983). Six formal properties of two-dimensional anisotropic visual filters: Structural principles and frequency/orientation selectivity. *IEEE Transactions on Systems, Man, and Cybernetics, SMC-13,* 882–888.

Fukushima, K. (1984). A hierarchical neural network model for associative memory. *Biological Cybernetics, 50,* 105–113.

Grossberg, S. (1982). *Studies of the mind and brain: Neural principles of learning, perception, development, cognition and motor control.* Boston: Reidel.

Harvey, L., & Gervais, M. (1978). Visual texture perception and Fourier analysis. *Perception and Psychophysics, 24,* 534–542.

Hebb, H. (1949). *The organization of behavior.* New York: John Wiley.

Hubel, D., & Wiesel, T. (1977). Functional architecture of macaque monkey visual cortex. *Proceedings of the Royal Society of London, B, 198,* 1–59.

Hummel, R., & Zucker, S. (1983). On the foundations of relaxation labelling processes. *IEEE: PAMI, 3,* 267–287.

Julesz, B. (1981). Textons, the elements of texture perception and their interactions. *Nature, 290,* 91–97.

Julesz, B. (1984). Toward an axiomatic theory of preattentive vision. In G. Edelman, W. Gall, & W. M. Cowan (Eds.), *Dynamic aspects of neocortical function.* (pp. 585–611). New York: John Wiley.

Kohonen, T. (1977). Associative memory - A system theoretic approach. New York: Springer.

Liu, Z. Q. & Caelli, T. M. (1986). An adaptive prefilter matching technique for detection of signals in non-white noise. Applied Optics, 25, 1622–1626.

Marr, D., & Mildreth, E. (1980). Theory of edge detection. *Proceedings of the Royal Society of London, B, 207,* 187–217.

Richards, W. (1979). Quantifying sensory chananels: Generalizing colorimetry to orientation and texture, touch and tones. *Sensory Processes, 3,* 207–229.

Rodieck, R. (1973). *The vertebrate retina: Principles of structure and function.* San Francisco: Freeman Press.

Rosenfeld, A., & Kak, A. (1976). *Digital picture processing.* New York: Academic Press.

Torre, V., & Poggio, T. (1986). On edge detection. *IEEE Transactions on Pattern Analysis and Machine Intelligence, PAMI-8(2),* 147–163.

Wilson, H., & Bergen, J. (1979). A four mechanism model for spatial vision. *Vision Research, 19,* 19–32.

Witkin, A. (1983). Scale-space filtering. *Proceedings of the Eighth International Joint Conference on Artificial Intelligence, 2,* 1019–1022.

Yuille, A., & Poggio, T. (1986). Scaling theorems for zero crossings. *IEEE Transactions on Pattern Analysis and Machine Intelligence, PAMI-8(1),* 15–25.

4
The Measurement of Binocular Disparity

Michael R. M. Jenkin
Allan D. Jepson

University of Toronto

INTRODUCTION

Current theories of stereopsis involve three distinct stages (Marr, 1982; Marr & Poggio, 1976, 1977; Grimson, 1981). First, the two images of a stereo pair are processed separately to extract monocular features. One common choice of feature is the presence of a zero-crossing in a bandpassed version of the image. Second, the monocular features in one image are matched with corresponding features found in the other image. In practice this second stage cannot be expected to produce only the correct matches, and a third stage must be considered in order to remove the incorrect matches ("false targets"). There are therefore three main issues in the design of such a traditional algorithm for stereopsis, namely, (a) the choice of image features, (b) the choice of matching criteria, and (c) the way false targets are avoided or eliminated.

In this paper we introduce a different approach. We propose that symbolic features should not be extracted from the monocular images in the first stage of processing. Rather we examine a technique for measuring the local phase difference between the two images. This essentially combines the first two stages of the traditional approach. A third stage is still necessary to remove false targets (which now arise as errors in the relative-phase measurements by roughly an integer multiple of 2π). The benefits of the new approach are discussed below.

There are several factors involved in the choice of suitable image primitives. We consider these factors from the traditional viewpoint of extracting symbolic monocular features. Of primary importance is that an extracted feature can be expected to correspond to a particular property of a surface in the scene being viewed, and that this surface property is likely to produce the same type of feature in both images. In other words, we wish to use features that can be expected to produce reliable information about *surface structure* once they have been correctly matched. It is also important that matches should occur sufficiently often to provide a fairly dense description of the disparity. While algorithms are available for filling in expected values of disparity given only sparse data (Terzopoulos, 1983), it is obviously preferable to have denser data. Finally, the choice of image features strongly effects the options available for obtaining matches and eliminating false targets. This has been expressed very clearly by Marr (1982, p. 127):

> The basic problem to be overcome in binocular fusion is the elimination or avoidance of false targets, and its difficulty is determined by two factors: the abundance of matchable features in an image and the disparity range over which the matches are sought. If a feature occurs only rarely in an image, the search for a match can cover quite a large disparity range before false targets are encountered, but if the feature is a common one or the criteria for a match are loose, false targets can occur within quite small disparities.

In brief, the features should (a) correspond to surface properties, (b) produce matches that are fairly dense, and (c) have distributions over a typical image such that false matches can be relatively easily avoided or eliminated.

The constraints on the density and on the ease of eliminating false targets are in direct opposition. In particular, the number of possible matches in a given region increases exponentially with the density of a given symbolic feature. Therefore, the problem of finding the correct match can be expected to become rapidly more difficult as the density grows.

The approach we take here does not rely on symbolic features to measure the disparity, and therefore we avoid this density/interpretation difficulty trade-off. Instead of using symbolic features, the entire output of various bandpass channels (applied separately to each image of the stereo pair) is used. Operators are constructed to use this output to compute the local phase difference between the two images. The current operators rely on a nonlinear feedback control which is derived from a signal-processing tool called the *phase-lock loop* (see Gardner (1979) for an excellent survey).

We are not alone in suggesting that the construction of a denser disparity representation should be considered. For example, Mayhew and Frisby (1981) have shown that zero-crossings alone are not enough to account for human stereopsis. They suggest that (at least) peaks and zero-crossings should be extracted as monocular primitives. A similar suggestion is made by G. Poggio and T. Poggio (1984, p. 382):

> . . . one should not overlook the possibility that the primitive measurements used for stereo matching may be far more dense and specific than just location and parameters of edges. Many different measurements at each point of the image, for instance various types of derivatives, may provide a rich and robust description of each image, suitable for matching.

However, we are unaware of any other work suggesting that the initial stage of symbolic feature extraction can and should be discarded. This idea is consistent with the general approach of using direct measurement techniques to extract image primitives such as orientation and velocity (see, for example, Fleet & Jepson (1986)).

In the following sections, we present the basic theory of the measurement approach. The actual use of these measurements in a complete algorithm for stereopsis is beyond the scope of the current paper. However, it is important to note here how these disparity measurements satisfy the constraints discussed above, which suggests that a complete algorithm is possible.

First we note that the local structure of bandpassed versions of an image tend to correspond to properties of objects in the scene, and not on the details of the imaging process. In fact, zero-crossings of such signals were chosen for precisely this reason (see Marr, 1982). Here we are using the entire local structure of the signal to extract relative-phase information. So the case for zero-crossings carries over to the disparity measurements.

The disparity measurements can also be expected to be dense since the operators are designed to be insensitive to the particular form in any one image alone. Roughly speaking, only two nontrivial bandpass signals are needed to extract relative-phase. The positions of features, such as zero-crossings, in either image separately is irrelevant.

Finally, we consider the third constraint, namely, that false targets can be avoided. The key here is to note that relative-phase information can, of course, only be computed modulo 2π. In particular, false targets arise in a phase measurement approach when an inappropriate integer multiple of 2π is used in the computed relative-phase. The key to avoiding or eliminating many false targets is to note that a good prediction for the distance between possible matches is given by the wavelength associated with the peak frequency of the bandpass filter being used. This is precisely the same information used by Marr and Poggio (1977) (about the expected spacing of zero-crossings) to develop a coarse to fine matching strategy (also see Grimson (1981)). For our purposes here we need only state that a similar strategy could be used on the results of the relative-phase measurements provided by our approach. The development of such a strategy is the subject of current research.

We conclude that relative-phase measurements have the potential of doing a significantly better job than the extraction and matching of symbolic features. They should provide denser information with no extra cost in removing false targets. The noise sensitive task of extracting symbolic features at an early stage of processing would be eliminated.

Later we will consider some novel consequences of the technique that are a result of the nonlinear feedback control. The operators are shown to be capable of tracking disparities that vary slowly in time. In addition they exhibit an interesting form of hysteresis (i.e. history-dependent behavior). Finally, some possible connections to biological vison systems will be discussed.

THE BASIC APPROACH

In this section we present the basic computational approach used to measure local disparity information. To illustrate the proposed technique in its simplest form we consider one-dimensional sinusoidal signals here. We note, however, that only relatively minor modifications of this basic technique will be needed in order to deal with more general two-dimensional images.

In particular, we let $I_l(x)$ and $I_r(x)$ be the left and right "images", where

$$I_l(x) = A \sin(\omega_l x + \theta_l); A, \omega_l > 0, \tag{2.1a}$$

$$I_r(x) = B \sin(\omega_r x + \theta_r); B, \omega_r > 0. \tag{2.1b}$$

Note that since different perspectives in the left and right views can alter the spatial frequencies of the observed patterns, we do not assume that $\omega_l = \omega_r$. However, we do assume that $\omega_l \approx \omega_r$ (i.e. $|\omega_l - \omega_r|/(\omega_l + \omega_r) << 1$), and in practice we ensure this by considering two bandpassed versions of the left and right raw images.

One way to define a local phase difference between two bandpass signals is to consider matching "features" such as peaks and zero-crossings. For I_l and I_r as above, this basically amounts to matching the arguments of the two sinusoids, modulo 2π. Thus, we define the local phase difference to be

$$\phi(x) = \lfloor(\omega_l - \omega_r)x + (\theta_l - \theta_r)\rfloor \in [-\pi, \pi). \qquad (2.2)$$

Here $\lfloor\theta\rfloor$ denotes the principal part of the angle θ, which is obtained by adding an integral multiple of 2π to θ so that the result lies in the interval $[-\pi, \pi)$. Note that this mod operation produces a discontinuous function, $\phi(x)$, whenever $\omega_l \neq \omega_r$. Moreover the discontinuities correspond to boundaries of intervals in x where the matching of peaks and zero-crossings between the two images is one-to-one.

The local disparity, $d(x)$, is the distance the images must be shifted with respect to each other so that I_l and I_r agree up to a multiplicative constant (B/A in the above example). This definition leads to the following form for the local disparity,

$$d(x) \equiv \frac{1}{\bar{\omega}} \phi(x), \quad \bar{\omega} \equiv 1/2 \, (\omega_l + \omega_r). \qquad (2.3)$$

With this definition it follows that the left image shifted to the right by $\frac{1}{2} d(x)$, namely $I_l(x - \frac{1}{2} d(x))$, is a constant times the right image shifted by the same amount to the left, namely $I_r(x + \frac{1}{2} d(x))$. Positive disparity, therefore, corresponds to objects that are *further* from the fixation point of the two eyes.

In order to extract $\phi(x)$ we consider the point-by-point multiplication of the left and right images, that is

$$I_l(x)I_r(x) = AB \sin(\omega_l x + \theta_l) \sin(\omega_r x + \theta_r)$$

$$= \frac{AB}{2} [\cos((\omega_l - \omega_r)x + (\theta_l - \theta_r)) - \cos((\omega_l + \omega_r)x + (\theta_l + \theta_r))].$$

By low-pass filtering this product with a filter, L, having a high frequency cut-off below $\omega_l + \omega_r$, we obtain

$$P(x) \equiv L^*(I_l(x)I_r(x)) = \frac{ABK}{2} \cos((\omega_l - \omega_r)x + (\theta_l - \theta_r)) \qquad (2.4)$$

$$= \kappa \cos(\phi(x)), \quad \kappa \equiv \frac{ABK}{2}.$$

Here $\phi(x)$ is as in (2.2) and $K = K(\omega_l - \omega_r)$ is the sensitivity of the low-pass filter to the frequency $\omega_l - \omega_r$. This result is encouraging since

$P(x)$ depends only on the desired local phase difference and a scale factor involving the product of the amplitudes. However a *local* technique is needed to disambiguate the amplitude and the relative phase information inherent in $P(x)$. The technique must be local since we cannot expect bandpassed images to be well approximated by (2.1a,b) over intervals of roughly the length of one wavelength of $P(x)$,

that is, over lengths $\dfrac{2\pi}{|\omega_l - \omega_r|}$. Therefore it is not possible to estimate the scale factor by examining the behavior of $P(x)$ over one of its wavelengths. Instead it is necessary to obtain an estimate for $\phi(x)$ based on information available over no more than a few wavelengths of the base frequency $\bar{\omega}$ $\left(\text{i.e. } \Delta x \approx \dfrac{2\pi}{\bar{\omega}}\right)$.

A suitable local technique can be obtained by the inclusion of a specific shift in one (or both) of the images before the pointwise product is calculated. For example, we define

$$P(x, s) \equiv L^*[I_l(x - \tfrac{1}{2}s)\, I_r(x + \tfrac{1}{2}s)] = k\cos(\bar{\omega}s - \phi(x)), \tag{2.5}$$

with $\bar{\omega}$ as in (2.3). Now, to obtain an approximation for ϕ near a given point x_0, the value of s is adjusted so that a particular local feature of $P(x, s)$ occurs at $x = x_0$. Suitable local features of P are, of course, features that are independent of the product AB, and include zero-crossings and local extrema. Here we chose to avoid seeking local extrema of $P(x_0, s)$, which can be fairly sensitive to noise. Instead we consider a simple feedback control which provide a value $s_\infty(x_0)$ such that

$$P(x_0, s_\infty(x_0)) = 0, \tag{2.6a}$$

$$\frac{\partial P}{\partial s}(x_0, s_\infty(x_0)) > 0. \tag{2.6b}$$

For the moment assume that such a $s_\infty(x_0)$ is computed. Then we see from (2.5) that (2.6) is satisfied if and only if

$$\lfloor \bar{\omega} s_\infty(x_0) - \phi(x_0) \rfloor = -\frac{\pi}{2}.$$

Therefore the local relative phase difference is given by

$$\phi(x_0) = \lfloor \bar{\omega} s_\infty(x_0) + \frac{\pi}{2} \rfloor, \tag{2.7}$$

which can be computed given $s_\infty(x_0)$ and the mean frequency $\bar{\omega}$.

The basic idea behind the control loop is best illustrated with the following simple scheme. Consider $s(t;x_0)$, which is defined to be the solution of the differential equation

$$\frac{ds}{dt} = -P(x_0, s(t;x_0)), \ t > 0, \tag{2.8a}$$

$$s(0;x_0) = s_0. \tag{2.8b}$$

Here s_0 is some initial guess for the local phase difference at x_0. The meaning of (2.8a) is made clear in Figure 1, where the arrows indicate the size and direction of $\frac{ds}{dt}$. Roughly speaking, equation (2.8a) simply states that if $P(x_0, s)$ is positive then s should be decreased, and if $P(x_0, s)$ is negative than s should be increased. Except for the unstable situations where the initial guess $s(0;x_0)$ precisely satisifies (2.6a) but with the inequality (2.6b) reversed, the solution $s(t;x_0)$ of (2.8) converges to a solution of (2.6) as $t \to \infty$. We refer to this limit as $s_\infty(x_0)$.

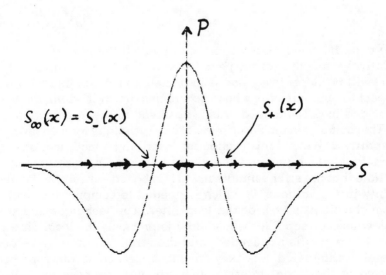

Figure 1.   Response of $P(0, s)$, and direction of flow, given a step-edge input.

If the mean frequency $\bar{\omega}$ is known, then the relative phase can be computed from (2.7) and the limit, $s_\infty(x_0) = \lim_{t \to \infty} s\,(t;x_0)$. In addition it follows from (2.3) and (2.7) that the local disparity $d(x_0)$ satisfies

$$d(x_0) = s_\infty(x_0) + \frac{\pi}{2\bar{\omega}} = s_\infty(x_0) + \frac{\bar{\lambda}}{4}. \tag{2.9}$$

In practice $\bar{\omega}$ might be approximated by ω_0, the peak frequency of the bandpass filter used to compute $I_l(x)$ and $I_r(x)$. The error in the disparity calculated using (2.9), with $\bar{\omega}$ replaced by ω_0, is then

$$\Delta d = \left(\frac{1}{\omega_0} - \frac{1}{\bar{\omega}} \right) \frac{\pi}{2} \approx \frac{1}{4} \left(\frac{\bar{\omega} - \omega_0}{\omega_0} \right) \lambda_0. \tag{2.10}$$

where $\lambda_0 = \dfrac{2\pi}{\omega_0}$ is the wavelength corresponding to the peak frequency, and $\dfrac{\bar{\omega} - \omega_0}{\omega_0}$ is the relative displacement of $\bar{\omega}$ from the peak frequency. For example, if the bandwidth of the bandpass filter is less than an octave, then the relative displacement of $\bar{\omega}$ is bounded by $\frac{1}{2}$, and (2.10) implies

$$|\Delta d| \lesssim \frac{1}{8} \lambda_0. \tag{2.11}$$

We note, however, that for images in which the power spectrum is relatively smooth near the peak frequency ω_0, the appropriate value of $\bar{\omega}$ will be closer to ω_0 than is indicated above. In this case we can expect to obtain results whose accuracies are well within the bound provided in (2.11). Indeed, this is observed in the next section.

This completes the description of one technique for measuring the disparity d. We end this section by describing a modification of this technique which provides the disparity to a greater accuracy.

Note that the error estimate in (2.11) is based entirely on the uncertainty in the value of $\bar{\omega}$, which is needed to compute the disparity from the limit, $s_\infty(x_0)$, of the loop filter. The technique illustrated below uses a loop filter whose limit is precisely the local disparity. The value of $\bar{\omega}$ is not needed, and the error produced by its use is thereby eliminated. The key to this approach is an appropriate choice of the initial filtering. In particular, suppose the left and right images I_l and I_r are obtained using a zero-phase and a $-\dfrac{\pi}{2}$

-phase filter, respectively. For example, sine and cosine Gabor filters could be used (see (3.8)). Then the right image becomes

$$I_r(x) = A \sin(\omega_r x + \theta_r - \pi/2). \tag{2.12}$$

Forming $P(x, s)$ as in (2.5a), with $I_l(x)$ as in (2.1a), we find

$$P(x, s) = k \sin(\bar{\omega}s - \phi(x)). \tag{2.13}$$

Now the control scheme (2.8) provides $s(t;x_0)$ such that $s(t;x_0) \to s_\infty(x_0)$ as $t \to \infty$, where $s_\infty(x_0)$ satisfies

$$\lfloor \bar{\omega}s_\infty(x_0) - \phi(x_0) \rfloor = 0. \tag{2.14}$$

Finally, we see that if $\bar{\omega}s_\infty(x_0) \in [-\pi, \pi)$ then (2.2), (2.3), and (2.14) imply that $s_\infty(x_0) = d(x_0)$. Therefore the control scheme converges to the local disparity, and the precise value of $\bar{\omega}$ is not needed.

PHASE-SENSITIVE DEMONS

The general form of the phase measurement scheme discussed above is given in Figure 2. We summarize this scheme below. The raw images, I_l^0 and I_r^0, are first sent through bandpass filters B_l and B_r, respectively, to produce

$$I_v(x) = \int_{-\infty}^{\infty} B_v(x - z)I_v^0(z)dz, \text{ for } v = l, r. \tag{3.1}$$

The left and right bandpass images are shifted by s, and the phase detector,

$$P(x,s) = \int_{-\infty}^{\infty} L(x - z) (I_l(z - s/2)I_r(z + s/2))dz, \tag{3.2}$$

is applied. Here L is a low-pass filter. The response of the phase detector is fed into the loop filter, which for now we take to be

$$\frac{ds}{dt} = -P(x,s), \text{ for } t > 0. \tag{3.3a}$$

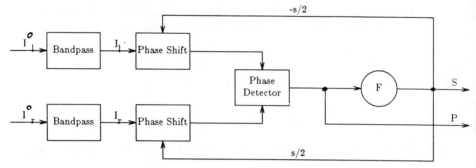

Figure 2.   Circuit diagram for the phase-sensitive demon.

It is also convenient to introduce an initial guess, say

$$s(0; x) = s_0. \tag{3.3b}$$

The value of s computed in (3.3) is fed back to the shift module. The loop is said to be *locked* when $P(x, s) = 0$, and in this case we set $s_\infty(x) = s$. We refer to this mechanism as a *phase-sensitive demon*.

We imagine the initial stereo images decomposed into a multiple spatial scale representation, perhaps through the use of a Burt pyramid (Burt, 1981), or the DOLP transform (Crowley & Stern, 1984). Also there is a similar hierarchical scheme that can be used to construct Gabor-like receptive fields (Fleet & Jepson, 1986). The filters B_l and B_r are taken to be members of such a family, and we assume that their output is resampled appropriately. For each spatial scale there is a demon applied at each sample point x and, like all good demons, they are taken to be independent of other demons at different x's or different scales. In particular, we emphasize that the demons are simply collecting *local* disparity information. A globally consistent interpretation of the computed local disparities is the job of either a subsequent level of processing or a network connecting demons to their neighbors. Either way we do not discuss globally consistent interpretations in this paper.

In this section we consider the application of several different demons, each of the above form, to the step-edge input

$$I_l^0(x + d/2) = I_r^0(x - d/2) = \begin{cases} 1, & \text{for } x > 0; \\ 0, & \text{for } x \le 0. \end{cases} \tag{3.4}$$

This step-edge has disparity d, as can be seen from the discussion immediately after equation (2.3).

The first specific form of demon we consider is described next. For simplicity we take $B_l = B_r = B$, with $B(x)$ the one-dimensional difference of Gaussians (DOG) filter given by

$$B(x) = \frac{1}{\sqrt{2\pi}\,\sigma_c} e^{\frac{-x^2}{2\sigma_c^2}} - \frac{1}{\sqrt{2\pi}\,\sigma_s} e^{\frac{-x^2}{2\sigma_s^2}} \tag{3.5}$$

It is useful to use σ_c and $\rho \equiv \sigma_s/\sigma_c > 1$ as the independent parameters for $B(x)$. Then ρ controls the bandwith of the filter $B(x)$ and σ_c can be used to adjust the scale. The peak frequency of $B(x)$ is given by

$$\omega_0 = \frac{2}{\sigma_c} \left(\frac{\ln(\rho)}{\rho^2 - 1} \right)^{1/2} ,$$

and $B(x)$ passes the frequency $\omega_0/2$ to a significant degree. Recall that the low-pass filter L, which occurs in the phase detection stage, must have a high frequency cut-off below $\omega_l + \omega_r$. For our purposes here we can take w_l, $\omega_r \gtrsim \omega_0/2$, and therefore the cut-off of L must be below ω_0. For simplicity we take L to be the Gaussian

$$L(x) = \frac{1}{\sqrt{2\pi}\,\sigma_l} e^{\frac{-x^2}{2\sigma_l^2}} , \tag{3.6a}$$

which has the half amplitude cut-off

$$\omega_c = \frac{\sqrt{2\ln 2}}{\sigma_l} . \tag{3.6b}$$

Therefore the condition on the cut-off frequency for the low-pass filter (i.e. $\omega_c < \omega_0$) becomes

$$\sigma_l > \sigma_c \left(\frac{(\rho^2 - 1)\ln(2)}{2\ln(\rho)} \right)^{1/2} . \tag{3.7}$$

For example, it is common to take p to be about 1.5 (see Marr & Poggio (1977)), for which (3.7) provides $\sigma_l > \sigma_c$.

The response of the phase detector for the step-edge stimulus (3.4) is given in Figure 3 for the disparity $d = 0$. Other values of the disparity produce precisely the same surface shape, but the surface is translated so that the peak at $(x, s) = (0, 0)$ is moved to $(x, s) = (0, d)$. The flow defined by the loop filter (3.3a) is sketched in Figure 4 (cf. Figure 1). In particular, for any starting point s_0 below the line $s = s_+(x)$, the loop causes s to converge to the zero of $P(x, s)$ at which

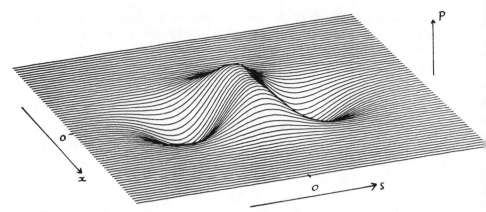

Figure 3. Response of $P(x, s)$ for a step-edge input having zero disparity.

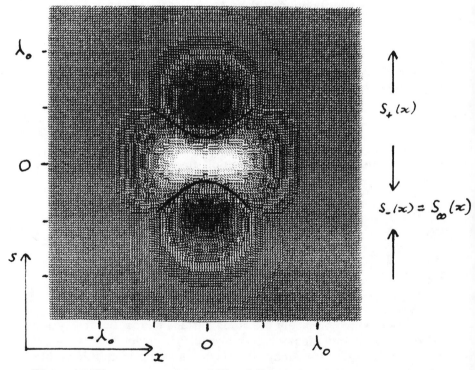

Figure 4. The zero-crossings of $P(x, s)$ (heavy lines). Here lighter (darker) points indicate positive (negative) values of P.

(2.5b) is satisfied, that is to $s_\infty(x) = s_-(x) < 0$. However, for starting values taken larger than $s_+(x)$, the loop diverges.

There is a complementary way to view this convergence behavior. Instead of thinking of the initial guess as varying, with a fixed stimulus disparity, it is more useful to think of the initial guess as fixed and consider the demon applied to various stimulus disparities. Because a change in the stimulus disparity simply shifts the response of the phase detector up or down the (x, s) plane, the two views are equivalent (at least for the loop filter considered in (3.3a)). The results stated in the previous paragraph indicate that the loop defined by (3.3a) and (3.3b), that is with the *fixed* initial guess s_0, converges whenever the disparity, d, of the stimulus satisfies

$$d < s_+(x) + s_0.$$

For larger values of the stimulus disparity, the demon diverges.

In practice, however, the convergence behavior is not quite this simple. First, in the absence of a signal in the bandpassed images $P(x, s)$ is identically zero. If the demon ever reaches such a point it stays there, that is, it locks. Since these points don't have any significant meaning as far as the underlying signal is concerned, we refer to this event as a *false lock*. For the step-edge stimulus of disparity d, the implemented demon will exhibit false locks whenever $|d|$ or $|x|$ is too large. Secondly, the discrete implementation of the loop filter can have much more complicated dynamics than the differential equation (3.3). This can cause the implemented demon to fail to converge for initial guesses that are within the convergence region of the demon discussed above. However, the basic behavior will be preserved, namely that the demon will converge for a sufficiently good initial guess. In particular, there exist bounds $\Delta d_+ > 0$ such that the implemented demon converges whenever the stimulus disparity satisfies

$$d \in (s_0 - \Delta d_-, s_0 + \Delta d_+). \tag{3.8}$$

This interval is called the *lock-in region*.

Several of the difficulties noted above can be dealt with through appropriate modifications of the loop filter (3.3). For example, the number of false locks can be reduced (but not eliminated) by replacing (3.3a) with

$$\frac{ds}{dt} + \epsilon[s - s_0] = -P(x,s), \tag{3.9}$$

where $\varepsilon > 0$ is a small constant and s_0 is called the *home* disparity value. The additional term involving ε causes the loop filter to lock on $s = s_0$ in the absence of any contribution from $P(x, s)$. Therefore regions in which $P(x, s)$ vanishes will not produce false locks except in cases for which P vanishes at $s = s_0$. The problem with the divergence of the loop filter (3.3a) for disparities larger than $s_+(x)$ can also be partially avoided. To see how, recall that the phase $\phi(x)$ can only be computed to within an integral multiple of 2π. By inspection of (2.7) we see that it is also natural to limit the range of s to lie in the interval $[-\pi/\omega_0, \pi/\omega_0]$. It is convenient to use $\lfloor s \rfloor$ to denote this principal part of s. The loop filter (3.3a) is then modified by making the shift "wrap-around", that is

$$\frac{ds}{dt} = -P(x,s), \text{ for } t > 0, \tag{3.10a}$$

$$s \leftarrow \lfloor s \rfloor, \text{ for } t > 0, \tag{3.10b}$$

$$s(0;x) = s_0. \tag{3.10c}$$

If s ever leaves the interval $[-\pi/\omega_0, \pi/\omega_0)$ its value is reset back in this interval by (3.10b). It is now easy to show that, for all disparities in the range

$$d \in [s_0 - \pi/\omega_0, s_0 + \pi/\omega_0) \tag{3.11}$$

(except for the one unstable point $d = s_0 + s_+(x)$), the loop filter (3.10) drives s to the desired point $s_\infty(x) = s_-(x)$. We emphasize, however, that this strong convergence property need not be satisfied for other loop filters involving wrap-around, or even for discrete time-step approximations to (3.10).

The lock-in regions for the two loop filters (3.3) and (3.10) are sketched in Figure 4. Notice that neither region contains the other. A second difference in the performance of the two demons is that the mod operation (3.10b) anchors the demon to the interval of disparity $[-\pi/\omega_0, \pi/\omega_0)$. This demon will never correctly lock to a step-edge stimulus having a disparity outside of this interval (although false locks are possible). On the other hand the loop filter in (3.3) provides no such anchor and, as we see below, the demon is free to track a slowly varying disparity well outside its lock-in region.

Next we briefly consider the accuracy of the computed disparity for the case of the step-edge input with $d = 0$ (other values of d will be computed to the same absolute error). We assume that the initial bandpass and the lowpass filters are as described in (3.5) and (3.6) above. We take the DOG parameters to be $\sigma_c = 1$, $p = 1.5$, and we set

$\sigma_l = 1.1\sigma_c$. If the stimulus disparity $d = 0$ is within the lock-in region for the loop filter used, then the loop converges to the limit $s_\infty(x) = s_-(x)$. For the demon at $x = 0$, this limit is given by $s_\infty(0) \approx -1.36$. The disparity can then be approximated using the peak frequency $\omega_0 = 1.14$ in (2.9). This provides $d(0) \approx 0.02$, which is well within the tolerance given in (2.11) (in fact, it represents a 0.3% error relative to the the wavelength λ_0). The accuracy is due, at least in part, to the fact that the step-edge stimulus has a smooth power spectrum, and therefore ω_0 is a good approximation to the average frequency in the bandpassed stimulus. However, as x is varied away from zero the accuracy decreases, so that at $x = \pm\pi/2$ the computed disparity has an error of roughly 25% relative to λ_0 (see Figure 4).

We end this section with a sketch of a phase-sensitive demon that uses initial filters B_l and B_r having a relative phase shift of $\pi/2$. As we show below, the resulting demon is much less sensitive to the position of the step-edge than the demon discussed above. The initial filtering is provided by the (slightly modified) Gabor pair

$$B_l(x) = \sin(\omega_0 x)\ \frac{1}{\sqrt{2\pi}\ \sigma_0}\ e^{\frac{-x^2}{2\sigma_0^2}}, \tag{3.12a}$$

$$B_r(x) = [\cos(\omega_0 x) - \gamma]\ \frac{1}{\sqrt{2\pi}\ \sigma_0}\ e^{\frac{-x^2}{2\sigma_0^2}}. \tag{3.12b}$$

Here the constant γ is set to $e^{-\omega_0^2\sigma_c^2/2}$, which is precisely the value needed to make the filter $B_r(x)$ bandpass. This pair is conveniently characterized by the peak frequency, ω_0, and the ratio of the corresponding wavelength to the diameter of the Gaussian, namely

$\beta \equiv \dfrac{\pi}{\omega_0\sigma_0}$. The parameter β is related to the bandwidth of the Gabor

filters, and the condition that this bandwidth is less than an octave is

$$\beta < \frac{\pi}{2\sqrt{2\ln 2}} \approx 1.33. \tag{3.13a}$$

Given such as β, we take $L(x)$ to be the Gaussian given in (3.6a) with the high-frequency cut-off, ω_c (see 3.6b)), taken to be ω_0 or smaller. This is equivalent to requiring that

$$\sigma_l > \frac{\sqrt{2\ln 2}}{\omega_0}. \tag{3.13b}$$

Finally, the remainder of the demon is taken from (3.2) (the phase-detector) and (3.3) (the loop filter without wrap-around).

In order to easily compare the results with the previous demon we chose $\omega_0 = 1.139$, which is the same peak frequency as before. Then (3.13b) becomes $\sigma_l > 1.0$, and we chose this value to be two. The response, $P(x, s)$, of the phase detector for the step-edge stimulus (3.4), with disparity $d = 0$, is given in Figure 5. Again, when d is varied the response profile of the phase detector is simply translated along the s-axis in the (x, s)-plane. The flow defined by the loop filter is sketched in Figure 6. For an initial guess $s_0 = 0$, the lock-in region of the demon is sketched in Figure 6. For an initial guess $s_0 = 0$, the lock-in region of the demon is

$$d \in (s_-(x), s_+(x)).$$

Here $s_\pm(x)$ are two curves of points sketched in Figure 6 at which $P(x, s) = 0$. These values are roughly given by

$$d \in (-3.5, 3.5) \approx (-1.3\lambda_0, 1.3\lambda_0).$$

For values of d within this lock-in region, the loop filter converges, and $\lim_{t \to \infty} s(t;0) \equiv s_\infty(0) = d$. In particular, the initial Gabor pair cause

the loop filter to converge directly to the desired disparity, which eliminates the need for approximating the disparity through the use

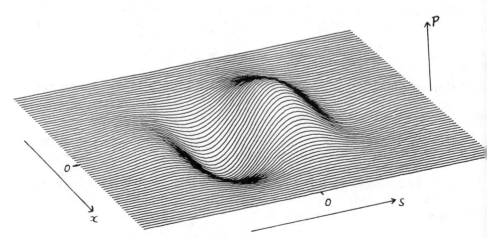

Figure 5.   Response of $P(x, s)$ constructed using Gabor filters.

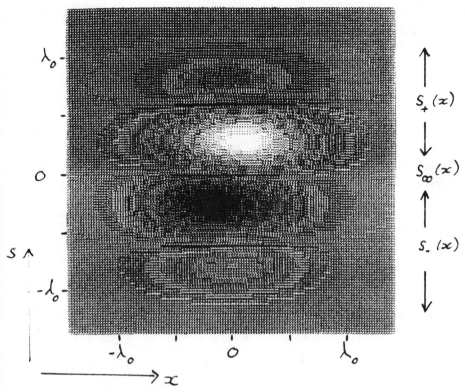

Figure 6. The zero-crossings of $P(x, s)$ (heavy lines) constructed using Gabor filters. Here lighter (darker) points indicate positive (negative) values of P.

of equation (2.9) and an estimate for $\bar{\omega}$. Furthermore, the limit is accurate to within 3% of λ_0 for *all* values of x within $\lambda_0/2$ of zero. This is a striking contrast to the performance of the demon based on DOG filtering, and is apparent in the difference between Figures 4 and 6.

TRACKING BEHAVIOR AND HYSTERESIS

We saw in the previous section that a particular demon, with a given initial guess, can have a finite lock-in region. Here we show that demons which are not explicitly anchored to a particular disparity range (i.e. which use, for example, a loop filter without wrap-around) can be pulled to disparities well beyond the initial lock-in region. In particular, we show that if the stimulus disparity is slowly varied in time, a demon can track this disparity outside of this region. The response of the demon is therefore seen to be history dependent, that is, it exhibits *hysteresis*.

In order to keep the necessary analysis at a minimum, we consider left and right images of the form (cf. (2.1a) and (2.12))

$$I_l(x,t) = A \sin(\omega_l x + \theta_l(t)); \, A, \omega_l > 0, \tag{4.1a}$$

$$I_r(x,t) = B \sin\left(\omega_r x + \theta_r(t) - \frac{\pi}{2} \right); \, B, \omega_r > 0. \tag{4.1b}$$

Here I_l and I_r can be thought of as idealizations of bandpassed left and right views of a textured surface. The phase shift of $-\pi/2$ in I_r could be introduced through the use of a pair of Gabor filters such as those described in (3.9). The relative phase difference of the left and right images is defined earlier ignoring the phase shift introduced by B_r. That is, we define

$$\phi(x, t) = [(\omega_l - \omega_r)x + \theta_l(t) - \theta_r(t)]. \tag{4.2}$$

In particular, the local phase difference ϕ depends on both x and t.

In this section we consider the following simple temporal behavior for ϕ:

$$\phi(x,t) - \phi(x,0) = \begin{bmatrix} 0 & \text{for } t \leq 0; \\ vt & \text{for } t \in (0,T); \\ vT & \text{for } t \geq T. \end{bmatrix} \tag{4.3}$$

This corresponds to a surface initially at disparity $d(x, 0) = \phi(x, 0)/\bar{\omega}$. Then the surface moves at a constant rate in disparity until it stops at time T. Note that (4.3) only describes the temporal behavior of the local phase *difference*, and not the complete motion of the surface. In particular, the texture could be moving across the images as well as changing in disparity and this would lead to precisely the same phase difference signal.

In using (4.1) we are ignoring the variation of the spatial frequencies ω_l and ω_r with changes in depth, and the nonzero bandwith of the initial filters B_l and B_r. While these effects change the quantitative analysis of a demon, they do not change the qualitative behavior discussed below. The important assumption here is that nontrivial local phase information is available at a particular spatial location x, say, over the duration of the motion. This assumption is also satisfied for step-edge stimuli if one of the "eyes" tracks the step-edge.

For I_l and I_r as in (4.1), the loop equation (3.3) (with $P(x, s)$ as in (3.2)) becomes

$$\frac{ds}{dt} = -P(x,s) = -\kappa \sin(\bar{\omega}s - \phi(x,t)), \text{ for } t > 0, \tag{4.4a}$$

where κ is as in (2.4), ϕ is as in (4.2), and $s \equiv s(t;x)$. Moreover, we assume that the loop is locked at $t = 0$, so that

$$s(0;x) = \frac{1}{\bar{\omega}} \phi(x,0) \equiv d(x,0). \tag{4.4b}$$

In order to analyze (4.4a,b) we assume that

$$|\bar{\omega}s - \phi(x,t)| << \frac{\pi}{2}, \text{ for } t \geq 0. \tag{4.5}$$

This assumption is certainly valid for $t = 0$, since we have assumed (4.4b). Furthermore, it follows from the results quoted below that (4.5) is satisfied whenever the rate of change of the disparity, $|v|$, is taken sufficiently small.

Given (4.5) the loop equation (4.4a) can be linearized. That is, we can approximate the solution of (4.4) by

$$s(t;x) \approx d(x,0) + u(t), \tag{4.6}$$

where $d(x,0) = \frac{1}{\bar{\omega}} \phi(x,0)$ is the initial disparity and u satisfies the *linear* problem

$$\frac{du}{dt} = C(vt - u), \ t \in (0,T); \tag{4.7a}$$

$$\frac{du}{dt} = C(vT - u), \ t \geq T; \tag{4.7b}$$

$$u(0) = 0. \tag{4.7c}$$

Here $C = \kappa\bar{\omega}$, with κ as in (2.4) (in particular, C is a positive constant). The solution of (4.7) is

$$u(t) = vt - v/C \ (1 - e^{-Ct}), \text{ for } t \in (0,T); \tag{4.8a}$$

$$u(t) = vT - [vT - u(T)] \ e^{-C(t-T)}, \text{ for } t \geq T, \tag{4.8b}$$

which is sketched in Figure 7. First note that for v sufficiently small, assumption (4.5) is satisfied for all $t \geq 0$. Therefore it follows that (4.6) provides a reasonable approximation for the computed disparity $s(t; x)$. From Figure 7 it can be seen that the loop lags behind the true disparity by an amount that approaches v/C. After the movement in depth stops at time T, the demon once again attains a perfect lock.

A similar analysis can be applied to the case of a single step-edge stimulus, such as the one considered in (3.4) but with a time-varying disparity. Note that the left eye tracks this stimulus (i.e., the left image doesn't change in time), so demons near $x = 0$ will obtain nontrivial phase information for the duration of the motion. The behavior of such a demon is the same as that sketched in Figure 7. In particular, the demon can be pulled well beyond its initial lock-in range of $[-\pi/\omega_0, \pi/\omega_0)$, and can then maintain a lock outside this range. This is precisely the hysteresis effect mentioned at the beginning of this section.

In fact the demon discussed above is free to wander infinitely far away from its initial guess, s_0, in pursuit of a slowly varying disparity. This is in contrast to a demon with the loop filter (3.10) (i.e., with wrap-around), which is anchored to a specific interval of disparities and cannot track disparities outside this interval. The remaining

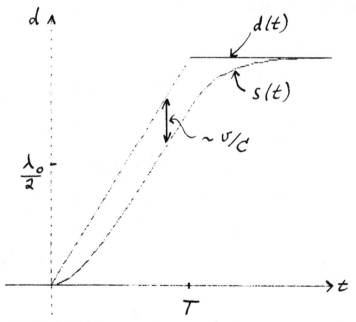

Figure 7.   Tracking behavior of the phase-sensitive demon. Here $d(t)$ is the stimulus disparity, and $s(t)$ is the response of the demon.

loop filter discussed earlier, namely (3.9), exhibits an interesting mix of these two behaviors. As we show below, it is capable of tracking a slowly varying step-edge stimulus outside of its lock-in region. However, it eventually drops out of lock when the disparity of the step-edge becomes too far from its home disparity, s_0.

The necessary analysis is similar to the one leading to (4.8) above, and we only sketch the argument below. Consider the stimulus to be the step-edge given in (3.4), with a time-varying disparity given by

$$d(t) = \begin{cases} 0 & \text{for } t \leq 0; \\ vt & \text{for } t > 0. \end{cases} \qquad (4.9)$$

Here we take the velocity, v, to be positive, although we note that the same behavior is observed for negative v. Suppose the phase detector is constructed using the Gabor pair (3.12) and the Gaussian low-pass filter (3.6a) with (3.13b) satisfied. The phase detector response for $t = x = 0$ is sketched in Figure 8. For our purposes here we need only note that $P(0,s)$ satisfies

$$P(0,s) \geq Cs \text{ for } s \leq 0. \qquad (4.10)$$

Here $C \equiv \dfrac{\partial P}{\partial s}(0,0) > 0$, and therefore the line $y = Cs$ is tangent to the

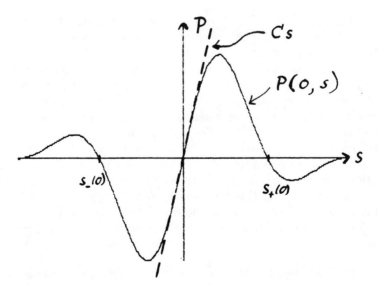

Figure 8.   The relation between the strength, C, of the disparity match and the response $P(0, s)$.

curve $y = P(0,s)$ at $s = 0$ (see Figure 8). This definition for C is essentially the same as that used in (4.7) above.

The loop filter for our demon is obtained from (3.9), where we take the home disparity, s_0, equal to zero. When the phase detector is applied to the moving stimulus its response is given by $P(0,s - d(t))$. Therefore, the equation of motion for the demon is

$$\frac{ds}{dt} + \epsilon s = -P(0,s - d(t)), \text{ for } t > 0, \tag{4.11a}$$

$$s(0) = 0. \tag{4.11b}$$

Through the use of (4.9) and (4.10) it now follows that

$$\frac{ds}{dt} + (C + \epsilon)s \le Cvt, \tag{4.12a}$$

so long as

$$d(t) \equiv vt \ge s(t). \tag{4.12b}$$

This differential inequality, along with the initial condition (4.11b), is easily solved. We obtain

$$s(t) \le b(t) \equiv vC \left(t/\tau - [1 - e^{-t/\tau}]/\tau^2\right), \tag{4.13}$$

where $\tau = \epsilon + C$. Since v, ϵ, and C are all positive, it follows from (4.13) that $b(t) \le d(t)$. In particular, (4.12b) is satisfied for all $t > 0$, and therefore (4.13) must be satisfied for all $t > 0$.

The bound obtained in (4.13) implies that the demon is slowly falling further behind the disparity of the step-edge. Eventually the demon will lag by roughly $2\pi/\omega_0$, or more. The phase detector response for such a large lag is essentially zero, that is, the step-edge has escaped from the demon. The demon then returns to home (zero), according to (4.11a) with $P \equiv 0$. The results of a simulation are given in Figure 9, for three values of ϵ. The demon with the smallest value of ϵ (namely, ϵ_1) successfully tracked the stimulus. The increasing size of the lag during the stimulus motion is apparent in Figure 9, and if the motion continued the demon would eventually drop out of lock. This occurs in Figure 9 for the demon with $\epsilon = \epsilon_3$, the largest value used.

Finally, we remark that $\frac{\partial P}{\partial s}(x,s_\infty(x))$ (which corresponds to the quantity C used above) is related to the *strength* of the disparity

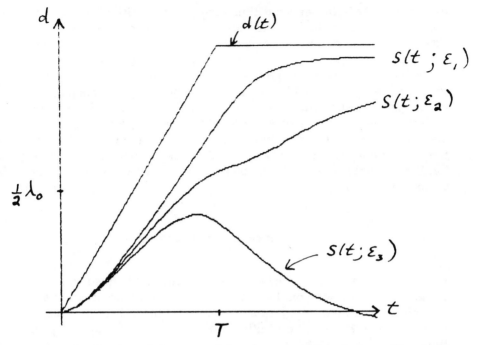

Figure 9. Tracking behaviour of the phase-sensitive demon with a home value of zero. Three different attachment strengths are used, with $0 < \epsilon_1 < \epsilon_2 < \epsilon_3$.

match at $d = s_\infty(x)$. For example, for sinusoidal input such as that given in (2.1), this quantity is proportional to the product of the amplitudes, AB (see (2.4)). Similarly, for step-edge input, C is proportional to the square of the size of the step. Moreover, the tracking error of the demon is related to the strength of the match. In particular, for larger values of C, the bound $b(t)$ for $s(t)$ provided in (4.13) becomes larger and closer to the true disparity $d(t)$. Since the bound provides a good approximation for the response of the demon before the lag becomes too large (the bound is actually the solution of a linearized equation), the demon tracks stronger disparities more accurately and over larger distances. A similar effect is seen with the demon considered in (4.4), for which the tracking error, v/C, is also inversely proportional to AB.

This effect may turn out to be highly desirable when networks of these demons are investigated. The job of the network would be to obtain a globally plausible model for the disparity, that is an attempt is made to remove any false targets. False targets (these are not equivalent to false locks) are matches obtained between two different scene features, and therefore the disparity they produce has nothing to do with depth in the scene (see Marr & Poggio (1976)). To remove these matches, we could consider a cooperative scheme in which the disparity signals produced in the neighborhood of a particular demon affect the home value, s_0, of the demon. Changes in this home value would have different effects on the demon, depending on the strength of its current lock. Demons locked to stronger disparity signals will stick more tightly to these signals than others, and therefore would be more likely to survive similar changes in the home value. This is precisely the sort of behavior one might hope for. Experiments of this nature, however, are beyond the scope of the current paper.

A RANDOM-DOT STEREOGRAM

In this section we briefly consider the application of phase-sensitive demons to two-dimensional stimuli. The necessary modifications to the scheme outlined in Figure 2 are minimal. In particular, we consider a demon based on an initial two-dimensional DOG bandpass filter

$$B(\vec{x}) = \frac{1}{2\pi\sigma_c^2} e^{\frac{-|\vec{x}|^2}{2\sigma_c^2}} - \frac{1}{2\pi\sigma_s^2} e^{\frac{-|\vec{x}|^2}{2\sigma_s^2}} \tag{5.1}$$

Here $\vec{x}$ is a two dimensional vector, and $|\vec{x}|$ is its Euclidean length. Again we use σ_c and $\rho \equiv \sigma_s/\sigma_c > 1$ as the independent parameters for $B(\vec{x})$. The low-pass filter L in the phase detector is taken to be the two dimensional Gaussian

$$L(\vec{x}) = \frac{1}{2\pi\sigma_l^2} e^{\frac{-|\vec{x}|^2}{2\sigma_l^2}} \tag{5.2}$$

As we did earlier, we take the cut-off frequency for the low-pass filter to be below the peak frequency of the DOG in (5.1). This condition is the same as in (3.7), and, when ρ is taken to be 1.5, it becomes

$$\sigma_l > \sigma_c. \tag{5.3}$$

The phase detector is defined using only *horizontal* shifts of the left and right images. That is, we define P by

$$P(\vec{x},s) = \int_{-\infty}^{\infty} \int_{-\infty}^{\infty} L(\vec{x} - \vec{z}) \, (I_l(z_1 - s/2,z_2)I_r(z_1 + s/2,z_2))dz_1dz_2. \quad (5.4)$$

The response of the phase detector is fed into the loop filter, which we again take to be

$$\frac{ds}{dt} = -P(\vec{x},s), \text{ for } t > 0. \qquad (5.5a)$$

(In the implementation we did not simulate (5.5a) directly. Since the image was not time dependent, we could use a much faster scheme to find the same limit.) It is also convenient to introduce an initial guess, say

$$s(0;\vec{x}) = s_0. \qquad (5.5b)$$

Finally, the value of s computed in (5.5) is fed back to the shift module. A separate loop could also be set up for the vertical disparity, with a phase detector defined as in (5.4) but with the shift applied to the second argument (the vertical direction) of I_l and I_r. But that is beyond the scope of the current paper.

The demon described above was tested on the random-dot stereogram given in Figure 10. There the dots are 8 pixels square. The center 64 by 64 region has been displaced by 8 pixels (the width of one dot) to produce a disparity of -8. The demon parameters were taken to be $\sigma_c = 3$, $p = 1.5$, and $\sigma_l = 1.1\sigma_c$. The results are given in Figure 11, where we use dark points to indicate the spatial position of demons that are locked to within ± 1 pixel of a given disparity. Three disparity planes are shown, corresponding to disparities $+8$ (top), 0 (middle), and -8 (bottom). Note that there are no locks in the surround region of the bottom plane, which is the appropriate response. There is only one demon responding (incorrectly) with disparity zero in the center region, while there are many demons reporting the correct value zero in the surround. Also, from the plane at disparity $+8$ we see that, as expected, false matches still occur. The results are therefore consistent with the theory discussed in the preceding sections.

We are currently investigating the behavior of a two-dimensional demon based on the phase-detector which uses the Gabor initial filters. We expect that the results will provide denser responses than are shown in Figure 11.

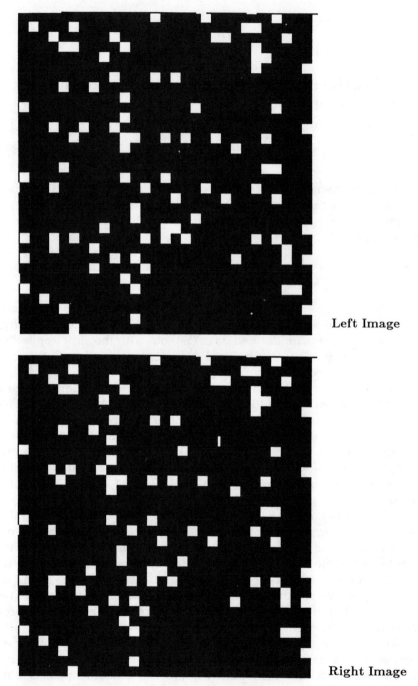

Left Image

Right Image

Figure 10. The random-dot stereogram used.

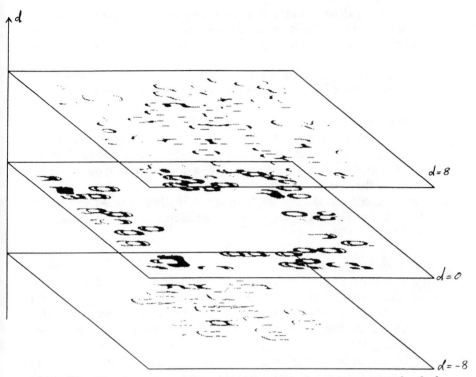

Figure 11. The demon response strengths at the three disparities $d = 8$, $d = 0$, and $d = -8$.

CORRESPONDENCES WITH BIOLOGY

One of the underlying motivations of the model for disparity detection that we are proposing was computational; a model was desired that would operate without the need for complex monocular features that may or may not be accurately and consistently extracted from a scene. In addition, the notion of point-to-point correspondences as the underlying computational process did not seem to simplify the task of stereopsis. As has been suggested (Jenkin & Kolers, 1986) correspondence as a token-to-token matching process seems ill-suited to stereopsis.

From a biological viewpoint a number of physiological and psychological results have guided our model. We begin first with the physiological results, which are listed below:

1. In the cat (Barlow, Blakemore, & Pettigrew, 1967) and the monkey (Poggio & Fischer, 1977) there exist classes of neurons which are

binocularly driven and which are tuned to very narrow regions of relative inter-ocular disparities. These *tuned excitatory neurons* are described by Poggio and Fischer as giving (p. 1392)

. . . excitatory responses over a narrow range of depth about a fixation distance often with inhibitory flanks nearer and farther.

2. In the monkey (Poggio & Fischer, 1977) there exist a class of neurons which respond to stimuli which have disparities nearer than the current fixation distance, and another class that responds to stimuli further away. These are referred to as *near and far neurons*, respectively.
3. There is also evidence for neurons selectively tuned to the 3-d direction of motion, which are called *looming detectors* (Beverly & Regan, 1973).

Our detectors probably do not correspond to the tuned excitatory neurons, but we now have the pieces from which to build a suitable model. For example, the tuned excitatory neuron may respond on a positive going, zero-crossing (as s is varied) in the phase-detector response $P(x, s)$. Similarly, the near and far neurons may be based on sampling the sign of $P(x, s)$. Finally, looming detectors detectors might be made by using velocity selective filters B_l and B_r in the initial stages of filtering. Note that the use of $P(x, s)$ in these applications effectively eliminates the unwanted structure of each image in a stereo pair, and the response depends only on relative properties of the two stimuli.

Three of the physiological results which have guided our work are:

1. Stereopsis operates independently from spatial frequency tuned channels (Julesz & Miller, 1975).
2. Binocular fusion exhibits significant hysteresis, with fusion being initially attainable only for fairly small disparities, but being maintained by tracking over a much wider range (Fender & Julesz, 1967).
3. Binocular fusion in humans appears to have a cooperative element. This is exhibited by the pulling effect induced by neighboring disparities (Julesz & Chang, 1976).

In terms of our demons, it is necessary to provide them with band-pass input (see the discussion after equation (2.1)), which at least is consistent with number 1 above. Also, the fact that they can exhibit hysteresis raises a new option. Marr and Poggio (1977) point out that the hysteresis may not be due to a cooperative algorithm, as had been suggested by Julesz (1971, p. 220). Instead they suggest the presence of a memory in the form of the 2½-d sketch. In this theory the sketch is continually updated, and new values for the disparities are sought near the values predicted by the sketch. The phase-locking model can be viewed as a model for how the disparities might be sought given a predicted value, and how the predicted value could be updated, *with or without* a 2½-d sketch. Finally, we note that a cooperative mechanism is very appealing for reducing the number of false targets. The way such a mechanism can be built around the phase-locking demons is an interesting area for further research.

REFERENCES

Barlow, H. B., Blakemore, C., & Pettigrew, J. D. (1967). The neural mechanism of binocular depth discrimination. *Journal of Physiology, 193,* 327–342.

Beverly, K., & Regan, D. (1973). Evidence for the existence of neural mechanisms selectively sensitive to direction of movement in space. *Journal of Physiology,* Vol. 235, pp. 17–29.

Burt, P. J. (1981). Fast filter transforms for image processing, *Computer Graphics and Image Processing, 16,* 20–51.

Crowley, J. L., & Stern, R. M. (1984). Fast computation of the difference of low-pass transform. *IEEE Transactions on Pattern Analysis and Machine Intelligence, 6*(2), 212–222.

Fender, D., & Julesz, B. (1967). Extension of Panum's fusional area in binocularly stabilized vision. *J. of the Optical Society of America, 57,* (6), 819–830.

Fleet, D. J., & Jepson, A. D. (1986). On the hierarchical construction of orientation and velocity selective filters. Submitted to *IEEE Transactions on Pattern Analysis and Machine Intelligence.*

Gardner, F. (1979). *Phaselock techniques,* (2nd ed.). New York: John Wiley and Sons.

Grimson, W. E. L. (1981). *From images to surfaces.* Cambridge: MIT Press.

Jenkin, M. R. M., & Kolers, P. A. (1986). Some problems with correspondence, research in biological and computational vision. TR-86-10, University of Toronto, April.

Julesz, B. (1971). *Foundations of cyclopean perception.* Bell Telephone Labs., Inc.

Julesz, B., & Chang, J. J. (1976). Interaction between pools of binocular disparity detectors tuned to different disparities. *Biological Cybernetics,* V. 22, 107–120.

Julesz, B., & Miller, J. E. (1975). Independent spatial-frequency-tuned channels in binocular fusion and rivalry. *Perception, 4,* 125–143.

Marr, D. (1982). *Vision.* New York: Freeman Press.

Marr, D., & Poggio, T. (1976). Co-operative computation of stereo disparity. *Science, 194,* 283–87.

Marr, D., & Poggio, T. (1977). *A theory of human stereo vision.* A. I. Memo No. 451, MIT.

Mayhew, J. E. W., & Frisby, J. P. Psychophysical and computational studies towards a theory of human stereopsis. *Artificial Intelligence, 17,* 349–385.

Mayhew, J., & Frisby, J. (1980). The computation of binocular edges, *Perception, 9,* 69–86.

Poggio, G., & Poggio, T. (1984). The analysis of stereopsis, *Annual Reviews of Neuroscience, 7,* 379–412.

Poggio, G. F., & Fischer, B. (1977). Binocular interaction and depth sensitivity in striate and prestriate cortex of behaving rhesus monkey, *Journal of Neurophysiology, 40*(6), 1392–1405.

Terzopoulos, D. (1983). Multi-level reconstruction of visual surfaces. In A. Rosenfeld (ed.), *Multiresolution image processing and analysis.* Berlin: Springer-Verlag.

5

Natural Constraints on Apparent Motion*

Michael Dawson and Zenon W. Pylyshyn

Centre For Cognitive Science
The University of Western Ontario

INTRODUCTION: THE PROBLEM OF PERCEPTUAL UNDERDETERMINATION

The pattern of stimulation of the light receptors in our eyes—even if extended over time—greatly underdetermines what we see. A single pattern of retinal stimulation (called the proximal stimulus) could have been caused by any one of an infinite number of physical arrangements in the world (called distal stimuli). The proximal stimulus cannot uniquely determine a distal stimulus, because it does not preserve the full dimensionality of the physical world. The mapping from three-dimensional patterns to two-dimensional patterns is many-to-one and hence not uniquely invertible.

Despite the fact that retinal stimulation *geometrically* underdetermines interpretation of the physical world, we as perceivers are not aware of this. The visual system appears to invert the distal-to-proximal mapping, and it does so in a manner that typically (though not invariably) yields the *correct* interpretation of the proximal stimulus. A central problem of the study of visual perception is this: How does the visual system generate interpretations that accurately reflect the

* This research was supported by an NSERC postgraduate scholarship and by the York University President's NSERC Award, awarded to the first author, and by NSERC operating grant No. A2600 awarded to the second author.

structure of the world, given the noninvertibility of the proximal-to-distal mapping?

One very general answer to this question defines the natural computation approach to the study of vision. This approach begins with the assumption that perception is a problem of information processing. Such problems can be divided naturally into three different components (Marr, 1977, 1982). The first component is a theory of the computation. Such a theory provides an abstract description of the information processing problem and the general principles used to solve it. The second component provides an account of the algorithm (i.e., the specific procedures) used by a system to solve the problem. The third component is a description of how the visual algorithm is actually implemented in a biological device. In this chapter we shall be concerned primarily with Marr's level of "computational theory," and will examine how such a theory provides an answer to a particular problem of underdetermination, a problem that the visual system must solve in the course of constructing, among other things, a percept of apparent motion.

COMPUTATIONAL THEORIES OF VISION

Marr (1977, 1982) describes a theory of a computation as an account of what is being computed, and why. What this means becomes evident by considering perception as the process of solving problems of underdetermination. A theory of the computation required to solve such problems can be viewed as a statement of a selection rule that can be used to choose one interpretation of the proximal stimulus from the set of possible interpretations. Such a rule provides a characterization of relevant attributes of a distal stimulus. The rule is applied as follows: if some distal stimulus possesses the relevant attributes characterized by the selection rule, then this distal stimulus will be perceived in the way prescribed by the rule.

For example, Ullman (1979) has developed a theory of the computation of three-dimensional object structure from perceived motion. His theory can be described as the application of the following selection rule: if a dynamic proximal stimulus is compatible with the interpretation that arises from a set of features on the surface of a rigid three-dimensional object in motion, then adopt this interpretation in preference to others.

Why a particular computation is performed becomes evident by noting the characteristics of distal stimuli that these selection rules describe. These characteristics are general physical properties, and tend to be true most of the time. For instance, the proximal stimulus is almost always caused by a configuration of rigid objects whose surfaced properties (e.g., luminance, texture, depth) vary smoothly and continuously through space. Discontinuities in these properties (i.e., object boundaries or occluding edges) make up only a vanishingly small proportion of the proximal stimulus. Such very general physical properties of distal stimuli, which can be used to constrain problems of underdetermination, are called *natural constraints*. Researchers have found that natural constraints can be used to resolve many different problems of underdetermination in the analysis of visual stimuli (see, for example, Marr, 1982). Thus one reason that one particular inverse mapping (from a proximal to a distal pattern) is selected by the visual system is that, because of certain very general properties of our physical environment, this interpretation is highly likely to be veridical *in our kind of world.* If a physical property is almost always true of the world, then a visual system that exploits this property as a natural constraint will almost always compute the correct interpretation of a proximal stimulus.

The purpose of this chapter is to provide a specific example of how natural constraints can be applied to solve problems of underdetermination in vision. We consider various natural constraints that might be applied to a problem of underdetermination encountered in motion perception, called the *motion correspondence problem.* This problem is introduced in the next section.

THE MOTION CORRESPONDENCE PROBLEM

When watching a motion picture, an observer is being presented a sequence of static images of some dynamic scene. The moving elements in the scene change their locations between successive images. However, the observer is not aware of the frame-by-frame structure of the motion pictures—one static view is not seen to be suddenly replaced by another. Instead, continuous motion is seen over time. The percept is a visual illusion called *apparent motion.*

In generating apparent motion, the visual system must be capable of identifying an element in a location in one image (Frame I) and another element in a different location in the next image (Frame II) as constituting different glimpses of the same moving element. A motion correspondence match between a Frame I and a Frame II element is such an identification.

Identifying a visual feature in one frame as corresponding to the same distal element as a feature in another frame is complicated by the fact that there are, in general, many different sets of such correspondence matches that might be computed from a pair of frames (see Figure 1). In fact, if there are N elements in each frame of view, then there are $N!$ potential sets of motion correspondence matches. The problem of selecting one set of interframe matches from among the many logically possible ones is called the "motion correspondence problem" (Ullman, 1979). Since, in general, the N proximal features are not distinguishable (except for their location), the visual system must exploit additional rules or principles in order to solve the motion correspondence problem. An important goal of apparent motion research has been to discover the principles actually ex-

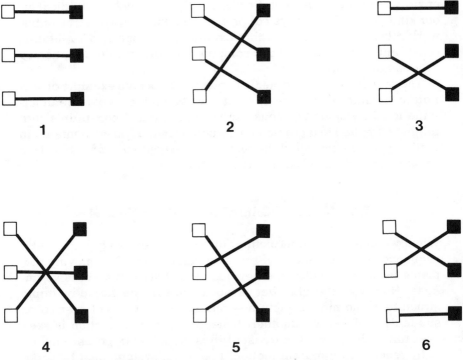

Figure 1. An example of the underdetermination of motion correspondence matches. In this figure, unfilled squares represent element positions in Frame I, filled squares represent element positions in Frame II, and solid lines represent motion correspondence matches between frames of view. In this 3-element example, the six possible sets of correspondence matches are illustrated. Only one of these sets (set 1) is computed by the human visual system.

ploited by the visual system to solve the motion correspondence problem (Attneave, 1974).

In the remainder of this chapter we will consider several putative natural constraints that could in principle help solve the motion correspondence problem. They are based on two distinct types of properties of stimuli. The first are figural properties of the set of visual features, that is, the 2-D geometrical patterns formed by the set of features. The second are properties of the movement of individual features, in relation to the movements of other features in the proximal stimulus.

Computing Motion Correspondence Using Image Matching

Many researchers have been concerned with the development of computer models capable of tracking moving objects in a visual scene (e.g., Aggarwal & Duda, 1975; Ferrie, Levine, & Zucker, 1982; Jain, Martin, & Aggarwal, 1979a, 1979b; Jain, Militzer, & Nagal, 1977; Price, 1985; Price & Reddy, 1977; Tsuji, Osada, & Yachida, 1979, 1980; Yalamanchili, Martin, & Aggarwal, 1982). In order for object tracking to be accomplished, motion correspondence matches must be computed. Typically those concerned with object tracking have done this by using some variation of an image matching procedure.

In image matching, motion correspondence matches are made between objects that have similar forms or similar image functions (i.e., similar 2-D intensity profiles). In such image-matching techniques the sampling rate of a scene is sufficiently high that drastic changes in object form will not have occurred between frames of view. Under such conditions, image matching usually produces the correct motion correspondence matches. In this sense, image matching procedures can be viewed as exploiting a natural constraint reflecting the figural integrity of objects in the world (i.e., frequently sampled 2-D projections of rigid slowly-moving 3-D objects, vary gradually in their figural properties). One might expect to find that the human visual system also uses figural information during the perception of apparent movement. However, it has been very difficult to confirm this expectation empirically. Kolers and his colleagues (Kolers, 1972; Kolers & Green, 1984; Kolers & Pomerantz, 1971; Kolers & Von Grunau, 1976) have shown that the visual system makes motion correspondence matches between objects that differ in shape or color. This research has provided many examples of apparent motion in which the visual system performs a correspondence match in a way that results in a perceived distortion of the shape of objects, even when there is an alternative way to make the correspondence that preserves the shape of the objects. This finding has been confirmed by several other researchers (e.g., Burt & Sperling, 1981; Navon, 1976; Ullman, 1979, chapter 2).

This evidence suggests that the visual system does not (or at least, need not) use figural information when motion correspondence matches are computed. However, there are reasons for being cautious about accepting this conclusion. A few recent studies have demonstrated that motion correspondence processing is sensitive to some figural information. Ullman (1980) found that there was a tendency for motion correspondence matches to be made between elements of similar size and orientation. Ramachandran, Ginsburg and Anstis (1983) demonstrated that motion correspondence matches can be made exclusively on the basis of similarity of low spatial frequency components of figures. High frequency information (i.e., oriented edges) did not appear to be involved, particularly when the frames of view of the display were alternated quickly. A similar finding is reported by Green (1986). Green found that motion correspondence matches appear to be made on the basis of the low-frequency components of stimuli, but that the orientation of these components and their intensity also played a role. Green suggests that the role of figural effects on motion correspondence processing have been difficult to demonstrate because researchers have tended to use stimuli composed of very high spatial frequency components (e.g., contour drawings of geometric shapes).

The evidence presented in this section indicates that the role of figural information in human motion correspondence processing has not been univocally established, despite the fact that image-matching has been used successfully by computer systems that perform object tracking (as mentioned above). It could be, as Green (1986) suggests, that researchers have not examined the correct set of figural properties. If motion correspondence matches are computed at a very early stage in low-level vision, then it would stand to reason that the only figural properties to affect correspondence processing would be those detected very early in vision (e.g., orientation, spatial frequency). This issue clearly deserves further empirical investigation.

Computing Motion Correspondence by Minimizing Element Velocities

Ullman (e.g., 1979) has studied the motion correspondence problem extensively by using the motion competition paradigm. In a motion competition display, two alternative opposing paths of motion are

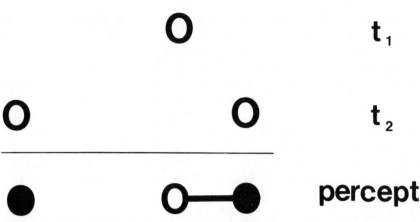

Figure 2. The motion competition paradigm. Subjects are asked to indicate the perceived direction of movement of the Frame I element. Typically this element is seen to move in the direction of the nearest Frame II element.

set into competition. This is done by replacing a central element in Frame I of an apparent motion display with two lateral Frame II elements, one to the left and one to the right of the Frame I element (see Figure 2). The Frame I element will be seen to move either to the left or to the right. Subjects are asked to indicate the perceived direction of motion of this display element. Of interest are the factors that influence which direction of motion is preferred by the human visual system. (Variations of this paradigm have been used by Chang and Julesz, 1985, to study the perception of motion in random-dot kinematograms, and by Burt and Sperling, 1981, to study perceptions of apparent movement in multiple frame displays.)

One of the strongest predictors of the perceived direction of motion in a competition display is interelement distance (i.e., the distance between an element in Frame I and potential corresponding elements in Frame II). All things being equal, the visual system prefers short motion correspondence matches over long motion correspondence matches (e.g., Burt & Sperling, 1981; Ullman, 1979, chapter 2): That is, if motion to the left involves a shorter interelement distance than motion to the right, then motion to the left will be preferred. This suggests that the visual system may exploit a "nearest neighbor" principle: all things being equal, a motion correspondence match should be created between a Frame I element and its nearest neighboring element in Frame II.

Is the nearest neighbor principle a natural constraint? Ullman (1979, pp. 114–118) argues that this is indeed the case. Consider a world in which all possible element velocities in 3-D occur with equal probability. When these velocities are projected onto a 2-D surface (e.g., the retina) the probability distribution of projected element velocities will be no longer be uniform. This is because the depth component of an element velocity in three-space is lost when it is projected onto a two-dimensional surface. For the projected motion, slower element velocities occur with much higher probability than faster element velocities. Since a preference for the nearest neighbor is equivalent to a preference for slowest 2-D velocities, it may be that the human visual system is exploiting the "natural constraint" that in 2-D, low velocities are more frequent that high velocities (assuming no systematic skew in the 3-D motion distribution).

The Maintenance of Element Integrity

Proximal visual features usually arise from physical features on the surface of solid objects. Thus, except for when these physical features become occluded by the movement of edges (either due to the rotation of the object in question or to different objects moving through the line of sight), visual elements do not appear or disappear from view. Since element occlusions are low probability events, at least in environments in which the density of edges is sparse in relation to the density of elements, this suggests a further constraint on solutions to the motion correspondence problem. Ullman (1979) calls this constraint the *cover principle*. According to this principle, in any solution to the motion correspondence problem each Frame I element must be matched to at least one Frame II element, and each Frame II element must be matched to at least one Frame I element. The cover principle is thus the claim that Frame I elements will not suddenly disappear, and that Frame II elements will not suddenly appear.

It might be noted here that, although the cover principle may be reasonable in most circumstances, there is reason to believe that the cover principle may not always hold for human vision: Specifically, it may not hold in cases where there is independent visual evidence for the presence of an occluding surfaces is an important variable in the perception of apparent motion. If the disappearance of an element cannot be accounted for by the presence of an occluding surface, then an attempt is made to cover the element with a motion correspondence match (i.e., if possible, the disappearance of the element from one location in the image will be accounted for by asserting that the element moved to a new location). If the disappearance of an element can be accounted for by the presence of an occluding surface, then a motion correspondence match is not computed for the element (i.e., it is not seen to move).

The evidence above suggests that the cover principle might only hold in the absence of visual evidence for occluding edges. Jenkin (1984) has developed a motion correspondence model where the problem of occlusion can be dealt with by noting the history of motion correspondence matches and element appearances and disappearances over a number of frames of view. Since we shall be adopting a different approach in the present paper, we present the nearest-neighbor principle and the cover principle simply to illustrate one particular well-developed model (due to Ullman, 1979) based on the natural constraint approach.

Minimal Mapping Theory

Ullman's minimal mapping theory demonstrates how the motion correspondence problem can be solved by minimizing element velocities and by applying the cover principle. In this model, a cost is associated with each possible match between a pair of elements. This cost is proportional to interelement distance. The greater this distance, the higher the cost of that match. Minimal mapping theory uses the following principle for solving the motion correspondence problem; the set of matches that has the lowest cost, and at the same time satisfies the cover principle, is the set of matches that is selected. Ullman (1979) has shown that this selection rule accounts for human interpretations of many apparent motion displays.

Minimizing Changes in Configuration

The purpose of our investigation of motion perception was to examine whether the human visual system might apply other natural constraints to the motion correspondence problem. In particular, we were concerned with one assumption that is built into Ullman's Minimal Mapping Theory, namely the assumption that the cost assigned to the match between any pair of elements is independent of the location of other display elements. This assumption seemed questionable to us, precisely because the perceptible world consists primarily of the surfaces of solid objects. To the extent that visual elements arise from physical features on the surface of solid objects, the movement of neighboring elements should reflect the coherence of surfaces. Thus such interdependencies could serve as natural constraints, and it would be reasonable to expect the visual system to be tuned to exploited such a constraint.

What types of interdependencies exist among the motion trajectories of elements lying on the surface of the same solid object? The coherence of matter ensures that neighboring points in three-space travel along similar 3-D trajectories. The projection of pairs of such neighboring points onto a 2-D plane preserves certain properties. If we had a general way of characterizing some properties of the 2-D motion of points projected from the surface of a solid object, we would have a general statement of a principle that could be exploited to constrain solutions to the motion correspondence problem. We would then have a principle by which the visual system could assign motion correspondence matches so as to minimize changes in the 3-D configuration of display elements, and thus consistent with the natural constraint of coherence of matter or rigidity of objects (c.f. Foster, 1978; Yuille, 1983).

Below we examine several approaches to this problem that have been pursued in recent years. We do this primarily in order to place our own proposal—which is rather simpler and more qualitative—in perspective in relation to these attempts.

Imagine viewing the arbitrary movement in 3D space of a rigid, transparent rod, and that four arbitrarily selected points on this rod are visible. In Frame I, these four points can be labeled A, B, C, and D (see Figure 3). In Frame II, these four points have been moved to different positions because of some arbitrary movement of the rod. The task of the visual system is to compute motion correspondence matches for this display—this is, to correctly assign the labels A, B, C, and D to the four points in Frame II.

One measure of the configuration of four points on a rigid straight object is the canonical cross ratio, which is defined as the following ratio of distances:

$$\text{Cross Ratio} = \frac{AD * BC}{AC * BD}$$

It can be proved that the value of this cross-ratio is preserved for any rigid movement of the rod about an observer, projected onto any plane (e.g., Cutting, 1986). In other words, the canonical cross-ratio is a measure of the configuration of the points that is not affected by arbitrarily moving the rod in 3D space, or by projecting this movement onto the 2D retina. Thus the visual system could solve the motion correspondence problem for this particular type of display by assigning motion correspondence matches between labeled rod

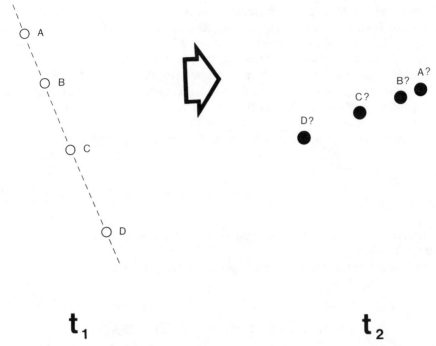

Figure 3.   The cross-ratio approach to the motion correspondence problem.
The labels A, B, C, and D in Frame II could be assigned in such a way that
the cross-ratio of the display in Frame II was equal to the cross-ratio of the
original display. This approach only works for the situation in which the four
points are assumed to be collinear, and undergo an arbitrary movement that
preserves their configuration.

points in such a way that the cross ratio was identical for the two
frames of view.[1]

One problem with the cross-ratio rule described above is that it
will only work for a very specific type of display. Yuille (1983) provides a more general formulation of a minimal configural change
rule. Imagine that Frames I and II of an apparent motion display
contain different views of some moving contour. Such a contour can
be represented as the set of unit vectors tangent to the contour at all
of its points (e.g., Dodwell, 1983). A motion correspondence solution

---

[1] This rule will only fail in the case in which there is an equal spacing between the
points, which produces two different solutions (one the reflection of the other) with
identical cross ratios. Both of these solutions are physically plausible, in the sense
that they preserve the configuration of elements.

could be described as an account of how the contour presented in Frame I was transformed into the contour presented in Frame II.

Yuille suggests that the desired transformation would be the one that minimizes the distortion of the figure as it moves. In other words, the desired transformation would be one that links each Frame I tangent vector to a Frame II tangent vector, such that differences between the matched vectors is minimized. Mathematically, this can be described as finding the set of motion correspondence matches that minimizes the following integral:

$$\int \left(\frac{\partial T}{\partial t} \right)^2 ds$$

where $\frac{\partial T}{\partial t}$ represents the measure of changes in tangent vectors with respect to time, taken at each point along the curve.

THE RELATIVE VELOCITY CONSTRAINT

The importance of Yuille's (1983) minimum transformation rule is that he was also able to prove that this measure bears a very strong relationship to another type of rule used to determine the motion of smooth contours. This rule is the smoothness principle of motion (Hildreth, 1983; Horn & Schunk, 1981). According to this rule, smooth motion is motion in which neighboring points on contours move in nearly identical fashions. Differences in velocity (called relative velocity) vary smoothly over the entire contour.

The relationship between a measure of figural variation and a measure of variation of relative velocities is very important. Hildreth (1983) demonstrated that the smoothness constraint is a natural constraint, because objects moving arbitrarily in three-dimensional space produce unique, smooth patterns of retinal movement. The relationship discovered by Yuille (1983) demonstrates that motion correspondence solutions that minimize figural variation over time also conform to this constraint.

These ideas can be used as the basis for a hypothesis about a particular constraint for the motion correspondence problem. A motion correspondence match can be viewed as a motion vector, because such a match is an assertion that a Frame I element has moved in a set direction at a set speed to occupy a new position in

Frame II. It is hypothesized that the visual system selects the set of motion correspondence matches that minimizes the relative velocities between neighboring display elements, in an attempt to minimize local changes in the configuration. Relative velocity is defined to be the difference between two motion correspondence matches (interpreted as motion vectors) after they have been centered to a common origin. This selection rule is called the relative velocity constraint (Dawson, 1986; Dawson & Pylyshyn, 1986).

The relative velocity constraint is very similar to Yuille's (1983) condition for minimizing a measure of figural distortion. In minimizing the relative velocities of neighboring elements, the selected set of motion correspondence matches will minimize changes in the relative positions of elements over time. The constraint is also very similar to Hildreth's (1983) smoothness constraint, because both attempt to minimize local differences in the same property (relative velocity).

However, it is important to point out that the present use of the relative velocity constraint is quite different from the smoothness constraint. First, the smoothness constraint was designed to be applied to image points that make up continuous contours, while the relative velocity constraint is designed to be applied to discrete display elements. Second, the smoothness constraint is typically used to solve a problem of underdetermination in motion which arises because only one component of the velocity of a contour is known (in the case of a smooth curve, only the component of velocity perpendicular to the tangent of the curve can be seen—this is called the "aperture problem"). Because of this indeterminacy, an infinite number of interpretations of 3-D movement are possible.

Using the relative velocity constraint to solve the motion correspondence problem among discrete elements is quite different since all possible motions of each discrete display element can be enumerated and the problem becomes that of selecting the most plausible one. Each element in Frame I is initially associated with a small number of complete velocity vectors (i.e., potential motion correspondence matches), and this initial state can lead to the computation of only a finite number of correspondence solutions.

The relative velocity constraint also differs from Yuille's (1983) measure of figural variation. Yuille's measure is computed over the entire continuous contour, and frames of view are assumed to be sampled at infinitesimally small time intervals. The relative velocity constraint is intended to be applied in a situation in which element locations are known, but no figural information (e.g., the orientation or curvature of elements) is available. It is also assumed that frames of view can be sampled at relatively large time intervals, corresponding to apparent motion displays.

A Computer Model and Some Empirical Results

Can relative velocity information be successfully used to constrain solutions to the motion correspondence problem? This question was investigated by developing a computer simulation model that was guided by the relative velocity constraint—that is, that attempted to minimize the relative velocity between neighboring points (Dawson, 1986; Dawson & Pylyshyn, 1986). Frames I and II of an apparent motion display are used to construct a labeled network. The nodes in the network correspond to Frame I elements, and labels attached to these nodes represent potential motion correspondence matches to be assigned to these elements (see Figure 4). Discrete relaxation labeling (Rosenfeld, Hummel & Zucker, 1976) was used to minimize the relative velocity between each node and its neighbors (i.e. its topological neighbors in the network). (A model similar to this, but which uses continuous relaxation labeling, is presented in Barnard & Thompson, 1980).

The process proceeds as follows. The average relative velocity is computed between each possible motion correspondence match and the other motion correspondence matches assigned to neighboring nodes. At each iteration, the motion correspondence match at each node with the highest relative velocity (relative to its neighbors) is discarded. This process is repeated until no more labels can be discarded from the network without leaving nodes unlabeled. The labels remaining in the network represent a solution to the motion correspondence problem.

The computer model generates the same motion correspondence matches as the human visual system for a wide variety of displays, including translations, expansions, and rotations in the plane (Dawson, 1986).

One interesting test of the relative velocity constraint involves the perception of rigid rotations in the plane of sets of elements. For many of these displays, there is more than one interpretation that preserves the configuration of elements (the rigidity of the configuration is not altered). However, the interpretations differ in terms of the total relative velocity between motion correspondence matches. If the relative velocity constraint is applied, then human observers should compute the motion correspondence matches that lead to minimum total relative velocity. If some other constraint is used, then other display interpretations should be selected.

Possible percepts:

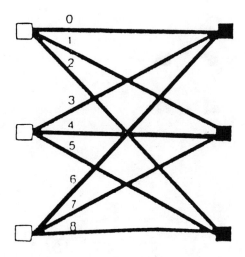

Representation of percepts in network:

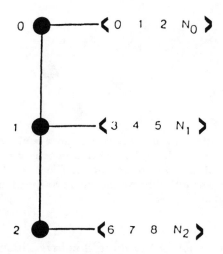

Figure 4. The representation of apparent motion displays in the computer simulation model. The potential motion correspondence matches are listed in the upper part of the figure. These potential matches are represented in the lower part of the figure as labels attached to nodes in a local network.

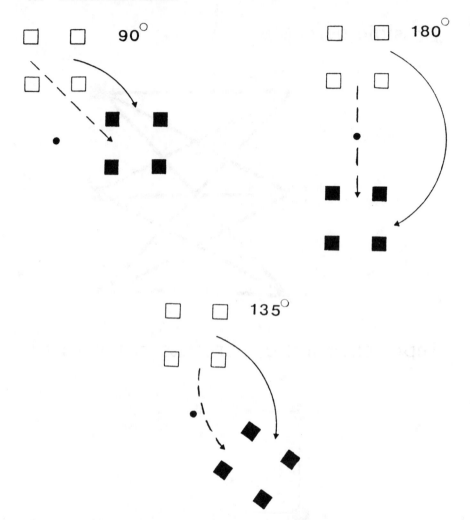

Figure 5. Alternative interpretations of square configurations of elements that undergo a rotation about the origin. The solid arrows indicate the circular rotation applied to the configuration. The dashed arrows indicate an interpretation consistent with the motion correspondence matches generated by the computer model.

For example, consider the three displays illustrated in Figure 5. All three displays can be described in terms of a geometric rotation of the four vertices of a square about the origin (0,0). Figure 5a can be described as a 90 degree circular rotation. If this is the pattern of motion seen, the scalar value of the total relative velocity of the display is 64 degrees of visual angle.[2] This pattern of motion is depicted with solid arrows. As equally plausible interpretation of this display is in terms of a rigid translation (unfilled arrows). This interpretation has a relative velocity of 0° visual angle. According to the relative velocity constraint hypothesis, this interpretation should be preferred. Figure 5b shows two possible interpretations for a display in which a geometric rotation of 180 degrees is applied. If this is the motion that is perceived, the total relative velocity of the solution is 128° visual angle. The alternative interpretation is of a translation with 0° total relative velocity. Figure 5c can be interpreted as a 135 degree rigid rotation (total relative velocity = 109.29° visual angle) or as a "hyperbolic" rotation in which the configuration travels in towards the origin and then rotates outwards (total relative velocity = 18.75° visual angle).

In all these cases, the computer simulation model generates the motion correspondence matches that lead to the lowest overall relative velocity.[3] The question of interest was whether human observers would respond in a like manner. This was tested experimentally as follows (details of these studies are reported in Dawson, 1986). Twelve introductory psychology students were presented a series of displays like those depicted in Figure 5. Geometric rotations of ± 45, ± 90, ± 135, and 180 degrees were used. Each student saw 20 instances of each display type (in random order), and was asked to describe the perceived motion by emulating it using a plexiglas model of the display, and by verbally describing the motion they saw. Their responses were coded into one of three different categories: geometric rotation, translation, or "hyperbolic" or inner rotation.

The results from this experiment are presented in Table 1. From this table it can be seen that in the vast majority of cases the motion

---

[2] The relative velocities given in the text were computed for a 2 degree square placed 4 degrees from the origin. These dimensions were used because they were used in the experiment that is reported. Total relative velocity was computed by summing the relative velocities of the motion of each point with the motion of its two nearest neighbours in the square configuration. The same pattern of relative velocities would emerge if the relative velocity of each point was taken with respect to all of the other points in the display.

[3] Although this procedure minimizes the vector differences in velocity between each possible correspondence match and its neighboring correspondence matches at each node, we have not proved that this procedure always results in a global minimum relative velocity over the network. In the specific cases that we have investigated empirically—described below—we have, however, determined that the global minimum does in fact correspond to the correspondence choices made by the algorithm.

Table 1. The results for the rotating squares experiment. Numbers indicate the proportion of response types for each amount of rotation. These proportions are based on 240 observations per display type (20 per subject). Numbers in parentheses indicate proportions predicted by the model.

ROTATION	RESPONSE TYPES		
	Circular	Translation	Hyperbolic
45	0.896	0.000	0.104
	(1.000)	(0.000)	(0.000)
90	0.158	0.821	0.020
	(0.000)	(1.000)	(0.000)
135	0.125	0.000	0.875
	(0.000)	(0.000)	(1.000)
180	0.017	0.983	0.000
	(0.000)	(1.000)	(0.000)
−135	0.158	0.000	0.842
	(0.000)	(0.000)	(1.000)
−90	0.129	0.846	0.025
	(0.000)	(1.000)	(0.000)
−45	0.954	0.008	0.038
	(1.000)	(0.000)	(0.000)

perceived by the subjects was the same as that predicted by the computer model. Thus it appears that the human visual system prefers the set of motion correspondence matches that minimizes relative motion, even when other sets of matches that preserve figural integrity are possible.

Context Effects on Apparent Motion

One of the major assumptions used in deriving the relative velocity constraint was that there are interdependencies between the movement of object parts. Is the human visual system sensitive to these interdependencies? The evidence described in the previous section provides indirect support for this assumption—human observers preferred sets of motion correspondence matches that minimized relative motion. However, a more direct test of the assumption is desirable, since this assumption is not accepted by all researchers. In particular, Ullman (1979) explicitly assumes that the perceived motion of one display element is independent of the perceived motion of other display elements.

Dawson (1987) explicitly tested this independence hypothesis using a variation of the motion competition paradigm. Subjects indicated the perceived direction of motion of a target element of a motion competition display in one of three context conditions. The first was a condition in which the only movement involved in the display was the perceived movement of the target competition display. The second and third conditions provided an additional movement pattern: a set of context elements moved either to the left or to the right. Half of the context elements moved above the target competition display, half moved below it. If the independence hypothesis was correct, then the presence of the moving contexts should not affect the perceived direction of movement of the target competition display. On the other hand, if the independence hypothesis is incorrect and the visual system is sensitive to interdependencies among the movement of elements, then the contexts should affect perceptions of the display. In particular, the relative velocity constraint predicts that a context moving to the right should increase the probability of seeing the target competition element move to the right. A context moving to the left should increase the probability of seeing the target competition element move to the left.

The results showed clearly that the presence of a moving context affected the perceived direction of movement in the competition display—and in the predicted direction. In addition, the effect of the moving context was greater the closer the context was to the target display, and the greater the number of elements in the context. These results show that the independence hypothesis, as adopted by Ullman (1979), is not correct. It appears that computation of motion correspondence matches is sensitive to patterns of relative movement. This finding is consistent with our assumption that a relative velocity constraint is being exploited by the human visual system.

However, additional results (also reported in Dawson, 1987), also show that the relative velocity constraint is not the only constraint used by the human visual system in solving the motion correspondence problem. Although the moving context increased the likelihood that the target would be seen to move in the minimal relative velocity direction, it did not invariably result in the target being perceived to move in the same direction as the context. A very short interelement distance on the side opposite to that of the direction of motion of the context could overcome the effect of context and result in the target being perceived to move in the opposite direction (i.e. toward the closer competition element). This suggests that the visual system might compute motion correspondence matches by simultaneously exploiting the relative velocity constraint and the minimum element velocity constraint.

CONCLUSION

This chapter has examined some candidate natural constraints that might be useful in constraining solutions to a specific problem of underdetermination in vision: the motion correspondence problem. It was shown that natural constraints based on maintaining the figural identity of objects (i.e., image matching constraints) do not appear to be exploited by the human visual system. The human visual system appears to solve the motion correspondence problem by applying natural constraints based upon the properties of the movement of objects. The work of Ullman (1979) has shown that, in many cases, the motion correspondence solution that minimizes the total distance that the target elements travel, and that prevents elements from appearing or disappearing, is preferred by the human visual system. Our own research has shown that the human visual system might also exploit a natural constraint that minimizes the (vector) difference between the velocities of neighboring elements. Further research is required to determine how constraints derived from the distribution of both scalar values of the velocity of elements, and from the relative velocities of neighboring elements, interact in the human visual system.

REFERENCES

Aggarwal, J. K., & Duda, R. O. (1975). Computer analysis of moving polygonal images. *IEEE Transactions On Computers, C-24*, 966–976.

Attneave, F. (1974). Apparent movement and the what-where connection. *Psychologia, 17*, 108–120.

Barnard, S. T., & Thompson, W. B. (1980). Disparity analysis of images. *IEEE Transactions On Pattern Analysis And Machine Intelligence, PAMI-2*, 333–340.

Braddick, O. (1980). Low-level and high-level processes in apparent motion. *Philosophical Transactions Of The Royal Society Of London, B290*, 127–151.

Burt, P., & Sperling, G. (1981). Time, distance, and feature trade-offs in visual apparent motion. *Psychological Review, 88*, 171–195.

Chang, J. J., & Julesz, B. (1985). Cooperative and non-cooperative processes of apparent movement of random-dot cinematograms. *Spatial Vision, 1*, 39–45.

Cutting, J. (1986). *Perception with an eye for motion.* Cambridge, MA: MIT Press.

Dawson, M. R. (1986). Using relative velocity as a natural constraint for the motion correspondence problem. U.W.O. Centre For Cognitive Science Technical Memorandum #27.

Dawson, M. R. (1987). Moving contexts do affect the perceived direction of movement in motion competition displays. *Vision Research, 27,* 799–809.

Dawson, M. R., & Pylyshyn, Z. W. (1986). Using relative velocity information to constrain the motion correspondence problem: Psychophysical data and a computational model. *Proceedings of the 6th Biennial Meeting of the Canadian Society for Computational Studies of Intelligence.*

Dodwell, P. C. (1983). The Lie transformation group model of visual perception. *Perception & Psychophysics, 34,* 1–16.

Ferrie, F. P., Levine, M. D., & Zucker, S. W. (1982). Cell tracking: A modeling and minimization approach. *IEEE Transactions On Pattern Analysis and Machine Intelligence, PAMI-4,* 277–291.

Foster, D. H. (1978). Visual apparent motion and the calculus of variations. In E. L. J. Leeuwenberg & H. F. J. M. Buffart (Eds.), *Formal theories of visual perception.* New York: Wiley.

Green, M. (1986). What determines correspondence strength in apparent motion? *Vision Research, 26,* 599–607.

Hildreth, E. C. (1983). *The measurement of visual motion.* Cambridge, MA: MIT Press.

Horn, B., & Schunk, B. (1981). Determining optical flow. *Artificial Intelligence, 17,* 185–203.

Jain, R., Martin, W., & Aggarwal, J. (1979a). Extraction of moving images through change detection. *Proceedings Of The Sixth International Joint Conference On Artificial Intelligence,* 425–428.

Jain, R., Martin, W., & Aggarwal, J. (1979b). Segmentation through the detection of changes due to motion. *Computer Graphics And Image Processing, 11,* 13–34.

Jain, R., Militzer, D., & Nagel, H. (1977). Separating non-stationary from stationary scene components in a sequence of real-world TV-images. *Proceedings Of The Fifth Joint International Conference On Artificial Intelligence,* 612–618.

Jenkin, M. (1984). Applying temporal constraints to the problem of time-varying imagery. *Proceedings of the Fifth Bennial Conference of the Canadian Society for Computational Studies of Intelligence,* 106–113.

Kolers, P. A. (1972). *Aspects of motion perception.* Toronto: Pergamon Press.

Kolers, P. A., & Green, M. (1984). Color logic of apparent motion. *Perception, 13,* 249–254.

Kolers, P. A., & Pomerantz, J. (1971). Figural change in apparent motion. *Journal of Experimental Psychology, 87,* 99–108.

Kolers, P. A., & Von Grunau, M. (1976). Shape and colour in apparent motion. *Vision Research, 16,* 329–335.

Marr, D. (1977). Artificial intelligence—a personal view. *Artificial Intelligence, 9,* 37–48.

Marr, D. (1982). *Vision.* San Francisco: W. H. Freeman.

Navon, D. (1976). Irrelevance of figural identity for resolving ambiguities in apparent motion. *JEP:HPP, 2,* 130–138.

Price, K. (1985). Relaxation matching techniques—A comparison. *IEEE*

Transactions On Pattern Analysis and Machine Intelligence, PAMI-7, 617–623.

Price, K., & Reddy, R. (1977). Change detection and analysis in multispectral images. *Proceedings Of The Fifth International Joint Conference On Artificial Intelligence*, 619–625.

Ramachandran, V. S., Ginsburg, A. P., & Anstis, S. M. (1983). Low spatial frequencies dominate apparent motion. *Perception, 12*, 457–461.

Rosenfeld, A., Hummel, R. A., & Zucker, S. W. (1976). Scene labelling by relaxation operators. *IEEE Transactions On Systems, Man, and Cybernetics, SMC-6*, 420–433.

Sigman, E., & Rock, I. (1974). Stroboscopic movement based on perceptual intelligence. *Perception, 3*, 9–28.

Tsuji, S., Osada, M., & Yachida, M. (1979). Three-dimensional movement analysis of dynamic line images. *Proceedings of the Sixth International Joint Conference On Artificial Intelligence*, 896–901.

Tsuji, S., Osada, M., & Yachida, M. (1980). Tracking and segmentation of moving objects in dynamic line images. *IEEE Transactions On Pattern Analysis And Machine Intelligence, PAMI-2*, 516–522.

Ullman, S. (1979). *The interpretation of visual motion*. Cambridge, MA: MIT Press.

Ullman, S. (1980). The effect of similarity between line segments on the correspondence strength in apparent motion. *Perception, 9*, 617–626.

Yalamanchili, S., Martin, V. N., & Aggarwal, J. (1982). Extraction of moving object descriptions via differencing. *Computer Graphics and Image Processing, 18*, 188–201.

Yuille, A. L. (1983). The smoothest velocity field and token matching schemes. MIT AI memo #724.

6

The Hypercomplex Neuron: Has Vision Been Thrown A Curve?*

Alan Dobbins,
Steven W. Zucker*
Max S. Cynader†
Computer Vision and Robotics Laboratory
McGill Research Centre for Intelligent Machines
McGill University

Inferring the functional role of visual cortical neurons has been closely linked to the stimuli used to study them. For moving bar stimuli, neurons typically have receptive fields that are oriented, velocity-selective, and to varying degrees, directionally-selective and binocular. In addition, hypercomplex (or end-stopped) neurons are selective for bar length: their response increases up to some length and decreases thereafter. On the basis of these stimuli it is usually held that the hypercomplex property is related to length or end-point detection. However, a different perspective is supplied by our modeling of the computations necessary for orientation selection—that local curvature estimates are required. We therefore propose that hypercomplex cells are involved in curvature estimation, and that the endstop property is a consequence of this curvature estimation. To support this proposition, we develop a mathematical model of end-inhibited simple, or simple hypercomplex (SH) cells, which captures their length-response behaviour, but which also exhibits curvature-selectivity. The model is consistent with recent discoveries about the local cortical circuits responsible for end-inhibition.

* We wish to thank Lee Iverson for programming wizardry. Research supported by NSERC grant A4470. DREA grant 09SC.97707-3-3800, and MRC grant MA6125 to the second author and by NSERC grant A9939 to the third author.

*Fellow, Canadian Institute for Advanced Research.

† Dept. of Opthalmology, University of British Columbia, Vancouver, B.C.

Physiologists usually define orientation selectivity as the differential response of neurons to stimulus orientation. From a computational perspective, however, one is interested in the inverse question: how can information about orientation be inferred from neural responses? A computational analysis should identify, in abstract terms, the components necessary for orientation selection, and should be verified by computer simulations. We are developing such a model based qualitatively on physiological constraints; and in this paper we develop one aspect of it to show how the abstract computations can be reduced to concrete, physiological terms. More generally, this suggests that the function of neurons in the early visual system can be more subtle than is normally assumed based on stimulus properties.

Orientation selection is part of the process of inferring contours from images. Mathematically such inferences can be characterized in two stages, the first one aimed at recovering local information, and the second one global (Zucker, 1986). We shall concentrate on the first, local stage, whose goal is to recover the trace of the curve, or the set of points in retinotopic coordinates through which the curve passes, together with the direction in which the curve is going (its tangent) and how the tangent is changing near each point (its curvature). We shall refer to such information collectively as a tangent field.

The job of inferring the tangent field is usually relegated to *simple* cells. The rationale for this is clear: for long, widely spaced, straight lines, such cells will respond maximally when centered on the line at the proper orientation. The intermediate responses from cells at improper orientations, or from just off the line, can be eliminated by a lateral inhibitory network that has been postulated to extend both spatially and across orientations (Blakemore, Carpenter, & Georgeson, 1970; Carpenter & Blakemore, 1973).

But orientation selection schemes like the one above cannot work in general (Zucker, 1985; Parent & Zucker, 1985), and they do not explain the available physiological data regarding cell size/layer distributions, the prevalence and variability in end-stopping, and local cortical circuitry. Computationally it can be shown that such networks respond inappropriately in the neighborhood of corners and curves, and lead to incorrect interactions between nearby curves (Fig. 1) Information about curvature is necessary in these situations, and we now show how recent work in visual cortical physiology can be interpreted to accomplish curvature estimation.

Hubel and Wiesel (1965) first discovered certain cells that respond best to stimulus bars of less than some critical length, and they

Figure 1. Illustration of the need for curvature information in orientation selection. Two arcs of constant curvature in which the vertical tangents to the two arcs are collinear. In simulations a large *simple* cell-like operator centered between them and which spans both is strongly activated. This would lead to an incorrect representation of the two distinct arcs as a single, "S"-shaped curve.

termed these cells hypercomplex. More recently, Gilbert, Wiesel and coworkers have begun to elucidate the cortical circuitry giving rise to such end-inhibition in Area 17 of the cat. Layer VI cells with long receptive fields apparently exert an inhibitory influence on similarly positioned and oriented cells with smaller receptive fields found in Layer IV, since selective inactivation of Layer VI cells abolishes end-stopping in Layer IV and superficial layer hypercomplex cells (Bolz & Gilbert, 1986). There is anatomical evidence for such local circuits as well (McGuire, Hornung, Gilbert, & Wiesel, 1984). Similarly, the striato-claustral circuit could also contribute to the synthesis of cells with end-inhibition, since the claustrum contains cells with long receptive fields and they project most densely to Layer IV of visual cortex (LeVay & Sherk, 1981; Sherk & LeVay, 1981).

Such physiological and anatomical data allow development of a model of *simple* hypercomplex (SH) cells that is sufficient for demonstrating their curvature selectivity. Briefly, the model consists of two similarly positioned and oriented *simple* cells of different size (Fig. 2).

The end-inhibited response is obtained as a weighted difference of the responses of the small and the large *simple* cells, appropriately rectified. In symbols, let R_S denote the response of the small

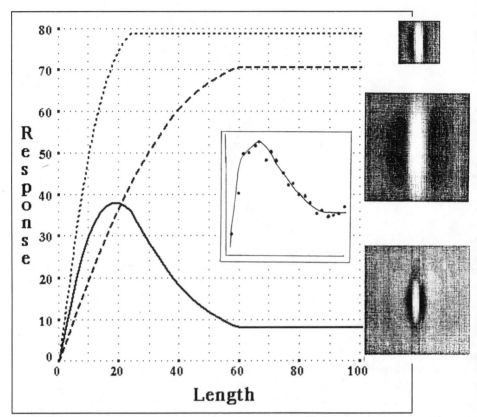

Figure 2. The right column depicts the receptive field (RF) profiles of the components of the simple type hypercomplex (SH) model and their linear combination. Regions lighter than the background are excitatory and darker regions are inhibitory. Binocular and temporal properties are not incorporated, allowing the retino-geniculo-cortical pathway to be collapsed into a spatial receptive field, and the cell's response to be modelled as a 2-D spatial convolution. Each *simple* RF may be modeled as a Gabor function (14) or, as in this case, a difference of Gaussians (DOG). The RF has a Gaussian weighting along the long axis and a DOG in the perpendicular direction. For each *simple* RF shown here the width ratio of the Gaussians comprising the DOG $(\sigma_{x2}/\sigma_{x1})$ is 2.5 and the aspect ratio (σ_y/σ_{x1}) is 4. The smaller RF (top) is 24 pixels long and the larger is 60 (middle). A linear combination of the two is also shown (three times the small one minus the large one, bottom). This representation corresponds to the SH model only when the result of both *simple* convolutions are positive. The large graph shows the response of the SH cell model to bars of different length centered on the RF and optimally oriented. The SH model (solid curve) and the component responses of the two *simple* cells ($c_S \cdot \phi(R_S)$ and $c_L \cdot \phi(R_L)$, the dotted and dashed curves, respectively) were computed for bars of width 2 pixels at 50

simple cell to a particular contrast pattern, and R_L the response from the *simple* cell with a large receptive field at the same position and orientation. In anatomical terms, these correspond to appropriate layer IV and layer VI cells, respectively. Letting $\phi(\cdot)$ denote a rectifying or clipping function which equals its argument when positive and is zero otherwise, the end-inhibited *simple* cell (SH) response is given as follows:

$$R_{SH} = \phi(c_S \cdot \phi(R_S) - c_L \cdot \phi(R_L))$$

where c_S and c_L are positive constants that normalize the area difference between the receptive fields. The nonlinear clipping in the model increases the stimulus-specificity by preventing a negative value for R_L from being flipped into a positive contribution to R_{SH}, the response of the SH cell. $\phi(\cdot)$ models the inability of spike trains to code negative numbers on a background of low spontaneous activity as is found in Area 17.

A length response curve obtained with the model is shown in Fig. 2, as are the component responses of the two *simple* cells. A length response curve obtained by Kato, Bishop and Orban (1978) is also shown for comparison in Fig. 2. By varying the size of the two receptive fields (within physiological limits) and the weighting constants it is possible to replicate any of the length response data found in the literature while maintaining position and orientation tuning.

Two further implications of the model are immediate, and both are consistent with the available data. First, note that end-inhibition is orientation-specific, with the inhibitory and zones having the same preferred orientation as the receptive field center (Hubel & Wiesel, 1965; Orban, Bishop, & Kato, 1979b). For a stimulus consisting of a short, optimally oriented bar in the receptive field center and a second bar of varying orientation in one of the end zones (Fig. 3), the reduction in response as a function of end zone stimulus orientation is similar to that found by Orban et al. (1979b). Secondly, observe that the region of strongest end zone inhibition is collinear with the strongest excitatory region of the receptive field center (Orban, et al., 1979b). The orientation-tuning curve with these model parameters had a width at half-height of 20 deg. corresponding to

---

lengths ranging from 2 to 100 pixels. c_S and c_L have the values 3 and 1, respectively. The difference of the two monotonic components is the characteristic nonmonotonic hypercomplex length-response. The small inset graph shows the off-response to flashed stimuli of varying length obtained from an SH neuron (adapted from Figure 4d. in (12)).

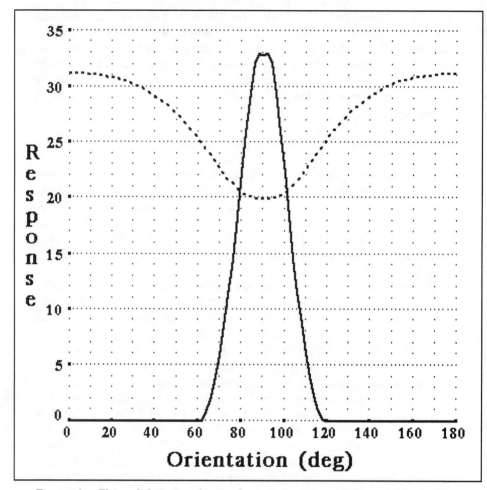

Figure 3. The solid curve shows the orientation tuning curve for a bar stimulus (dimensions: 24 by 2 pixels) evaluated at 90 equally spaced orientations. The dotted curve shows the response when an optimally oriented central bar (24 by 2 pixels) is accompanied by a bar (18 by 2 pixels) at one of 90 orientations in one of the two end zones. The response decreases as the orientation of the end zone stimulus approaches the optimal orientation of the RF.

one of the more orientation-selective SH cells described by Kato, Bishop and Orban (1978) and by Gilbert (1977).

Given that the model replicates the available length- and orientation-tuning curves, we now consider curvature. A curvature response curve can be obtained by evaluating the model over a series of cir-

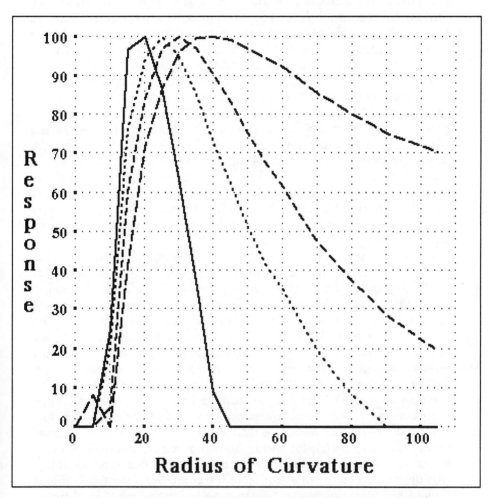

Figure 4. The result of convolving the SH model with semi-circular arcs of width three pixels and different radii to illustrate the sensitivity to curvature of the SH model. The curves are positioned so that the tangent to the curve at the point on the RF center is parallel to the long axis of the RF. The four curves are obtained with receptive fields of different size. In each case the large inhibitory receptive field is 60 pixels long and $c_L = 1$. The sizes and weights of the small receptive fields (size/c_S) are 16/5, 24/3, 36/1.8, and 48/1.32, respectively. This order corresponds to the order of the curves from left to right. Each model was convolved with 20 curves varying in radius from 5 to 100 pixels. For comparison the curves have been normalized to a peak value of 100.

cular arcs with increasing radius (Fig. 4). For a stimulus with a low radius of curvature, the curve passes through the excitatory region of the large receptive field to allow effective inhibition. As the radius of curvature increases, the stimulus becomes increasingly effective in driving the large receptive field and the inhibition increases.

The parameter values given above are consistent with those found physiologically. Assuming that the length of the SH cell discharge region approximates the length of the short receptive field, and that the discharge region plus end zones approximates the length of the long inhibitory receptive field, one can infer the ratio of receptive field lengths. Calculations based on data obtained in Table 1 of Orban, Bishop and Kato (1979a) imply a mean ratio of receptive field lengths of about 0.45. Our model, with the parameter values of Figures 1 through 3, has a length ratio of 0.4. The four sets of parameter values in Figure 4 have length ratios of 0.2, 0.4, 0.6, and 0.8.

One of the principal arguments against using curvature is mathematical. Recall that the tangent is the first derivative of the curve (with respect to arc length), and curvature is the second. Thus a common approach for estimating curvature is to first estimate the curve by, for example, fitting polynomials, and then to differentiate the result twice (de Boor, 1978). But this amplifies the errors in the initial fit substantially, leading to numerical instability. Our scheme works differently, however, avoiding the initial polynomial fit. Rather, the curvature estimates are obtained from the orientation estimates carried by the simple cells and the result is more stable.

Human psychophysical investigations also provide indirect support for the curvature-from-orientation estimation procedure. Timney and MacDonald (1978), for example, using adaptation and masking paradigms, found that curvature-specific adaptation (and masking) was dependent on the orientation of the curve but not attributable to orientation filter adaptation. Curvature is further implicated by results in dot pattern perception (Link & Zucker, in press).

To summarize, both local receptive field properties and neural circuit interactions are necessary to understand early visual function. In order to avoid incorrect interactions between cells stimulated by nearby but different curves, computational analyses indicate that curvature information is necessary. We have shown that the end-stopping property, which is normally associated with length or end-point detection, can also be explained as part of a curvature estimation process implemented by the local circuits underlying hypercomplex cells. Although we concentrated on end-stopped *simple* cells, end-stopping also occurs for *complex* cell types. We therefore think it likely that curvature plays a role in understanding their function as well. Furthermore, since the analog to curvature in space/time is acceleration, the role of such models in early optical flow should be investigated as well (Zucker & Iverson, 1986).

REFERENCES

Blakemore, C., Carpenter, R., & Georgeson, M. (1970). Lateral inhibition between orientation detectors in the human visual system. *Nature (London). 228*, 37–39.

Bolz, J., & Gilbert, C. D. (1986). Generation of end-inhibition in the visual cortex via interlaminar connections. *Nature, 320*, 362–365.

de Boor, C. (1978). *A practical guide to splines.* New York: Springer-Verlag.

Carpenter, R., & Blakemore, C. (1973). Interactions between orientations in human vision. *Experimental Brain Research, 18*, 287–303.

Gilbert, C. D. (1977). Laminar differences in receptive field properties of cells in cat primary visual cortex. *Journal of Physiology, 268*, 391–421.

Hubel, D. H., & Wiesel, T. N. (1965). Receptive fields and functional architecture in two non-striate visual areas (18 and 19) of the cat. *Journal of Neurophysiology, 28*, 229–89.

Kato, H., Bishop, P. O., & Orban, G. A. (1978). Hypercomplex and simple/complex cell classifications in cat striate cortex. *Journal of Neurophysiology, 41*, 1071–95.

LeVay, S., & Sherk, H. J. (1981). The visual claustrum of the cat I. Structure and connections. *Journal of Neuroscience, 1*, 956–80.

Link, N., & Zucker, S. W. (in press). Corner detection in curvilinear dot grouping. *Biological Cybernetics.*

Marcelja, S. (1980). Mathematical description of the responses of simple cortical cells. *Journal of the Optical Society of America, 70*, 1297–1300.

McGuire, B. A., Hornung, J. P., Gilbert, C. D., & Wiesel, T. N. (1984). Patterns of synaptic input to layer 4 of cat striate cortex. *Journal of Neuroscience, 4*, 3021–33.

Orban, G. A., Bishop, P. O., & Kato, M. (1979a). End-zone region in receptive fields of hypercomplex and other striate neurons in the cat. *Journal of Neurophysiology, 42*, 818–32.

Orban, G. A., Bishop, P. O., & Kato, H. (1979b). Dimensions and properties of end-zone inhibitory areas in receptive fields of hypercomplex cells in cat striate cortex. *Journal of Neurophysiology, 42*, 833–49.

Parent, P., & Zucker, S. W. (1985). Trace inference, curvature consistency, and curve detection. (Report No. 85-12R). Montreal: Computer Vision and Robotics Laboratory, McGill University.

Sherk, H. J., & LeVay, S. (1981). The visual claustrum of the cat III. Receptive field properties. *Journal of Neuroscience, 1*, 993–1002.

Timney, B. T., & Macdonald, C. (1978). Are curves detected by 'curvature detectors'? *Perception, 7,* 51–64.

Zucker, S. W. (1985). Early orientation selection: tangent fields and the dimensionality of their support. *Computer Vision, Graphics and Image Processing, 32,* 74–103.

Zucker, S. W. (1986). The computational connection in vision: early orientation selection, *Behaviour Research Methods, Instruments, and Computers, 18,* 608–617.

Zucker, S. W., and Iverson, L. (1987). From orientation selection to optical flow. *Computer Vision, Graphics, and Image Processing, 37,* 196–220.

II
Visual Routines

7

Curve Tracing Operations and the Perception of Spatial Relations*

Pierre Jolicoeur

University of Waterloo

In many situations the information that the human visual system must deliver to other systems (perceptual, motor, cognitive, and so on) involves specifying the relationship between elements in a visual scene, or the relationship between the observer and elements in the scene. Such information is required, for example, when we move through the environment, when we reach for objects, and when we engage in tasks involving complex pattern recognition such as reading. In all these examples, the visual system must integrate information across visual space and establish certain spatial relations between elements in that space.

How is the integration of information across diverse spatial locations achieved? The answer to this question is unlikely to be simple or based on a single mechanism. One approach designed to provide a general answer to the problem is to suppose that the visual system has a set of procedures or basic operations that can be invoked when the system computes a particular spatial relation between elements in a display (Jolicoeur, Ullman, & Mackay, 1986; Ullman, 1984). In this view, the visual system has at its disposal a limited set of basic operations each one of which can perform a very simple task. For example, we may have a basic operation to index or mark a particular point in visual space (Pylyshyn, 1988), and an operation that

---

* The work presented in this chapter has benefited from the collaboration of Shimon Ullman and Lynn Mackay. I thank Pia Amping for the artwork in Figures 1 and 2. I also acknowledge the contributions of Bruce Milliken, Doug Snow, Donna Henderson, Donald Clark, and Pat Hachey who helped set up the experiments, test subjects, and process the data.

determines whether a set of indexed locations are collinear (e.g., Uttal, Bunnell, & Corwin, 1970). A number of such operations concatenated appropriately could yield solutions to more complex problems. A particular ordering of basic operations designed to compute the answer to a general problem can be thought of as a visual routine (Ullman, 1984). By analogy to computer systems, basic operations are similar to basic machine operations (machine language), while visual routines are similar to short compiled programs or subroutines that can be invoked when required.

Progress toward a general theory of the perception of spatial relations is likely to follow detailed empirical and theoretical studies of the perception of simple relations. In the context of the visual routine framework, a detailed understanding of the postulated basic operations is a fundamental prerequisite for an explicit and complete general theory. The studies summarized in this chapter were designed to increase our understanding of one possible basic operation involved in the perception of spatial relations: *curve tracing*.

There is considerable evidence that contours and boundaries play a fundamental role in perception (e.g., Attneave, 1954; Biederman, 1985; Hubel & Wiesel, 1968). Furthermore, there are many visual tasks that require the integration of information that lies along a contour or is connected by a contour or boundary. The interpretation of geographic maps and of electronic circuits provide interesting situations in which information lying on, or connected by, contours must be integrated over large spatial regions. The following tasks are good examples: determining the altitude of a geographic location when reading a topographic map (see Figure 1); determining whether two lakes are connected by a river; determining whether one could drive from one city to another using a particular road; and deciding whether two electronic components are connected when reading an electronic circuit diagram (see Figure 2). Each one of the above examples can be thought of as a special case of a more general problem: determining whether two locations are joined by a continuous contour or boundary. A basic operation capable of following along or *tracing* a contour in a visual display would be useful to the visual system when solving this type of problem. The hallmark characteristic of such a tracing operation is that more time should be required to trace a longer curve distance than to trace a shorter distance. In this chapter I review recent experimental evidence in which more time is, in fact, required to solve spatial relations problems involving longer curve distances, which suggests that the visual system uses curve tracing operations to solve certain types of spatial relations problems. The experiments also explore how vari-

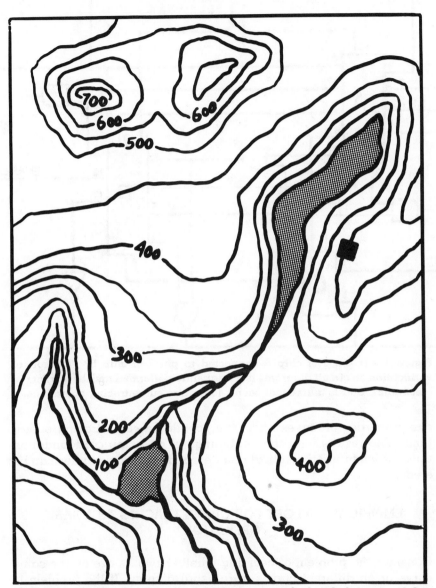

Figure 1.   What is the altitude of the location marked with the square? One way to answer this question is to trace the contour line from the square to the nearest altitude marking.

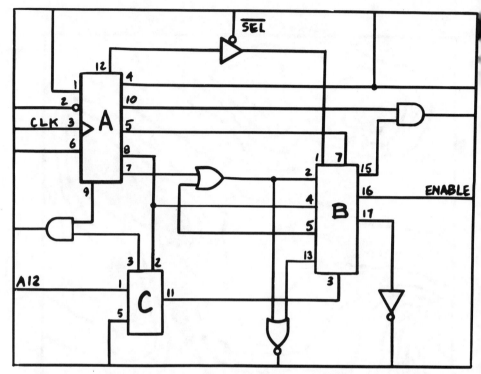

Figure 2. Is pin 5 of chip A connected to pin 7 of chip B? This sort of verification on circuit diagrams is a frequently performed task by workers in electronics, and is a task that seems to require curve tracing operations.

ous properties of curves affect the postulated curve tracing operations. The results form the beginnings of an empirical foundation upon which future theoretical and detailed empirical studies can be erected.

EMPIRICAL SUPPORT FOR CURVE TRACING AS A BASIC OPERATION

Consider the problem of deciding whether two Xs are on the same curve or on different curves in a visual display. If the problem is solved using a tracing operation, then, all else being equal, more time should be required when the Xs are joined by longer curves than when they are joined by shorter curves. In Experiments 1 and 2, an experimental paradigm is developed to test this prediction of the tracing hypothesis.

Experiment 1[1] A set of displays used in the present experiment

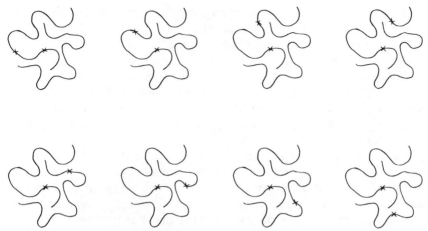

Figure 3.   A set of displays used in Experiment 1. The top row shows patterns used in "same" trials with curve distance increasing across displays from left to right. The bottom row shows patterns used in "different" trials. Reprinted from Jolicoeur, Ullman, & MacKay, 1986, by permission of The Psychonomic Society.

can be seen in Figure 3. In each display there are two nonintersecting curves and two Xs. These displays and others constructed in a similar way were shown to subjects in tachistoscopic presentations. The task was to decide, as rapidly as possible while keeping errors to a minimum, whether the two Xs were on the same curve or on different curves. In half of the displays the Xs were in fact on the same curve, as in the top row of displays in Figure 3, whereas in the other half the Xs were on different curves, as in the bottom row in Figure 3. One of the Xs was always located at the center of the display, at the location previously occupied by a fixation point that was present between trials but not during the exposure of the target display. The critical variable in the experiment was the distance *along the curve* that separated the Xs when they were on the same curve. Four different curve distances were used in trials in which the Xs were on the same curve. Distance along the curve was measured to the second X starting from the central X (i.e., the X located at the center of the display). One way to conceive of this measure is to

---

[1] More details on Experiments 1, 2, and 3 can be found in Jolicoeur et al. (1986). The other experiments are described in more detail in Jolicoeur and Milliken (1988), and Jolicoeur, Ullman, and MacKay (1988). Throughout this chapter, only the response time results from the experiments are reported. In all cases, the error rates either mirrored the response time results or they were not systematically associated with them, and thus the patterns of response times were not the result of speed-accuracy trade-offs.

imagine laying a string exactly on top of the curve and then cutting the string so that it begins at the central X and ends at the second X. The distance along the curve between the two Xs is the length of the string after it has been straightened to form a line.

Although the distance between the Xs along the curve was varied systematically, the physical distance between the Xs in the displays was held constant and equal in all cases. This control was achieved by positioning the Xs on an imaginary circle of fixed radius (1.8° of visual angle) concentric about the fixation point. Several different pairs of curves like the one shown in Figure 3 were used in the experiment. A fixation point was present between trials and was removed immediately prior to the onset of the target display, which remained in view for 250 msec. The results from the experiment are shown in Figure 4. The mean response time for trials in which the two Xs were on the same curve ("same" trials) increased monotonically with increasing curve distance. These results are as would be expected if the visual system's solution to the "two-Xs-on-a-curve" problem involved tracing the curve joining the two Xs. On the assumption that the increase in mean response time reflects curve tracing operations, the rate of curve tracing was about 48°/sec. As is evident in Figure 4, the average response time for "different" trials was relatively long, which is consistent with the notion that "different" responses were emitted after the curve going through the central X was traced without finding a second X. These tracing times would thus tend to be as long or longer than the longest "same" trial.

Experiment 2. One aspect of the results in Experiment 1 that was not expected was the nonlinear increase in response time with greater curve distance. Initially, it was thought that the nonlinearity could have been produced by the use of a relatively brief exposure duration (250 msec). This reasoning was based on the assumption that the representation used for tracing would persist for a brief time and then fade (Coltheart, 1980). Furthermore, tracing over a fading representation would affect longer distances more than shorter ones because more time would be required to trace over a longer curve distance, and the representation that was traced over could have faded more for long than for short distances. To investigate this possibility, the experiment was repeated with a longer exposure duration of the target curves (2500 msec), which should preclude the possibility that the representation that is traced would fade while it is processed.

The results from the present experiment are shown in Figure 5. Evidently, tracing over a fading representation is not the sole reason for the nonlinear results in the first experiment given that the present

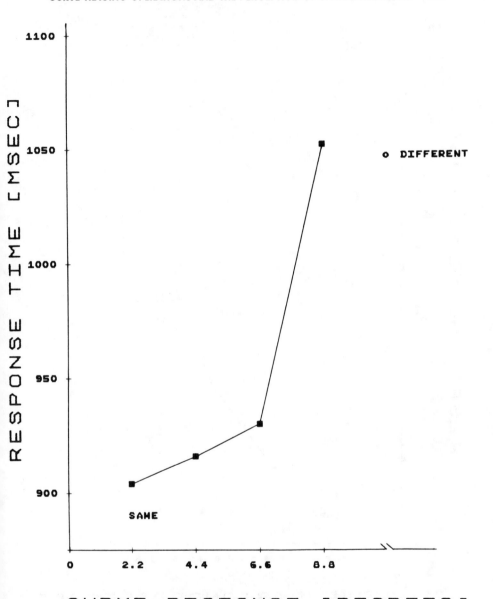

Figure 4.   Results from Experiment 1 (250 msec exposure duration).

results also departed significantly from linearity. However, the results provide a replication of the most important findings of Experiment 1, namely that response time increases monotonically with increasing curve distance. On the assumption that the slope of the

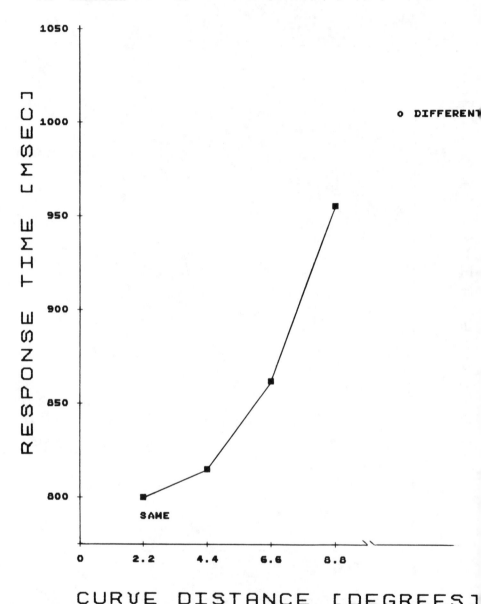

Figure 5. Results from Experiment 2 (2500 msec exposure duration).

distance effect reflects the rate of curve tracing, the rate in the present experiment was 43°/sec.

Overall, the main results from Experiment 1 and 2 are clear-cut: More time is required to respond to trials involving longer curve distances than those involving shorter distances. Taken together, the results suggest that subjects traced the curve joining the two Xs when they solved the "two-Xs-on-a-curve" problem. The findings in these experiments could not have resulted from superficial properties of the displays, such as the complexity of the curves or the placement of the Xs. Given that the same pair of curves was used equally often in each distance condition, longer distances did not involve more complex displays than shorter distances. Also, although the curve distance separating the Xs was varied systematically, the actual physical distance between Xs was constant and therefore this variable or any other associated with the physical positioning of the Xs (such as retinal eccentricity) cannot account for the main findings associated with curve distance.

The brief exposure duration used in Experiment 1 (250 msec) suggests that the tracing operations can operate internally, without eye movements. In fact, there was no statistical difference in the rate of curve tracing across Experiments 1 and 2, which suggests that subjects did not use eye movements to trace the curve even when they could have done so.

SOME PROCESSING PROPERTIES OF CURVE TRACING OPERATIONS

Does Curve Tracing Proceed in a Center-Out Self-Terminating Scan?

In Experiments 1 and 2, longer curve distances were associated with longer response times. One possible model of the underlying operations producing these results is that subjects scanned the central curve in the direction of the second X, starting at the central X, and responded "same" as soon as the scan of the curve reached the second X. This series of processing operations can be thought of as a center-out self-terminating scan. Experiments 3 and 4 examined this possibility.

Experiment 3. The displays used in the "same" trials of Experiments 1 and 2 were employed in modified form in the present experiment. In half of the displays, there was a small gap along the curve joining the two Xs. These displays were used in "gap" trials. In the other half, the joining curve was unbroken. These displays were used in "no-gap" trials. Example displays can be seen in Figure 6. The task was to decide whether or not the curve joining the two Xs contained a gap. The no-gap displays included four different curve distances, and thus these displays were identical to the displays used in "same" trials in the first two experiments. For gap trials with

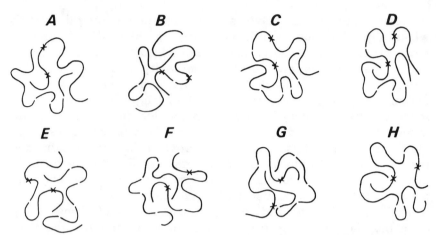

Figure 6. Example stimuli in Experiment 3 and 4. Display A shows a no-gap trial with a distance of 1 unit between the Xs. Displays B to H show gap trials. In B and C, the distance between the Xs is 1 and 2 units, respectively. In D and E, the Xs are 3 units apart; the gap is at the 1-unit mark in D and at the 2-unit mark in E. F, G, and H show trials with 4 units between the Xs; the gap is at the 1-unit mark in F, at the 2-unit mark in G, and at the 3-unit mark in H. This figure also shows the eight basic patterns of curves used in Experiments 1, 2, 3, and 4 (used without gaps in Experiments 1 and 2). Reprinted from Jolicoeur, Ullman, & MacKay, 1986, by permission of The Psychonomic Society.

one or two units of curve distance between the Xs, the gap was located half way along the joining curve between the Xs. In the case of displays with three units of curve distance separating the Xs, there were two possible locations for the gap, either at the one-unit location from the central X, or at the two-unit location. For displays with four units of curve distance between the Xs, there were three possible gap locations: at the one-unit, two-unit, or at the three-unit location as measured from the central X. Note that for gap trials with two, three, or four units of curve distance separating the two Xs, the gap was always located on the imaginary concentric circle used to locate the Xs themselves. Thus, as in Experiments 1 and 2, the physical distance between the central X and the gap was constant across all trials, and also the retinal eccentricity of the gaps was constant and equal to 1.8° of visual angle. In gap trials with one unit of curve distance separating the Xs, the gap could not be located on the imaginary circle and it was somewhat closer to the fixation point than 1.8°. The displays replaced a fixation point that was present between trials and remained in view until the subject made a response.

The results for no-gap and gap trials, averaging across gap locations for gap trials, are shown in Figure 7. Consider first the results for no-gap trials. These results are similar to those found in Experiments 1 and 2: Response times increased monotonically with increasing curve distance, as we would expect if the entire length of

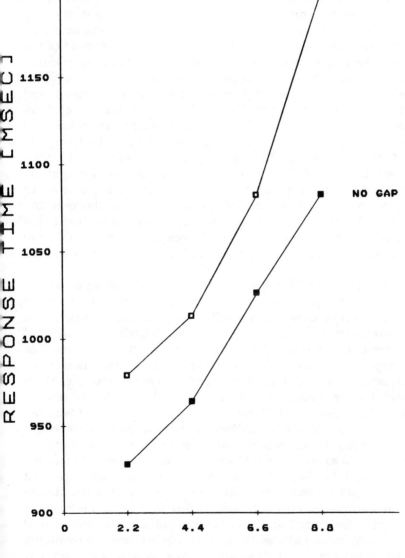

Figure 7.   Results from Experiment 3. The results for gap trials are averaged across gap locations.

the curve joining the two Xs had to be traced in order to ascertain that there was no gap. In fact, the increase in response times was more clearly linear than in the first two experiments, and the deviations from linearity were not significant.

Now consider the results for gap trials. As is evident in Figure 7, when the data are averaged according to the distance between the Xs, across trials that differed in gap location, the results are very similar to those found for no-gap trials. In fact, there was no statistical difference between the slopes of the linear component of the increase in response time with distance for gap and no-gap trials. This pattern of results is unexpected if the tracing process started at the central X and self-terminated as soon as a gap was encountered along the joining curve. According to this model, the slope of the distance function for gap trials should have a slope that is half that for the no-gap trials, given that the gap was located half way between the Xs, on average. The results plotted for each gap location are shown in Figure 8. The dotted lines in the figure highlight the results for gap trials in which the gap was located at the one-unit mark, as measured from the central X. Clearly, the results are inconsistent with the center-out self-terminating model. According to the model, the response times for gaps at the one-unit mark should have been the same for all trials, regardless of the distance between the Xs.

Experiment 4. In contrast with a search for two Xs along a particular curve in a display, the search for a gap does not appear to involve a self-terminating process. This result was surprising enough to lead to the present experiment. The subjects in Experiment 3 may not have realized that the curve tracing could self-terminate, given that all displays had two Xs on the target curve. In fact, the instructions were to respond "gap" if a gap was found *between* the two Xs, along the curve joining them. One possibility is that subjects always checked that a gap found along the central curve in fact was *between* the two Xs. Doing so would require tracing the entire length of the curve joining the Xs, which is apparently what subjects did.

In the present experiment subjects were told explicitly before the experiment that the two Xs would always be on the same curve and that they should respond "gap" as soon as a gap was found anywhere on the central curve. It was hoped that these instructions would lead to the adoption of a center-out self-terminating scan, if such a strategy makes efficient use of the basic processes available to the visual system. In all other ways, however, the procedure was identical to that used in Experiment 3.

The results for no-gap and gap trials, averaging across gap locations for gap trials, are shown in Figure 9. The results for no-gap

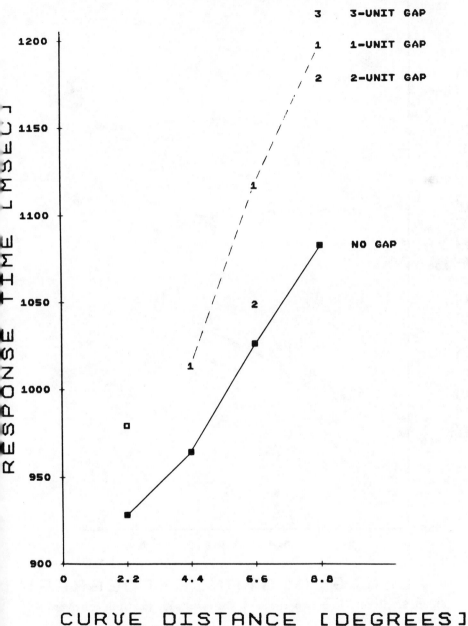

Figure 8.   Results from Experiment 3. The results for gap trials are shown
for each gap location. The plotted symbol represents the gap location in gap
trials.

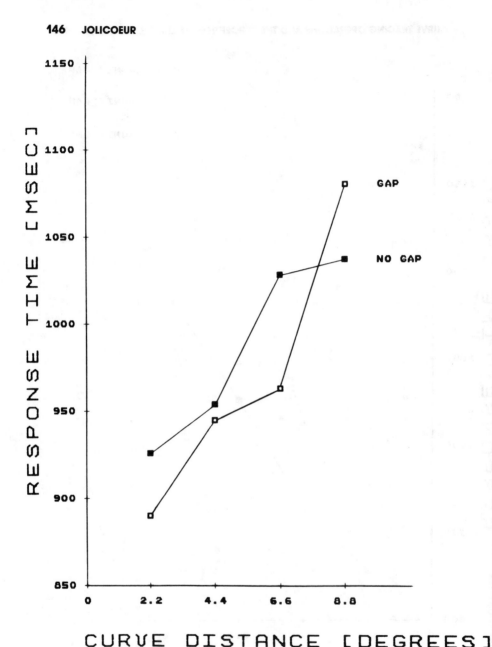

Figure 9.   Results from Experiment 4. The results for gap trials are averaged across gap locations.

trials again replicated the main results obtained in the first three experiments—response time increased monotonically as curve distance increased. As in Experiment 3, the results for gap trials also increased monotonically with increasing curve distance between the

*X*s. Also as in Experiment 3, there was no significant difference in the slope of the linear component of the increase in response time across distance between gap and no-gap trials. The results for each gap location in gap trials and those for no-gap trials are displayed in Figure 10. If a center-out self-terminating scan had been used, the results for gap trials in which the gap was at the one-unit mark should have been equivalent. Clearly, the results are inconsistent with the center-out self-terminating model.

The results of Experiments 3 and 4 suggest that curve tracing operations are used to detect the presence of a small gap in an otherwise continuous contour. However, one particular model of these tracing operations, the center-out self-terminating model, is inconsistent with the results. Although self-termination apparently did not occur in response to the target gap, it apparently did in response to the *X*s. In every case, the curve distance between the *X*s controlled response times. A model in which the curve between the *X*s was scanned exhaustively is more consistent with the results from gap trials. The exact reasons for this exhaustive scanning, as opposed to self-termination, remain unclear. In retrospect, the use of a gap as a target feature along the joining curve may not have been the most judicious choice. A more easily detected and indexed feature or object would perhaps have produced results consistent with the center-out self-terminating model. Experiments designed to test this conjecture are currently underway. To speculate a little further on this matter, it is possible that small gaps in an otherwise continuous curve are especially difficult for curve tracing operations to handle. In fact, a general curve tracing procedure should be able to gloss over such gaps, as when we interpret lines indicated by dashes and dots as political boundaries on geographic maps.

Are Curve Tracing Operations Spatially Serial or Spatially Parallel?

Experiment 5. The issue addressed in the present experiment is whether the curve tracing operations investigated in the first four experiments can take place on more than one curve simultaneously (spatially parallel processing), or whether these operations process a single curve at a time (spatially serial processing). Each display in the experiment had a circle and a dot at the center of the circle and one or more curves or paths starting at the central dot and terminating on the circle. Three variables were manipulated in the experiment. First, the curves were either all discontinuous (they all had a gap) or one of the curves did not have a gap and thus that path from

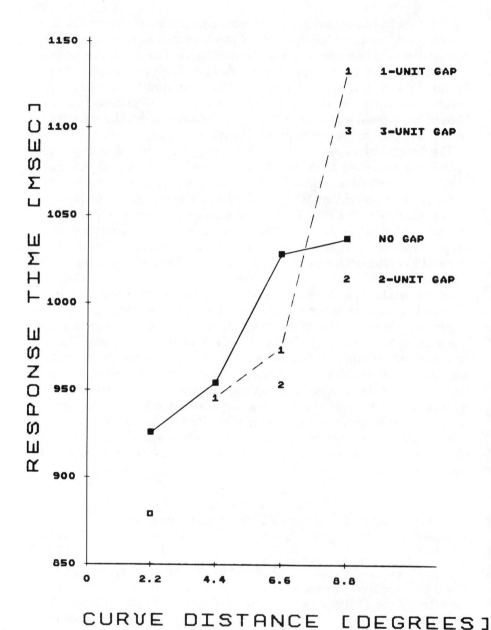

Figure 10.   Results from Experiment 4. The results for gap trials are shown for each gap location. The plotted symbol represents the gap location in gap trials.

the center to the circle was continuous. The task was to decide either that all the paths were discontinuous (all-gap trials) or that one of them did not have a gap (no-gap trials). In every display there was a random number of gaps positioned on the circle itself so that the gap/no-gap decision could not be made simply by detecting any gap in the display when there was only one path. The target gap had to be located on one of the curves linking the central dot with the circle. The second variable was the number of paths from the central dot to the circle: either one, two, or four. The length of the paths was also manipulated systematically, which was the third variable. The shortest path length was slightly longer than the radius of the circle. This path length was taken as the basic unit of curve distance (which was 3⅓° when displayed to subjects). Two other lengths were used: twice the shortest length (6⅔°) and three times the shortest length (10°). All the paths in any one display had the same length (either one, two, or three units). Example displays before the gaps were added can be seen in Figure 11. As in Experiments 3 and 4, the display remained in view until the subject made a response.

The results are shown in Figure 12. As in Experiments 1, 2, 3, and 4, response time increased as the length of the curve(s) to be traced increased, which is consistent with the notion that the paths were traced to decide whether one of them did not contain a gap. For both all-gap and no-gap trials, the number of curves also had a large effect: Response times increased as the number of curves was increased. These results suggest that curves were processed serially, one at a time, rather than in parallel. Three other findings support the serial model. First, as can be seen in Figure 12, the effect of path length was greater when there were more paths. If each path was processed one at a time, and if a longer path takes longer to trace, then the effect of path length should increase as more paths need to be processed because each additional path will take longer to trace. The second source of support for the serial model can be found in the difference in the slope of the path length effect between no-gap trials and all-gap trials. As shown in Figure 12, the effect of path length was smaller for no-gap trials than for all-gap trials. This is exactly what should have happened if subjects traced each path one at a time and did not trace new paths when they found a path that did not have a gap. That is, if the sequential processing of paths self-terminated when a continuous path was found. In this view, the effect of path length should be smaller in no-gap trials because, on average, fewer paths would need to be traced than in the all-gap trials, in which all the paths would presumably be traced in order to decide that none was continuous. The third piece of evidence supporting the

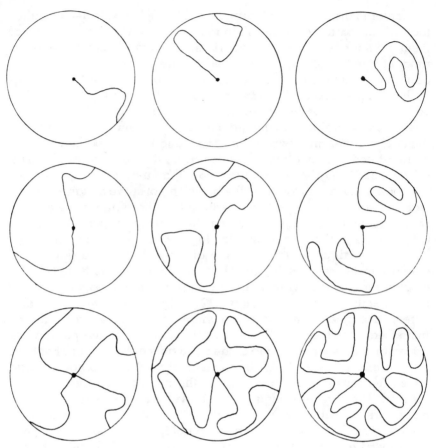

Figure 11.   Displays used in Experiment 5 (shown here without gaps).

serial model is that the effect of the number of paths was greater for all-gap trials than it was for no-gap trials (see Figure 12). This result is consistent with the serial model if all the paths were traced in all-gap trials and only some paths were traced in no-gap trials, as would be expected in a self-terminating process considering each path one at a time.

The serial model appears to provide a reasonable account for all the major findings in the present experiment. Thus, it appears that curve tracing operations proceed in a spatially serial fashion, at least when the goal of the system is to find a continuous curve in a set of discontinuous ones.

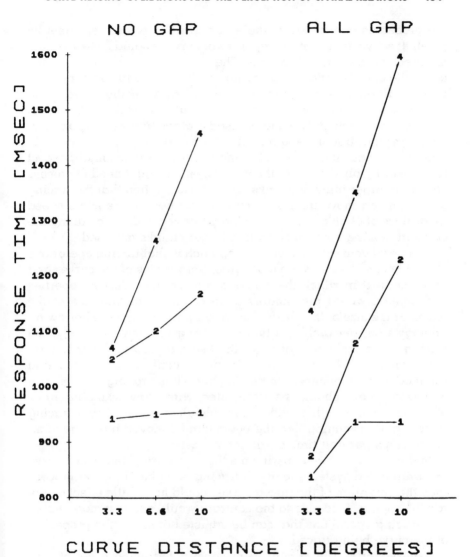

Figure 12.   Results from Experiment 5.

THE EFFECTS OF CURVE CHARACTERISTICS ON TRACING OPERATIONS

Does Curvature Affect the Rate of Curve Tracing?

Insight into the nature of tracing operations can be gained by studying the effects of various curve properties. Suppose that curve trac-

ing processes operate at a single fine grained scale (i.e., pixel by pixel). If so, various global properties of curves should have virtually no effect on the rate of tracing. The process would simply march along the curve from one local region to the next until some termination condition was met (such as finding an X on the curve). One advantage of such a process is that it would be easy for the visual system to implement. One major disadvantage of pixel-by-pixel processing is that tracing would tend to be a relatively slow process. In order to increase the speed of tracing beyond what one might expect from pixel-by-pixel tracing, the visual system would need to process curves in larger units or chunks. On the assumption that the tracing operation can move from one chunk to the next at the same speed regardless of chunk size, the tracing process could march along the curve at much greater speed when larger chunks are used.

A general consequence of this approach is that tracing operations should become sensitive to more global properties of the curves and of the context in which the curves are presented. Curve properties and context would presumably have an effect on the size of the chunk or the size of the "beam" used to scan the curve, to borrow an analogy from research on attention. Using the beam analogy, the tracing process should not move the beam a greater distance than the diameter of the beam, in order not to confuse the curve being scanned with other curves in the display. Thus, tracing a curve with a narrow beam should be associated with slow scanning rates whereas tracing with a wide beam should result in rapid tracing rates, again assuming that the beam can be moved the same number of steps per unit time at all beam sizes.

In the following two experiments the curvature of the traced curve is manipulated systematically. If tracing is a pixel-by-pixel process then the curvature of the traced curve should have little effect on the rate of tracing. Evidence to the contrary would suggest more elaborate tracing operations that can be adjusted or set by the properties of the curve being traced.

Experiment 6. The stimuli were constructed using a set of concentric circles in which the radii of the circles increased by a fixed amount, which ensured that the curves had equidistant neighbors at all curvatures, as can be seen in Figure 13. On the original stimuli from which slides were made later, the smallest circle had a radius of 5 mm and the increment in radius was 5 mm; the largest circle had a radius of 100 mm. Each display had two dots that were either on the same circle or on adjacent circles. The task was to decide whether the dots were on the same circle or on different circles, as rapidly as possible while keeping errors to a minimum. The dots were placed

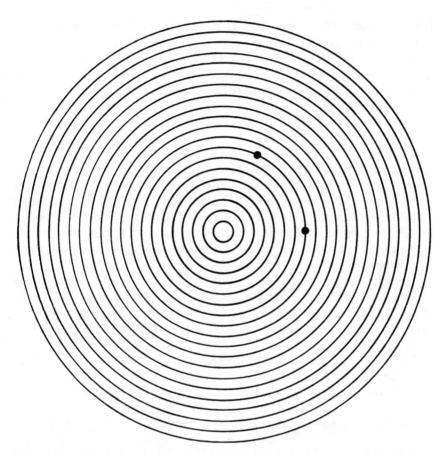

Figure 13. Set of concentric circles used to construct the stimuli for Experiments 6 and 7, and two dots yielding an example "same" trial.

on circles with different radii, and therefore on circles with different curvatures (a greater curvature for a smaller radius). Three sets of trials were created using the concentric ring display. These radii were 20 mm, 40 mm, and 60 mm. In addition, a display with parallel lines (infinite radii or zero curvature) was also used. The parallel line displays were always displayed such that the lines were oriented at 45° from vertical; both obliques were used equally often. For each curvature three sets of trials were created by varying the curve distance between the dots. Three distances were employed: 15 mm, 30 mm, and 45 mm, on the original stimuli. The stimuli were later displayed such that the curve distance between the dots was 2°, 4°, and 6° of visual angle, and such that one of the dots was at the center of the display. The target stimuli were preceded by a fixation point presented for 1000 msec. The display remained in view until the subject made a response.

It was expected that if curvature had an effect, it would act to slow the rate of curve tracing for curves with greater curvature. In the present experiment there was a confounding between the physical distance between the dots and curvature. For displays with zero curvature (parallel lines) the physical distance was the greatest, for any given curve distance. As the curvature of the target circle increased, the physical distance between the dots decreased, for any given curve distance. In the present paradigm I would expect that target dots closer to the fovea would produce faster response times than dots further from the fovea, if any difference was associated with retinal eccentricity. Thus, if anything, the confound between physical distance (which also corresponded with retinal eccentricity) and curvature should act to produce results in the opposite direction from what was expected for the curvature manipulation.

The results for "same" trials, which are of greatest interest, are shown in Figure 14 (similar results were also obtained for "different" trials). The results have two principal characteristics. First, response time increased approximately linearly with increased curve distance for all curvature conditions, which suggests that the circles (or lines) were traced in order to decide whether the dots were on the same circle (or line) or on different ones. Second, the distance effect was increasingly large as the curvature of the traced circle was increased. As expected if curve tracing operations do not take place at a pixel level, the rate of tracing was slower for circles with greater curvature.

Experiment 7. The present experiment is a replication of Experiment 6 in which the exposure duration of the displays was 180 msec. The purpose of the experiment was to discover whether the effects of curvature would be found when eye movements could not occur. If the results mirror those found in Experiment 6, we will have good evidence that the adjustments to the rate of curve tracing occur rapidly within an internal set of tracing operations.

The results are shown in Figure 15. Although the results are slightly more noisy (as would be expected because of the brief exposure duration and also because fewer subjects were tested), they replicate the main aspects of the results in Experiment 6. Response time increased approximately linearly for all curvatures, and the effect of distance was greater as the curvature increased.

Together, the results of Experiments 6 and 7 demonstrate that the rate of curve tracing is not constant across all applications of tracing operations. This is to be expected if tracing operations can be adapted to the particular characteristics of the displays that are pro-

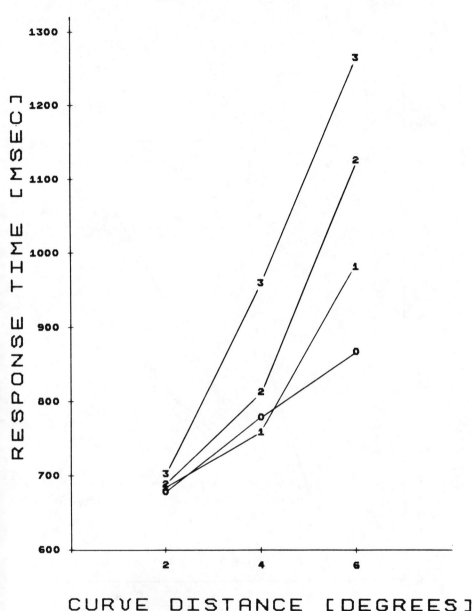

Figure 14.   Results from Experiment 6.

cessed. Pixel-by-pixel processing would not be expected to exhibit the sort of dependence on curvature demonstrated in Experiments 6 and 7. Additional evidence that tracing rate is affected by display properties is provided in Experiment 8.

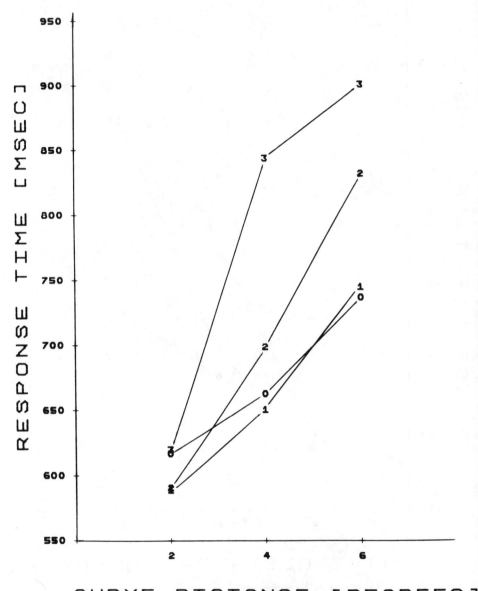

Figure 15.   Results from Experiment 7.

Is the Rate of Curve Tracing Affected by the Proximity of Distractor Curves?

Experiment 8. The results of Experiments 6 and 7 suggest that the rate of curve tracing is influenced by properties of curves such as curvature. One interpretation of these results is that curves with greater curvature must be processed in smaller chunks so that the tracing processes will not lose track of the curve. Another display variable that would be expected to affect the grain size of tracing operations is the proximity of other irrelevant curves in the display. In the experimental paradigm we are using, tracing operations serve as an identity operator whose purpose is to establish that two symbols in fact lie on one and the same curve. If another curve is brought in close proximity with the target curve, tracing the target curve would be expected to become more difficult because of a danger in confusing the distractor curve with the target curve. In order to avoid a tracing error in which the wrong curve became the object of processing, I expected that tracing operations would need to focus more closely on the target curve, as though the diameter of the beam used in the tracing had to be narrowed. If so, the rate of tracing should be slower when a distractor curve is potentially more confusable with the target curve than when it should be less confusable.

In the present experiment a test of the above prediction was provided by using target curves similar to those used in Experiments 1 and 2. However, in contrast with the displays in these experiments, the distractor curve in the present experiment intersected the target curve at several points. Figure 16 shows the type of arrangement used in Experiments 1 and 2 and in the present experiment. As in Experiments 1 and 2, the task was to decide whether two Xs were on the same curve or on different curves. The displays remained in view until the subject made a response.

The results are shown in Figure 17. As in earlier experiments, response time in "same" trials increased linearly as curve distance was increased. Furthermore, response time in "different" trials were relatively long, indicating that these responses may have been given when no second X could be found on the target curve. The most important finding, however, concerns the rate of tracing in the present experiment compared with that found in earlier experiments. The scale on the response time axis of the graph (ordinate) indicates that tracing was much slower for the present displays than in previous experiments. To illustrate this point more clearly, the results from "same" trials in the present experiment were compared with the results from no-gap trials in Experiment 3, in which the displays

Figure 16.   Example stimulus used in Experiment 8 shown at the top and a previous stimulus at the bottom.

also remained in view until the subject made a response. These means are shown in Figure 18. Clearly, the effect of the intersecting distractor curve was to cause a dramatic decrease in the rate of tracing. Thus, the results suggest that curve tracing does not usually take place at a pixel-by-pixel level, although the conditions of the present experiment may have produced this sort of tracing, which would result in the very slow tracing rates that were observed.

Is Tracing Horizontal or Vertical Lines Faster Than Tracing Oblique Lines?

Experiment 9. The visual system often processes properties of stimuli in relation with other stimuli in the visual field (e.g., Johansson, 1950) or in relation with a perceptual frame of reference (e.g., Corballis, Nagourney, Shetzer, & Stefanatos, 1978; Corballis,

Zbrodoff, & Roldan, 1976; Corcoran, 1977; Duncker, 1929; Feldman, 1985; Jolicoeur, 1988; Rock, 1973). In the present experiment subjects traced lines oriented in different directions with respect to the perceptual frame of reference. The main comparison of interest was

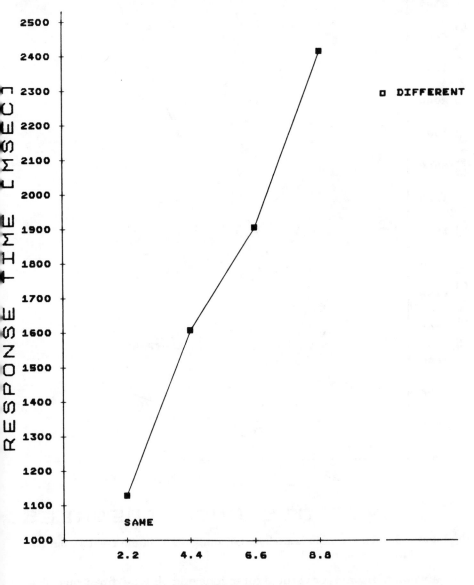

Figure 17.   Results from Experiment 8.

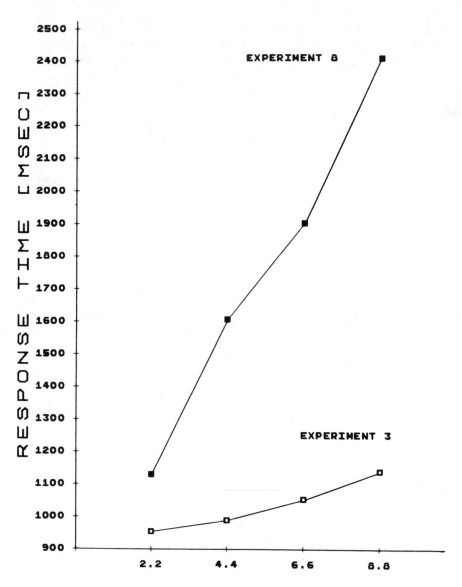

Figure 18.   Results from Experiment 8 compared with the results from no-gap trials in Experiment 3.

between horizontal or vertical directions and oblique directions. The displays were similar to those used in the zero curvature condition in Experiments 6 and 7. In these earlier experiments, however, the parallel line displays were always presented such that the lines were at oblique tilts from the vertical (clockwise and counterclockwise tilts were used equally often). In the present experiment, performance with displays at ±45° was compared with performance with displays at 0° or 90°. In each display two dots were either on the same line or on different lines, and the distance between the dots was varied systematically. In all displays, the two dots were placed on a circle of fixed radius concentric on the fixation point. Thus, in all conditions, the dots were equally far from the fixation point regardless of the curve distance between them or of the orientation of the lines in the display.

The results for "same" trials can be seen in Figure 19. Again, the effect of distance between the dots provides converging evidence in favor of the tracing hypothesis. In addition, oblique trials were slower overall than horizontal/vertical trials. Furthermore, the rate of tracing was slower for oblique trials than for horizontal/vertical trials. These differences between oblique and horizontal/vertical directions have been replicated in a number of other more recent experiments.

The results of the present experiment are consistent with earlier work of Church and Church (1977), in which spatial scanning processes were studied in the context of chess play. Church and Church asked a subject to decide whether a black king was in check or not in the presence of a single white attacking piece. The white piece could be a bishop, a queen, or a rook. The number of squares separating the white piece from the black king on the chess board was manipulated systematically. More time was required to decide that the king was in check as the number of squares between the two pieces was increased. Furthermore, the rate of increase with each additional square was much greater when the attack was along a diagonal (bishop or queen) than when it was along a parallel (rook or queen). It is likely that their task required some form of spatial scanning operation that may be similar to the curve tracing under study in the present study. In the Church and Church study, however, there were some variables covarying with the horizontal/vertical versus oblique comparison, which could have slowed the oblique scans relative to the horizontal/vertical scans. For example, scans along parallels may have been easier because they were parallel with the squares on the board whereas scans along diagonals did not have any lines parallel with the direction of the scan. The present experiment provides a more convincing demonstration that oblique spatial scanning (curve tracing in our case) is slower than horizontal/vertical scanning.

The differences between oblique and horizontal/vertical trials suggest that tracing takes place within a frame of reference whose

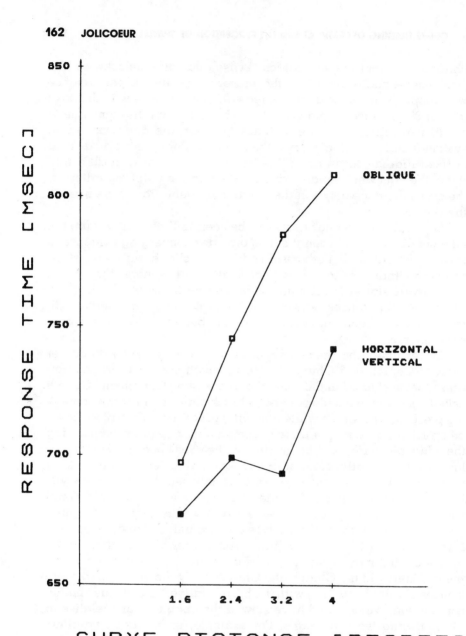

Figure 19. Results from Experiment 9.

directions are not completely isotropic (Appelle, 1972). More recent research investigated whether the frame of reference used for curve tracing is predominantly retinal or predominantly gravitational (or environmental). The results suggest a predominantly retinal influence (Jolicoeur & Milliken, 1988).

SUMMARY AND CONCLUSIONS

A decision tree shown in Figure 20 summarizes the principal conclusions from the present study. Each node in the figure can be thought of as a decision between two alternatives and the tree represents the combined results of the experiments described earlier. However, the tree is not strictly hierarchical in that the answer to one question does not necessarily depend on the answer to an earlier one.

Monotonic Distance Effect

The first node in the tree is simply whether the integration of information along a common contour, boundary, or curve is affected by the length of curve linking the critical information. The answer to this question is an unambiguous yes. All the experiments in the present study support this conclusion. The monotonic distance effect, which is often linear, provides support for the notion that curve tracing operations are used to achieve this type of spatial integration.

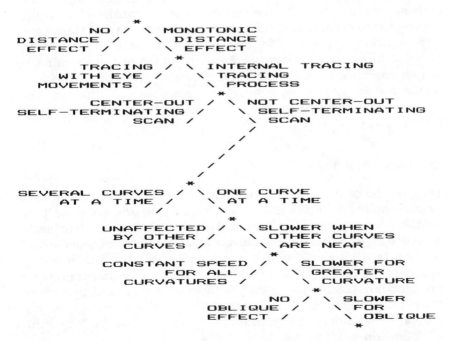

Figure 20.   Decision tree summarizing the main results from the present study.

Internal Tracing Process

At the second choice point in the tree, the issue is whether curve tracing is an internal process or whether it is external, in the sense of requiring eye movements to track along the curve. If the process is essentially an external one, then it would not be surprising that more time would be required to track the curve; presumably more eye movements would be required to trace longer distances. The brief exposure durations in Experiment 2 (250 msec) and in Experiment 7 (180 msec) precluded the use of eye movements to perform the task. Nonetheless, the results showed strong and systematic effects of curve distance. I conclude that curve tracing operations can take place internally, without eye movements, and that the distance effects in the present experiments reflect a spatial limitation of these operations: the rate of tracing is finite and thus longer distances take longer to trace.

Not Center-Out Self-Terminating Scan

Experiments 3 and 4 addressed the issue of whether tracing in the first two experiments always started from the central X, tracked the curve in the direction of the second X, and self-terminated upon arriving at the second X. The results were not consistent with the center-out self-terminating scan, when subjects were searching for a gap. I tentatively conclude that the center-out self-terminating model is incorrect. However, as discussed at the end of Experiment 4, this conclusion could reflect the choice of target in the experiments: a small gap along a curve. It is possible that other types of target would yield different results. This possibility is under investigation.

One Curve At A Time

The next choice in the tree is whether tracing operations can process several curves simultaneously or whether they are confined to a single curve at a time. The results of Experiment 5 in which subjects searched for a continuous path in a context of paths containing a gap conform to what would be expected from a serial self-terminating search of the paths. Thus, I conclude that curve tracing operations are probably confined to a single curve at a time and that curve tracing operations, in the context of the present experiments, are probably not automatic (Posner & Snyder, 1975; Schneider & Shiffrin, 1977; Shiffrin & Schneider, 1977).

Slower Tracing Rate When Other Curves Are Near And Slower Tracing Rate For Greater Curvature

The rate of curve tracing is affected by the proximity of a distractor curve (Experiment 8). Also, the curvature of the traced curve is important: The rate of curve tracing is slower for curves with a high degree of curvature and fastest for straight lines (Experiments 6 and 7). The dependence of tracing operations on these display properties allows us to argue against a model of tracing in which curves are tracked by rapidly moving the locus of a processor from one small local region to the next, that is in a pixel-by-pixel fashion. The results are more consistent with a more complex but more flexible system in which curves are traced at different scales depending on the nature of the display. When other curves can potentially be confused with the target curve or when the curvature of the target curve is high, the system selects a finer scale at which to perform the tracing—the operations become more local and, in the limit, possibly pixel-by-pixel. As a consequence of the finer grain of processing, however, the tracing rate becomes slower. In general, such a flexible arrangement allows the system to attain much greater tracing rates than would be possible if tracing was always performed at the finest possible grain.

One interpretation of the tracing results described in this study is that curves are first parsed into smaller parts or chunks, and that the distance effect in fact reflects an increase in the number of chunks as distance is increased (see Mahoney & Ullman, 1988). For example, Hoffman and Richards (1984) argue that curves are parsed into parts that begin and end at extrema of negative concavity along the curve. The hypothesis that curves are processed as a collection of parts and that the number of parts increases as curve distance is increased seems entirely plausible. There are a number of results in the present study, however, which are difficult to explain with this approach. For example, some displays (i.e., in Experiments 6, 7, and 9) required tracing straight lines. Response time increased linearly with the length of the line to be traced. Given that lines have constant curvature (i.e., zero curvature everywhere) they presumably are parsed into a single part—or perhaps they are parsed according to some other scheme (for example, by dividing the line using salient landmarks as boundaries, such as the dots or Xs in the experiments). In other cases, the curvature was nonzero, but was constant (Experiments 6 and 7) and again would presumably not be broken into different parts. Nonetheless, greater curve distances took longer to trace than shorter distances. Thus, the notion that curves are parsed into parts is not a sufficient account of the distance effects found in this study.

Slower Tracing For Oblique Lines

There is ample evidence from previous work that the vertical direction and occasionally the horizontal direction enjoy a special status in a number of perceptual tasks (e.g., Attneave & Curlee, 1977; Attneave & Olson, 1967; Attneave & Reid, 1968; Corballis et al., 1978; Corballis et al., 1976; Palmer, 1980; Palmer & Bucher, 1981; Rock, 1973; Wiser, 1980). Curve tracing operations are also affected by the orientation of the traced curve with respect to the perceptual frame of reference: Tracing is slower for oblique than for horizontal/vertical trajectories. These results are consistent with the notion that horizontal and vertical directions may be privileged orientations in the perceptual frame of reference within which tracing occurs. More recent work suggests that the perceptual reference frame is aligned more closely with retinal directions than with environmental/gravitational directions, which was investigated in an experiment that compared tracing performance with the head upright with performance with the head tilted (Jolicoeur & Milliken, 1988). The tendency to align the perceptual reference frame with retinal rather than with gravitational/environmental directions suggests that some components of the tracing operations may operate on relatively early representations in the visual system.

CONCLUSIONS

The visual system is often required to compute spatial relations between elements in space, as when we reach for objects under visual guidance, when we walk, drive, reach for objects, and in many other tasks. One type of spatial relation concerns the integration of information from diverse spatial locations that lie on a common curve. The results suggest that the visual system tracks along curves, contours, and boundaries in visual displays in order to integrate information linked by a common curve. Several tasks (such as reading maps, circuit diagrams, or graphs) performed routinely by people in real occupations appear to require this type of spatial integration. The experiments reported here suggest that tracing operations are complex, flexible, fast, and rapidly adapted to the particular characteristics of the curves to be traced. It is hoped that the present results will form part of an empirical foundation upon which we can build a detailed computational theory of the perception of spatial relations.

REFERENCES

Appelle, S. (1972). Perception and discrimination as a function of stimulus orientation. *Psychological Bulletin, 78*, 266–278.

Attneave, F. (1954). Some informational aspects of visual perception. *Psychological Review, 61*, 183–193.

Attneave, F., & Curlee, T. E. (1977). Cartesian organization in the immediate reproduction of spatial patterns. *Bulletin of the Psychonomic Society, 10*, 469–470.

Attneave, F., & Olson, R. K. (1967). Discriminability of stimuli varying in physical and retinal orientation. *Journal of Experimental Psychology, 74*, 149–157.

Attneave, F., & Reid, K. W. (1968). Voluntary control of frame of reference and slope equivalence under head rotation. *Journal of Experimental Psychology, 1*, 153–159.

Biederman, I. (1985). Human image understanding: Recent research and a theory. *Computer Vision, Graphics, and Image Processing, 32*, 29–73.

Church, R. M., & Church, K. W. (1977). Plans, goals, and search strategies for the selection of a move in chess. In P. W. Frey (Ed.), *Chess skill in man and machine*. Berlin: Springer-Verlag.

Coltheart, M. (1980). Iconic memory and visible persistence. *Perception & Psychophysics, 27*, 183–228.

Corballis, M. C., Nagourney, B. A., Shetzer, L. I., & Stefanatos, G. (1978). Mental rotation under head tilt: Factors influencing the location of the subjective reference frame. *Perception & Psychophysics, 24*, 263–273.

Corballis, M. C., Zbrodoff, N. J., & Roldan, C. E. (1976). What's up in mental rotation? *Perception & Psychophysics, 19*, 525–530.

Corcoran, D. W. J. (1977). The phenomena of the disembodied eye or is it a matter of personal geography? *Perception, 6*, 247–253.

Duncker, K. (1929). Uber induzierte Bewegung (ein Beitrag zur Theorie optisch warigenommener Bewegung). *Psychologishe Forschung, 2*, 180–259.

Feldman, J. A. (1985). Four frames suffice: A provisional model of vision and space. *The Behavioral and Brain Sciences, 8*, 265–289.

Hoffman, D. D., & Richards, W. A. (1984). Parts of recognition. *Cognition, 18*, 65–96.

Hubel, D. H., & Wiesel, T. N. (1968). Receptive fields and functional architecture of monkey striate cortex. *Journal of Physiology* (London), *195*, 215–243.

Johansson, G. (1950). *Configurations in event perception*. Uppsala: Almquist and Wiksells Boktryschkeri AB.

Jolicoeur, P. (1988). *Orientation congruency effects on the identification of disoriented shapes: Implications for reference frame models and pattern rotation models*. Submitted.

Jolicoeur, P., & Milliken, R. B. (1988). What's up in curve tracing and in mental extrapolation. Submitted.

Jolicoeur, P., Ullman, S., & Mackay, L. (1986). Curve tracing: A possible basic operation in the perception of spatial relations. *Memory & Cognition,* *14,* 129–140.

Jolicoeur, P., Ullman, S., & Mackay, L. (1988). Curve tracing properties. Submitted.

Mahoney, J. V., & Ullman, S. (1988). Image chunking. Defining building blocks for scene analysis. In Z. W. Pylyshyn (Ed.), *Computational processes in human vision: An interdisciplinary perspective.* Norwood, NJ: Ablex.

Palmer, S. E. (1980). What makes triangles point: Local and global effects in configurations of ambiguous triangles. *Cognitive Psychology, 12,* 285–305.

Palmer, S. E., & Bucher, N. M. (1981). Configural effects in perceived pointing of ambiguous triangles. *Journal of Experimental Psychology: Human Perception and Performance, 7,* 88–114.

Posner, M. I., & Snyder, C. R. (1975). Attention and cognitive control. In R. L. Solso (Ed.), *Information processing and cognition.* Hillsdale, NJ: Erlbaum.

Pylyshyn, Z. W. (1988). Here and there in the visual field. In Z. W. Pylyshyn (Ed.), *Computational processes in human vision: An interdisciplinary perspective.* Norwood, NJ: Ablex.

Rock, I. (1973). *Orientation and form.* New York: Academic Press.

Schneider, W., & Shiffrin, R. M. (1977). Controlled and automatic human information processing: I. Detection, search, and attention. *Psychological Review, 84,* 1–66.

Shiffrin, R. M., & Schneider, W. (1977). Controlled and automatic human information processing: II. Perceptual learning, automatic attending, and a general theory. *Psychological Review, 84,* 127–190.

Ullman, S. (1984). Visual routines. *Cognition, 18,* 97–159.

Uttal, W. R., Bunnell, L. M., & Corwin, S. (1970). On the detectability of straight lines in visual noise: An extension of French's paradigm into the millisecond domain. *Perception & Psychophysics, 8,* 385–388.

Wiser, M. A. (1980). *The role of intrinsic axes in the mental representation of shapes.* Unpublished doctoral dissertation, Massachusetts Institute of Technology.

8

Image Chunking Defining Spatial Building Blocks for Scene Analysis*

James V. Mahoney† and Shimon Ullman

Massachusetts Institute of Technology

INTRODUCTION

Visual judgements about the properties and spatial relations of objects are the crux of our interactions with our surroundings. We perceive and conceive of the world in terms of objects and configurations, which we recognize, handle, navigate by, and reason about. Our primary source of information about these spatial entities and relations is vision. The visual system makes this information available in a manner that leaves the subjective impression of immediate, complete, effortless awareness. For example, you might look up from this text for a moment, reach for your cup, and take a drink, with hardly a thought. For that matter, vision transparently discerns the words and phrases you are reading, while your conscious thoughts are focused on their meaning.

Visual processing begins with an image array of pointwise measurements of light intensity. The physical entities in terms of which we conceive of our surroundings may have widely varying spatial extent, so they are not, in general, explicitly described in the image

---

* This paper describes research done within the Artificial Intelligence Laboratory at the Massachusetts Institute of Technology. Support for the A.I. Laboratory's artificial intelligence research is provided in part by the Advanced Research Projects Agency of the Department of Defense under Army contract number DACA76-85-C-0010, in part by DARPA under Office of Naval Research contract N00014-85-K-0214.

† Author's current address: Xerox Palo Alto Research Center.

Figure 1.   Three prominent blobs.

array or any other pointwise scene descriptions derived from it. Moreover, meaningful scene entities may appear in a very wide range of shapes and configurations, and what is meaningful may depend on the task at hand. Therefore the problem of making the relevant components of a scene visually distinct is complex from a computational standpoint, and it is quite remarkable that this is normally achieved in human vision in what seems an instant.

For example, consider Figure 1. We immediately perceive three large, striking blob shapes amidst an irregular background of curves. The speed with which the human visual system can locate and describe the outstanding blobs in figures like this does not depend noticeably on the total length of curves in the figure. (See also Figure 2.) Consider also the ease with which we can often solve connectivity-related problems, such as those in Figure 3. The capacities that these schematic examples illustrate—locating and isolating relevant figures rapidly—serve in all realms of visually guided activity. For example, the task of finding the largest spoon in Figure 4 seems to require no effort.

The problems of visually-guided interaction with the physical world impose various requirements on visual processing organization. One of the most crucial requirements is speed. This paper explores the design of computational processes that could approach the time performance of the human visual system in analyzing visual spatial information, particularly in regard to the extraction of meaningful scene components. The rate of execution of the pro-

Figure 2.   Three prominent blobs.

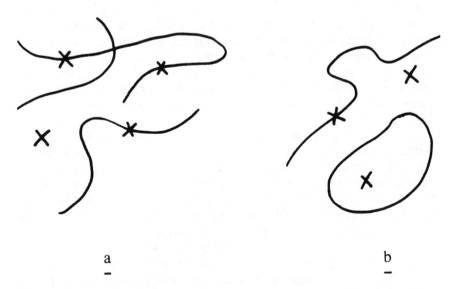

a

b

Figure 3.   (a) Are there two "X"s on the same curve? (b) Are there two "X"s inside the same closed curve?

Figure 4. Find the largest spoon in this picture.

cesses supporting spatial analysis in the human visual system has been tentatively estimated on the basis of the results of psychophysics and current knowledge of the neurological structures in the visual system. Current estimates suggest that the number of basic computational steps devoted to extracting a scene component is normally in the tens (Edelman, 1985; Shafrir, 1985). Assuming that these estimates are around the correct magnitude, and ignoring the detailed derivation of them, how can visual processing be organized to be so rapid?

This chapter presents a framework for the fast extraction of scene entities, based on a simple, local model of parallel computation. An image chunk is a subset of an image that can act as a unit in the course of spatial analysis. A parallel preprocessing stage constructs a variety of simple chunks uniformly over the visual array. On the basis of these chunks, subsequent serial processes rapidly locate relevant scene components and rapidly assemble detailed descriptions of them. The next section introduces visual routines, Ullman's proposal for the organization of visual processes leading to the perception of shape properties, and spatial relations (Ullman, 1984). Included in this discussion are the considerations that suggest a two-stage processing framework, in which the first stage is bottom-up and spatially uniform and the second is spatially-focused and demand-driven. Later sections present computational motivations for defining spatially extended primitives in early vision. The remaining sections summarize the results of our study of the required representations (Mahoney, 1987; Shafrir, 1985).

SPATIAL ANALYSIS BY VISUAL ROUTINES

The General Requirements of Spatial Analysis

Ullman (1984) formulated the problem of visually analyzing the spatial properties and relations of scene entities in terms of the following three general requirements: (a) *abstractness*—the capacity to establish computationally abstract properties and relations; (b) *open-endedness*—the capacity to establish a large and extensible variety of properties and relations; and (c) *complexity*—the ability to cope efficiently with the computational complexity involved.

A property or relation is said to be abstract if (a) its support is so large that it would be prohibitively expensive to detect the property or relation using a straightforward application of template-matching; and (b) the set of instances of the property or relation contains *regularities* that can be captured by an efficient computation.[1] Many properties and relations of fundamental importance to vision are abstract in the above sense, and the abstractness requirement implies that a visual system must employ computations for capturing the regularities inherent in these properties and relations. Notable examples of abstract relations are *connectivity* and its close variants, such as "inside/outside" (Figure 3), "same-curve" (Figure 3), etc. The notion of a "two Xs inside the same closed curve" template-matching detector is implausible—the support of the connectivity relation is the entire input, so a different template would be required for every possible case.

The variety of potentially useful properties and relations is open-ended. It is inconceivable that a detector for every possibly relevant property or configuration could be predefined. For example, the task of Figure 5 would require a "small circle nearest the third-largest circle" detector! The open-endedness consideration implies that the

---

[1] Intuitively, the *support* of a spatial predicate is that subset of an input upon which the predicate "really depends" (Minsky, 1969). For the purposes of this discussion, *template-matching* between plane figures can be defined as the cross-correlation between the figures.

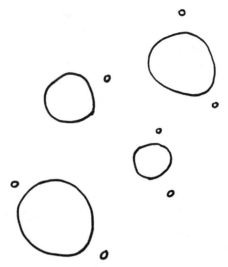

Figure 5. Find the small circle nearest the third-largest circle.

processes for detecting different properties or relations must use common machinery.

The complexity requirement addresses the fact that the computations for capturing abstract properties and relations, such as the connectivity relation, may be quite complex from a computational standpoint, and the implementation of them may be expensive. Moreover, the inputs to these computations are not constrained—relevant scene entities may appear with a very wide range of shapes, and they may occur with any spatial extent, at any location, and in any number across the visual field. The complexity considerations, too, imply that the processes for detecting a property or relation at different locations must use common machinery.

To summarize, all meaningful scene entities, and all their potentially relevant properties and relations, cannot be detected at once, due to combinatorial explosion in required computational resources. Instead, visual processes must be *spatially focused* and *goal driven*, locating.and assembling scene entities, and computing their relations and properties as the need arises. These processes must meet the exacting time-performance requirements imposed by the goal of interacting with an active, changing world.

It is, however, possible and useful to detect certain simple, local, viewer-centered properties and features in a manner that is bottom-up, spatially uniform, and parallel. The very first processes of vision can make this sort of local information available to subsequent focused processes. Research in low-level vision has elaborated the computation of essentially pointwise properties and features, such as intensity changes, depth, surface orientation, texture, and so on. The processes that have been studied create pointwise primitive spatial elements—tokens characterizing depth, orientation, motion, and so on, at a point. (For discussions of these low-level vision processes, (see Marr, 1982; Horn, 1986; & Barrow, 1978, for example.)

Visual Routines

Ullman's proposal for meeting the requirements of spatial analysis is that properties and relations should be established in two stages, by the goal-driven application of visual routines—sequences of basic operations drawn from a fixed set—to a set of base representations that are created in a bottom-up, spatially uniform, parallel manner. New routines are assembled to establish newly specified properties or relations. Routines for establishing different properties and relations, or applications of a routine to different locations, share the machinery implementing the elemental operations they use.

The goals of the study of visual routines are (a) to establish a theory of what spatial properties and relations are useful in the context of spatial reasoning, object recognition, and so on; (b) to determine a set of basic operations that makes it possible to robustly compute these properties and relations, and to specify the visual routines involved; and (c) to devise efficient implementations of the basic operations and related machinery.

Ullman proposed a partial set of basic spatial operations, including region coloring; boundary tracing and coloring; location marking with respect to spatial reference frames; shift of a spatial processing focus; and "indexing"—processing shift to a salient location. These choices were motivated mainly on grounds of potential usefulness in establishing a wide variety of relations. Figure 6 illustrates one possible routine for establishing an instance of the inside/outside relation. The task is to determine whether there is an "X" figure inside a closed curve. The procedure is as follows:

Until an "X" inside a closed curve is found, or Step 1 fails:

1. *Shift* the processing focus to the location of an unseen "X".
2. *Mark* the location of this "X" seen.
3. *Color* the white region that includes the processing focus.
4. *Shift* the processing focus to a location at the periphery.
5. If the periphery location is not colored, stop—the most recently visited "X" is inside a closed curve; otherwise "uncolor" colored locations and repeat.

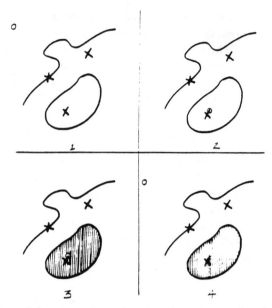

Figure 6. Part of the execution of a visual routine to decide if there is an "X" inside a closed curve.

Step 1 involves indexing to certain local features characteristic of an "X" figure, such as terminations. Further processing is required at this step to establish that the figure at the location shifted to is indeed an "X". This routine is based on the simplifying assumption that all boundaries in the input can stop the spread of coloring. For some tasks, it is necessary to first single out relevant boundaries, and allow only these to stop coloring spread. The shift to a periphery location is assumed to be a basic operation.

Basic Operations

This subsection explores the basic operations Ullman proposed in somewhat more detail. It is possible to divide them into the following three main families: *activation* operations; *selection and shift* operations; and *reference frame and marking* operations.

Activation operations. Tracing, coloring, and one form of location marking may be viewed as instances of a fundamental operation in the analysis of spatial information referred to here as *activation*. Activation is the operation of *singling out (uniquely labeling) a set of locations* or primitive spatial elements in the visual array, so that subsequent operations may be applied specifically to the distinguished set. The various activation operations differ according to the criteria that they apply in generating the distinguished set of locations or spatial elements. Region and curve coloring operations activate a set of locations connected to a given starting location. Curve tracing operations activate a set of locations connected to a given starting location, subject to a restriction on the number of neighbors each element of the set may have.

Selection and shift operations. Spatial analysis requires the capacity to apply certain operations at selected locations. To accomplish activation, for example, it is necessary to begin tracing or coloring at one or more locations of the relevant set. In the example routine above to determine whether an "X" is inside a closed curve, coloring begins from the location of the "X" and a decision is based on the knowledge of where coloring started. Therefore, the study of visual routines must account for the processes for (a) selecting the location at which a given operation is to be applied, and (b) shifting the point of application—the focus of processing—to that location.

Indexing is defined to be a shift of the processing focus to a salient location. The discussion of indexing in this paper is mainly concerned with the processes for determining saliency. Various workers have demonstrated pop-out effects in visual search tasks involving arrays of letters and similar stimuli (Treisman, 1980; Julesz, 1981). These demonstrations indicate that local features such as color, curvature, line terminations, and so on can serve for direct indexing, as long as the figure of interest is distinguished from irrelevant figures by a single one of these features. Figure 7 illustrates that global properties of an *extended* figure may support direct indexing as well.

Reference frame and marking operations. Reference frames are implicit in many useful spatial properties and relations, such as "above," "left-of," "vertical," "clockwise," and so on. Spatial analysis therefore requires the capacity to employ and manipulate frames of reference. One general application of reference frames is in establishing orientation. For example, the interpretation of a shape may depend on the reference frame with respect to which the shape is described. Figure 8 (a) contains a famous example of a figure for which different assignments of the "forward" direction lead to different interpretations (from Jastrow, 1900; see also Rock, 1984). Figure 8 (b) illustrates the *frame effect*: the direction attributed to the arrow depends on which of the enclosing rectangles is attended to.

A reference frame also serves as a system for describing and recording location. The location of a spatial entity is not usefully described in absolute retinocentric terms. It is more useful to de-

Figure 7. Figures are indexable based on prominence in global properties.

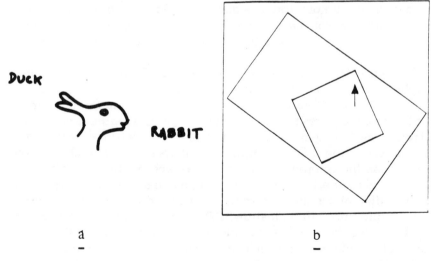

Figure 8. (a) Interpretation of a shape involves the assignment of a reference frame. (b) An illustration of the frame effect: which way the arrow seems to point depends on which enclosing rectangle one attends to.

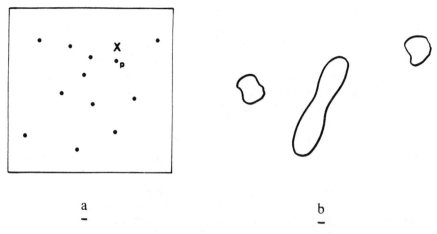

a

b

Figure 9.   (a) Describe the position of the point p. (b) Are the two compact blobs on the same side of the elongated blob?

scribe the position of one scene component in relation to others. This requires the capacity to anchor a coordinate system at the retinal location of a given scene component. The operation of marking a location for later reference involves generating and storing a descriptor of a scene location with respect to a particular assignment of an internal reference frame. In Figure 9 (a), the point p may be described as "immediately below" the "X" or "in the upper-right" of the enclosing square (from Ullman, 1984). In the former case, the internal frame is anchored on the "X" figure; in the latter it is anchored on the enclosing square. The task of Figure 9 (b) involves *aligning* a reference frame with the elongated blob.

PARALLEL MODELS OF COMPUTATION FOR VISION

Visual processing begins with a set of essentially pointwise descriptions of the scene in terms of intensity, color, texture, motion, depth, surface orientation, intensity boundaries, and so on. From an abstract point of view, the problem of extracting meaningful entities from these pointwise spatial representations is closely related to the graph theoretic problem of *connected component labeling*. Because sequential connectivity algorithms have running times linear in the number of elements in the relevant component, a lot of research has gone into parallel connectivity algorithms. A number of connectivity algorithms have been proposed whose complexity is poly-log (a polynominal in the logarithm of the problem size) for a polynomial (and therefore feasible) number of processors. The best known connectivity algorithms have $O(\log N)$ time-complexity (Shiloach, 1982; Lim, 1986; see also Cook, 1983; Hirschberg, Chandra, & Sarwate, 1979). At the completion of a run of such an algorithm, each connected component of elements in the input is, in effect, uniquely labeled, so that it is possible to determine in constant time whether any two given elements are "in the same region," or "on the same curve," for example.

Every poly-log parallel connectivity algorithm we know of assumes a model of computation in which (a) any processor can communicate in a single time step with any other processor, and (b) it is possible to associate unique identifiers with processors, and a processor may store such an identifier, or transmit it to another processor. We will refer to the class of models providing these two capabilities as *global parallel models*. The global parallel models that have been proposed for connectivity differ mainly with respect to the conventions for arbitrating *read* and *write* conflicts. The $O(\log N)$ algorithm of Shiloach and Vishkin (1982) assumes a synchronous computing model in which processors communicate via a common random access memory; concurrent reads from a location and concurrent writes to a location are allowed. Lim (1986) recently described connectivity algorithms with a best case performance of $O(\log N)$, based on a model restricted to exclusive read and exclusive write.

It is costly and difficult to implement global, direct communication between all processors in a large set. For moderately large sets of processors, truly direct communication is physically unrealizable. In practice, global communication is implemented by dynamically routing messages through high-dimensional networks (for example, see Hillis, 1985; BBN, 1985; and Gottlieb, Grishman, Kruskal, McAuliffe, Rudolph, & Snir, 1983). The duration and/or reliability of this delivery process depends on the pattern of message traffic through the network at a given time. For example, message delivery may be severely hampered when many processors simultaneously attempt to send a message to the same processor. As such, time-complexity figures that treat a message delivery cycle as a primitive computational operation involving "one" time step—as the $O(\log N)$ parallel connectivity figures do—cannot be taken at face value.

Ullman (1976) proposed a number of criteria to characterize what he called *simple local processes*, the primary ones being locality and simplicity. The locality criterion is a restriction on the range of the direct communication in the computation. The simplicity criterion is a restriction on the power of the processing elements. It is not feasible to build very large networks of powerful processors, owing to considerations of size and cost. We refer to models of parallel computation observing these restrictions as *simple, local parallel models*. The general properties of this class of models can constrain representations and algorithms, without regard to the details of any particular instance of the class. Importantly, *the capacity to establish and transmit unique processor identifiers is not compatible with the class of simple local parallel models*.

A simple, local model of computation suggests itself as a possible basis for exploiting parallelism in connectivity computations on two independent grounds. For low-level applications in machine vision systems, simple, local models may be preferable to global parallel models because they are less costly and simpler to build—for the same price, they can provide more parallelism. For the study of biological visual systems, the restrictions of simplicity and locality are consistent with what is known about biological information processing, so computations observing these restrictions are more likely to be revealing.

The goal of this research is to understand how parallelism can be exploited to make the inherently serial processes of scene analysis sufficiently fast, without relying on global parallel computation models. Our proposal is that complex scene entities can be rapidly extracted from pointwise initial representations using only simple, local computations, if a two-stage, parallel-then-serial processing organization is used, in which the first stage assembles simple, extended spatial building blocks. We begin by elaborating on the building-construction metaphor, for it helps to make some important ideas vivid.

PREFABRICATION—A METAPHOR FOR CHUNKING

If you had to build a house very quickly, you would assemble it from prefabricated sections, not individual bricks and boards. You would use prefabricated components specifically suited to the type of structure you were building; the structural elements required for a tropical bungalow are very different from those required for a bombshelter. Moreover, you would select the complement of component sections that entailed the shortest possible number of assembly steps. That is, your main objective would be *optimization* of the assembly process. The first consideration, *specialization*, also helps to optimize the assembly time—it seems natural that components designed with your particular application in mind will make the construction task go faster and more easily.

If you made a living selling prefabricated building supplies, you would offer simple components that could be easily combined into a wide range of more complex structures; they would be regular sections of wall or floor, for example. There are two reasons for this. First of all, it would be easier and more economical to construct such simple components. Moreover, more complex structures would have narrower *scope*, in the sense that they would fit into fewer customers' designs. The components would span a range of sizes and shapes just broad enough that most customers could come reasonably close to their goal of optimizing assembly time. These considerations of *economy*—providing simple components across a sensible range of sizes and shapes—strike a balance between demand and production costs. The optimization criterion must be traded off with those of scope and economy. It would be reasonable to ignore the occasional customer whose needs were outlandish. You would observe the specialization requirement by providing structural components tailored to each of the most common applications.

The power of prefabrication is that it exploits the parallelism inherent in the assembly process. The process of building a house is inherently serial due to the constraints of physics: the roof cannot be put up before the walls (or other supports), and the walls must follow the foundations. As such, if a house is built brick by brick, the first board in the roof must follow the last brick in the walls, and so on. This serialism, however, is not inherent at the level of individual bricks, but only at the higher level of walls, roofs, and so on, so *these abstract components may be concurrently pre-assembled*. The main problems of prefabrication, then, are (a) to determine, given a detailed description of a complex structure to be assembled, a decomposition into simpler substructures that can be economically pre-assembled in parallel; and (b) to assemble these simpler components.

IMAGE CHUNKS

An *image chunk* is defined as any subset of a pointwise representation of the scene that can act as a unit in the sense that applying some spatial operation to (or reading out some spatial properties of) the set of constituent elements requires minimal computational effort. Spatial operations on image chunks are the elements from which more complex spatial operations are composed. Coloring, for example, can generally be expressed as an iteration of the following parallel operation: *color every uncolored image chunk adjacent to a colored image chunk.*

The process of making a scene entity distinct on the basis of a pointwise initial description can be likened to an assembly problem. From this viewpoint, image chunking is closely analogous to pre-fabrication, and is governed by similar considerations. The motivation for image chunking is to optimize the run-time of the processes that extract scene components. To achieve this, image chunks must be specialized to particular basic operations—the representational requirements of indexing are different from those of tracing, for example. Combinatorics limits the range of distinct instances that a chunking process can generate, so these instances must be carefully chosen to provide the widest possible scope. In this context, scope refers to the class of inputs for which a particular class of image chunks supports efficient processing, and economy pertains to the hardware cost and time performance of the chunking process.

There are three main problems to be solved in the design of image chunks to support a particular basic operation. First, a computational definition of the basic operation must be formulated, in terms of its constituent operations. Second, image chunks must be devised that (a) would lead to the desired performance, and (b) are economically computable from a pointwise representation of the scene by spatially uniform, parallel processes. Finally, the processes for computing these chunks, and high-performance, chunk-based formulations of the associated basic operation, must be defined, implemented, and tested. All of this must be accomplished in the context of a realizable, affordable model of computation.

The remaining sections of this paper summarize the results of a study of image chunking in support of the most potentially time-consuming operations in the extraction of scene components—indexing, tracing, and coloring. Consideration of a wide variety of perceptual phenomena has led to a theory of image chunking that involves several novel low-level vision processes in a novel processing organization. As illustrated in Figure 10, the computation of image chunks can be divided into four principal stages: (a) the computation of local image properties from the intensity image; (b) the computation of local boundary information, which constitutes the primary input to chunking processes; (c) the computation of local prominence information, which augments the boundary information;

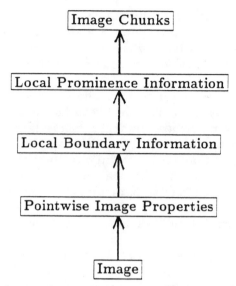

Figure 10. Image chunking and the organization of low-level vision.

and (d) the application of the chunking processes to this input. Chunks useful for indexing scene entities may be defined by dividing the boundary data into relatively isolated configurations of boundary locations. Chunks useful for curve tracing may be defined by dividing the boundary data into regions each containing a single segment of a boundary. Chunks useful for region coloring may be defined by dividing the boundary data into regions each containing no boundary locations. All of these classes of chunks are computable by simple, local processes.

The use of image chunks has implications for the computation of spatial properties and relations relevant to both the design of machine vision systems and the study of human vision. Image chunking raises questions regarding human visual perception at two levels. At the more general level, the hypothesis that image chunks are made use of may lead to novel interpretations of perceptual phenomena, without regard to the particular processes by which the chunks are generated. Some pertinent observations in this regard are made in the sections on chunking for indexing and for curve tracing. At a more detailed level, it may be possible to make hypotheses about how image chunks are generated in the human visual system, based on psychophysical studies and an understanding of the range of possibilities for computing the required image chunks. Future work may take up this intriguing problem.

LOCAL BOUNDARY INFORMATION

Many schemes for scene analysis are based on the assumption that meaningful scene components give rise to homogeneous image regions. Most *region segmentation* schemes, for example, are based on this assumption. Although uniform areas are often found in images, scene entities often *do not* give rise to image regions that are nearly uniform in any properties. There are several observations supporting this view: (a) the processes that often generate the surface markings of objects, for example, growth and accretion, generally do not operate in a globally uniform manner, and so do not give rise to globally uniform markings; (b) even when surface marking properties such as orientation, curvature, elongation, and size are uniform over the entire object surface, perspective projection does not preserve this uniformity; (c) similarly, changes in distance and orientation of the surface with respect to the viewer do not, in general, preserve uniformity. In general, therefore, extended scene entities, or their extended parts, cannot be *directly* individuated on the basis of region properties. A variety of perceptual examples demonstrate that distinct objects are compellingly perceived in human vision where no image property is uniform over the areas of the objects. An example in the case of a local orientation property is shown in Figure 11 (see also Muller, 1986). The Craik/O'Brien/Cornsweet brightness illusion illustrates a related phenomenon in the intensity domain (Cornsweet, 1970).

A more general characterization of the areas corresponding to

Figure 11.   Objects defined primarily by abrupt changes at their boundaries, not by uniform region properties.

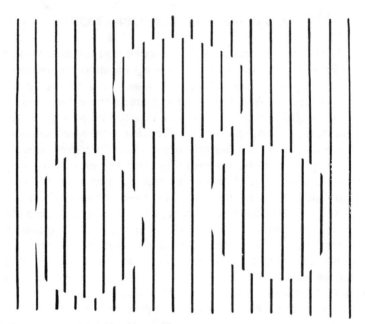

Figure 12.   Objects defined primarily by bounding alignment contours, not by uniform region properties.

scene entities in images is that they *vary only slowly* in some properties. For example, each hair in a coat of fur is necessarily similar in orientation only to neighboring hairs. Physical object boundaries—which include occlusion boundaries, sharp changes in surface orientation, and abrupt changes in surface material—almost always give rise to abrupt changes in image properties or to alignment contours (Figures 11 and 12). As illustrated in Figure 11, more abrupt changes at the boundary of an object give rise to a more definite perception of its shape and size.

The first step toward detecting scene entities is therefore to detect their boundaries. In the main, the relevant boundaries can be detected by simple computations sensitive to significant local difference or similarity in the properties of local image elements or surface marking elements (see Mahoney, 1987). The relevant properties include intensity, color, motion, size, and orientation.

LOCAL PROMINENCE INFORMATION

The information in images is normally highly redundant. Some of this redundancy arises at the level of the shape or arrangement of the scene's physical components. A variety of perceptual examples show that human vision copes very effectively with complex scenes when the scene is highly redundant at this level (e.g., Figures 1 and 2). Two observations about the natural world suggest that this *figural redundancy* occurs often enough in scenes to make computations dedicated to exploiting it very beneficial. First, uniform or slowly varying distributions of similar objects or surface markings are quite common. Second, it is common for an object with very different properties to occlude, or be embedded in, such a distribution of objects. Such a situation is referred to here as a *figure/background situation*. Spatially uniform, parallel computations for exploiting figural redundancy provide part of the solution to the difficult problem of extracting meaningful figures rapidly from complex scenes.

Local prominence information about local segments of boundary may be used to, in effect, *separate* boundaries of salient objects from other boundaries in parallel, so that chunking processes may be applied selectively to the former. It is shape properties that distinguish foreground figures from background figures in the figure/background situation. The distinguishing shape properties may be properties of local portions of the figure or properties of the entire

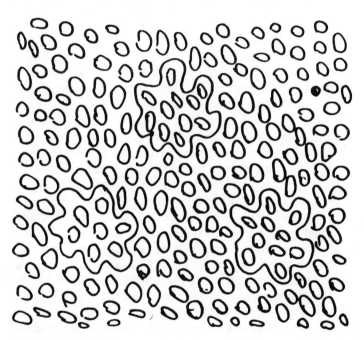

Figure 13.   The big blobs are not very prominent because local segments of their boundaries are not very different in curvature from neighboring background segments.

figure. Differences in global properties are often reflected locally. For example, when there is a substantial difference in size between a foreground entity and background entities, local boundary curvature may be substantially lower for the foreground entity than for the background figures, as in Figure 2. (In Figure 13, the big blobs were camouflaged somewhat by keeping the local curvature of their boundaries similar to that of the background blobs.) Local prominence may be detected by simple computations sensitive to local differences in the properties of local boundary segments, such as orientation, curvature, and length (see Mahoney, 1987).

CHUNKING FOR INDEXING

Indexing is defined to be a direct shift of processing to a salient location. It involves a spatially uniform, parallel process for selecting a single location for focused processing. Indexing can mediate a wide variety of general purpose intermediate visual processes, including object tracking, visual search, scanning and counting, simple spatial reasoning, and the initial stages of recognition. Because these processes are primarily concerned with objects, the criterion for saliency of a location should (a) reflect the likelihood that a meaningful object or object part is present at that location, and (b) reflect the probable importance of that object in relation to other objects in the field of view. As the objects of interest in vision are normally extended, such a saliency measure must involve the computation of global properties of extended subsets of the image, such as size, elongation, and orientation. Since possibly relevant objects may be distributed in the field of view in an unconstrained way, the measure must be applied at every location, so it must be a simple, not-too-costly computation. This means that the computation will yield only approximate values of object properties. It appears that crude measures are normally sufficient for identifying truly salient entities. (The occasional need to locate nonsalient objects may preclude the use of crude, rapidly available descriptors—and, hence indexing—leading to substantially slower performance.) It also appears that it would be appropriate to base the measure of saliency on *local* comparisons of object properties (see Figure 14). The final choice of a processing location must be based on *global* comparisons of the results of the saliency measure across locations. Note that the crude descriptors of scene components are important not only as *targets* for processing shifts, but also as *anchors* for spatial reference frames (for example, see Figure 9 (b).)

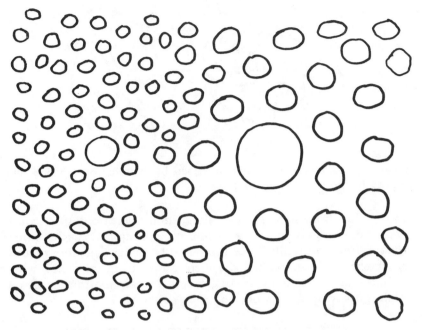

Figure 14.   Figural prominence is a local phenomenon.

Quite a variety of local configurations can indicate the presence of a meaningful, comparatively interesting spatial entity, for example curve junctions, chunks of free-space, local parallel structures (flow patterns), proximity groups, bilateral symmetry groups, and so on. Each kind of configuration is an indicator of saliency applicable in a particular class of situations. Perhaps the local configuration providing the most direct indication of a possible physical object is a *locally isolated* (or locally prominent) configuration of boundary locations (Mahoney, 1987). Local isolation is some measure of the distance of a given set of boundary locations to surrounding boundary locations. By this criterion, Figure 15 would be described as containing two objects; by the criterion of *connectivity* it contains four. Boundary configurations, and their degree of isolation, may be measured by a form of template matching—configurations of predefined shape, size, and orientation are *fit* to a boundary array, in a parallel, spatially uniform manner (Figure 16), using a measure of fit designed to express the degree of isolation. A predefined configuration used for this purpose is termed a *basic configuration*. Basic configurations are simple, general shapes, such as ellipses, bars, and arcs (see Figure 17). Useful attributes of the best fitting basic configura-

Figure 15. A figure containing four objects by the criterion of connectivity, and two by the criterion of isolation.

tion at a location, such as size, elongation and orientation, can be directly associated with the boundary configuration (the *figural chunk*) it fits. These attributes provide a basis for selecting processing locations. For example, the location of a figural chunk whose attributes are prominent in comparison to neighboring chunks, or match a given specification, might be selected.

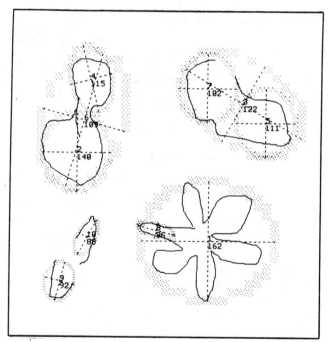

Figure 16. Figural chunks detected in a hand-drawn input, by crudely fitting ellipses.

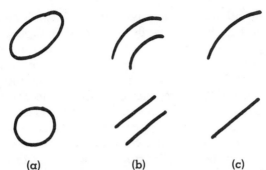

(a)   (b)   (c)

Figure 17.   Some useful basic configurations. (a) Ellipses. (b) Bars. (c) Arcs.

Thus, the two-stage spatial analysis framework supports two distinct levels of description of the objects in a scene. The figural chunk level is crude and limited, but immediately available and covers the entire image. The other level can be as detailed as necessary but is demand-driven, spatially focused, and potentially time-consuming. The detailed information must be spatially indexed by the crude information. The overall implication of this framework is that the speed and assurance with which a visual task can be accomplished depends on the level of description that it utilizes. Execution of a task will be much faster at the crude level, so the framework provokes a bias toward solutions at that level. Such solutions will apply under more restrictive conditions, though, than their detailed level counterparts.

Figural chunks make possible the immediate, parallel classification of global configurations of many elements. It is possible to create a summary description in which the distributions of chunk attributes in a global configuration is expressed consisely. Figures 2 and 14 give rise to an assured but qualitative sense of the overall arrangement of many similar items. This description may be augmented by a serial, detailed local description which moves from one location to another. The subjective impression of immediately available panoramic detail in such scenes may result from the effective integration of the immediate, parallel, global summary description with the incremental, detailed, local descriptions. The detailed description need not be performed exhaustively over the scene—if the summary description indicates that the scene has uniform or slowly varying structure, it is possible to extrapolate the local detailed description.

Another important implication of indexing on the basis of figural chunks is that a scene entity will be indexable only if it or its simple parts give rise to prominent figural chunks. Consider Figure 18—the long curves stand out. This phenomenon is independent of the number of curves in the input or their lengths. In Figure 19, however, a meandering curve that is more than twice as long as any other curve in the figure is effectively hidden. If it is assumed that connected and/or continuous curves are distinguished and described in low-level vision, say by curve tracing, then the discrepancy between these two figures is hard to account for. Processes amounting to curve tracing should have no more difficulty describing a winding curve than a straight one. In the view we propose, configurations of boundary locations that may be approximately described as *simple arcs* are individually described in early vision, but more complex curves are described only in terms of their simpler components. As such, the long lines in Figure 18 give rise to local descriptors that are outstanding in length, relative to the local descriptors generated for other lines. The longer curve in Figure 19, on the other hand, rise to no outstanding descriptors. Similar arguments apply to Figure 20, in which a curve twice as long as any of the others is hidden. In this case, the early descriptors of the figures express the properties of the small, compact region effectively occupied by each curve, not properties of the curves themselves.

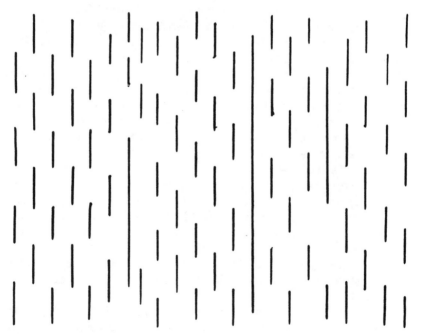

Figure 18.   Boundary prominence due to extent.

Figure 19. Long, meandering streaks are easy to hide.

Figure 20. One curve in this figure is twice as long as all the others.

CHUNKING FOR CURVE TRACING

Curve tracing has a tradition of pixel-at-a-time formulations (for example, see Duda, 1976; Pavlidis, 1982; & Rosenfeld, 1976). The run-time of pixel-at-a-time curve tracing processes is proportional to the number of curve pixels, and this is inadequate for a general purpose, real-time visual processor under reasonable restrictions on hardware cost (Edelman, 1985). The parallel algorithms applicable to this problem that have been proposed rely on global parallel models of computation. This section presents a new approach to exploiting parallelism in the operation of curve tracing. The basis of the approach is the observation that it is possible to trace a curve segment-at-a-time rather than pixel-at-a-time. A *curve chunk* is defined to be a region containing a single segment of a curve (Figure 21).[2] The internal operations of tracing can be applied to such regions as a whole, rather than to individual pixels. Curve chunks make it possible to trace curves in primitive steps that are much larger than a pixel, with the result that extended curves can usually be traced in a relatively small number of steps (Figure 22).

To reduce the number of iterations incurred by the tracing operation, the tracing process must simultaneously label a chain of curve elements—termed a *curve segment*—at a step, rather than a single element. The definition of a curve chunk is region-based owing to the combinatorics of this problem of labeling a curve segment in parallel. Any operation for simultaneously labeling all elements of a curve segment must be hard-wired. There are so many possible segments, however, that the cost of dedicated labeling hardware for each possibility would be prohibitive. This implies that the parallel labeling machinery must be shared across different curve segments. Perhaps the most straightforward way of achieving this sharing is to use machinery for labeling *an entire region* to label any curve ele-

---

[2] A curve is a set S of curve elements such that (i) for any two elements p, q in S, there is a path from p to q, and (ii) *either* all elements of S have exactly two neighbors in S *or* $|S| - 2$ elements of S have exactly two neighbors in S and two have exactly one. (Assume that $|S| \geq 2$. A path from an element $p = p_0$ to an element $q = p_n$ is a sequence of elements $p_0, p_1, \ldots p_n$ such that p_i is a neighbor of p_{i-1}, $1 \leq i \leq n$ [Rosenfeld, 1979].) A curve segment is a chain of curve elements. By "the curve segment contained in a region" we mean a maximal chain of curve elements included in the region. Thus the precise definition of a curve chunk is as follows. If the curve elements in a given region are viewed as nodes of a graph, and edges represent the neighborhood relation between curve elements, then *a curve chunk is a region such that* (a) *the set of curve elements in the region constitutes a chain (a connected graph of maximum degree 2)* and (b) *no curve element in the region has more than two neighbors, where neighbors outside the region are counted.*

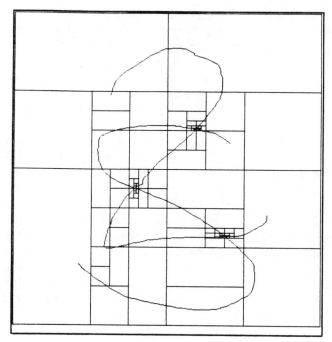

Figure 21. Result of dividing the input into large regions, called curve chunks, each containing one curve segment.

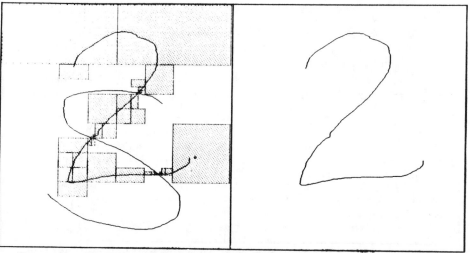

Figure 22. Result of tracing a curve in chunks. Each shaded region on the left was copied to the output in a single parallel operation.

ments falling within the region. A parallel operation for labeling a region may be used to label any curve segment that occupies the region provided nothing else occupies the region. Hence, the need for an essentially parallel labeling operation leads to a definition of a curve chunk as a region containing exactly one curve segment.

A graph may be used to represent curves at the chunk level in the following way. A node in the graph corresponds to a region in the input. Each node has an associated two-valued label which may take on the values *valid/invalid*. A region is said to be *vacant* if it contains no curve elements, and *valid* if it is vacant or if it contains exactly one curve segment. An edge in the graph corresponds to an instance of a curve continuity relation between two adjacent regions—it indicates that the curve segments in the two regions are connected. This representation of curves preserves the basic form of the standard serial curve tracing process. Let C be a variable that can be associated with a single curve chunk. A single valid node s is given to start, which is initially assigned to C.[3] Each curve chunk has an associated two-valued label which may take on the values *seen/unseen*, say. The basic form of a chunk-at-a-time curve labeling process is as follows:[4]

While C has a valid neighbor n labeled *unseen*:

1. label n and all the curve elements in the corresponding region *seen* (this is a parallel operation);
2. assign n to C;
3. repeat.

The effectiveness of a decomposition into curve chunks is measured in terms of the number of steps incurred in tracing, which is related to the size of the curve chunks—in general, the larger the curve chunks, the fewer tracing steps required. The strategy of *divide-and-conquer*[5] provides an efficient way of generating large curve

---

[3] In this paper, we do not discuss the processes that determine the starting chunk s. Generally, some independent process, such as indexing, will shift the processing focus to a location coinciding with or near an element of the relevant curve. A mechanism is required for accessing the chunk node s corresponding to the region containing the processing focus.

[4] This process stops when an invalid region is encountered. An extension is required to enable the tracing process to label the correct curve segment in an invalid region (i.e., a region containing more than one curve segment). A number of possibilities for coping with invalid regions are discussed in (Mahoney, 1987). One is to resort to pixel-level tracing within such a region.

[5] Curve chunking is open to a variety of implementation strategies. The scheme described here is one of several techniques developed and evaluated in (Mahoney, 1987).

chunks— large valid regions are constructed by combination of smaller regions that have previously been shown to be valid, subject to a simple rule. The main components of a divide-and-conquer curve chunking scheme are (a) an initial decomposition of the input into small curve chunks; (b) rules for combining valid regions into valid regions; (c) a pyramid representation and algorithm for building up valid regions by applying the rules.

The Initial Decomposition

The divide-and-conquer curve chunking process begins with an initial, direct decomposition of the input into curve chunks of small predetermined size. Regions defined in this initial decomposition are referred to as *elementary* regions. The algorithm for generating the initial decomposition is a parallel process over all elementary regions. The region-validity computation is simultaneously applied to each region and the corresponding *valid/invalid* label is assigned the appropriate value.

Mahoney (1987) develops a serial tracing method and a parallel feature matching method for determining validity of the small elementary regions. The serial validity check works by distinguishing and counting each of the curve segments in the region, using a pixel-level tracing process. The main requirement on the tracing process is that it must be capable of distinguishing two intersecting curve segments; this means the process must be able to maintain local curve direction. This is straightforward to implement using standard image analysis techniques.

The parallel method of establishing validity involves detecting certain combinations of two local features. The *border* of a region r is defined to be the set of locations in r that have fewer than eight neighbors in r. An *exit* of a region r is a connected set s of curve elements in the border of a region adjacent to r, such that some curve element e in s has a neighbor in r. A *termination* is a curve element (a) included in the region, and (b) having no more than one neighboring curve element in the region. The termination count T and the exit count E can be used to establish that a region contains a single curve segment (and no curve elements with more than two neighbors), assuming that the region contains no closed curve loops. With no loops, a region is guaranteed to be valid if $T + E \leq 2$, as shown in Figure 23. It is necessary to exclude loops because T and E cannot be used to distinguish between the cases in Figure 23 (a and b) and for example. For small regions, this "no loops" assumption is likely to be correct; when it is violated, tiny loops very near the relevant curve

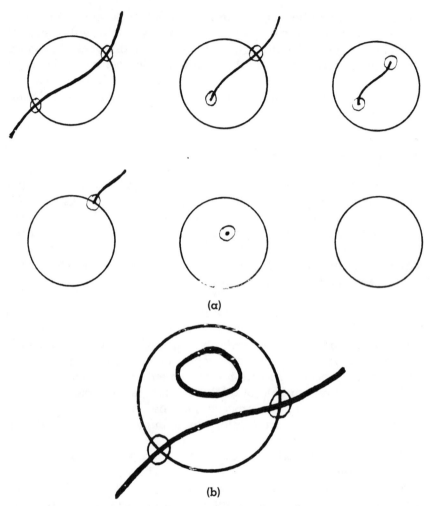

Figure 23. (a and b). Valid cases for the parallel validity check based on termination count and exit count.

may be incorrectly labeled, but for most applications this is likely to be harmless.[6]

The parallel validity check is implementable by simple, local processes. Terminations and exits are detectable by simple template matching. The simplest perceptron-like devices may be used to de-

[6] Mahoney (1987) develops a parallel validity check for regions of arbitrary size based on a simple modification of the parallel validity check for small regions. Restrictions on local curvature and the total range of local curve orientations are added to explicitly exclude loops.

tect the valid combinations of T and E (Minsky, 1969). A general counting mechanism is not required; it is only necessary to implement predicates of the form "There are exactly 2 terminations in the interior of the region," and so on. The outputs of these predicates can be combined by simple logical operations.

The Rule For Merging Regions

There is curve continuity between two adjacent regions $r1$ and $r2$ if there exist some curve elements $e1$ in $r1$ and $e2$ in $r2$ such that $e1$ and $e2$ are neighbors. If this condition is met, the common boundary between $r1$ and $r2$ is said to be *crossed*. *The union of two adjacent valid regions is valid if (a) either region is vacant; or (b) the common boundary is crossed.* (See Figure 24.)

The Pyramid Representation and Curve Chunking Algorithm

A pyramid representation is used to compute the decomposition of the input into curve chunks. Each node in the pyramid represents a region of the image. Associated with each node is a label which can take on the values *valid/invalid*. The size of the region represented by a node is determined by the node's level in the pyramid. The position of the region represented by a node is determined by the node's position in the array of nodes that are at the same level. The array of nodes at a particular level represents a regular tesselation of the input at some scale. A node is linked to a number child nodes in the level immediately below it. The children of a node in the pyramid represent the subregions of the region that the parent node represents. The base-level nodes represent the elementary regions in the initial decomposition. The pyramid as a whole can represent a

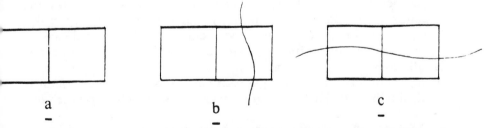

a   b   c

Figure 24.   Illustration of the rule for merging adjacent regions: (a) vacant regions are valid: (b) a vacant region and a valid, non-vacant region form a valid region; (c) two valid, non-vacant regions form a valid region if their common boundary is crossed.

large number of *irregular* tesselations of the input. A subset of the pyramid's nodes constitutes a description of a particular tesselation if (a) no two of the regions these nodes represent overlap, and (b) the union of these regions constitutes the entire input array. A two-sub-region-per-region pyramid scheme (termed a *binary image tree* by Parlidis, 1982) is adopted because it leads to the largest curve chunks—a failure to merge in a two-subregion scheme is less costly than in any other scheme—and it involves the simplest and most economical merging computations.[7]

The divide-and-conquer curve chunking process may be thought of in terms of three main processes. One process builds up valid regions. Another process identifies maximal valid regions. The third process establishes curve continuity between maximal valid regions. Valid regions may be built up in a single pass from the bottom of the pyramid to the top. For each node r at level l both of whose children are valid, the merging rule is applied to its subregions and the *valid/invalid* label is modified as appropriate. Maximal valid regions may be identified in a single pass from the top of the pyramid to the bottom. For each active valid node r at level l, every descendant node of r is marked non-maximal.[8] The curve continuity process is not dependent on the other two processes. The check may be applied in parallel for every pair of nodes in the pyramid that represent adjacent regions. For any given pair of regions, the curve continuity check involves determining whether their common boundary is crossed.

Performance of Tracing Based on Divide-and-Conquer Curve Chunking

The runtime of tracing using the curve chunks provided by the divide-and-conquer scheme depends on the sizes of the curve chunks into which the relevant curve is decomposed. Chunk size, in turn, depends on two factors. The primary factor is the geometry of the input curves—total curvature, local curvature, and incidence of curve crossings and points of curve proximity. (Taken by itself, curve length does *not* affect chunk size, so tracing time is scale independent.) The influence of curve geometry makes a precise performance analysis difficult. The secondary factor is the scheme by which the

[7] Mahoney (1987) defines a *symmetric binary image tree* because a divide-and-conquer process based on the standard binary image tree gives noticeably orientation-dependent results.

[8] A descendant node of r is any node that can be reached by following parent-child links downward from r through the pyramid.

regions to be checked for validity are defined. Each such scheme has different performance ramifications, and there is a wide, essentially unexplored range of possibilities. For that reason, the following paragraphs attempt to characterize the *general* aspects of the behavior of divide-and-conquer curve chunking, that is, the aspects that do not depend on the particular scheme used to define regions.

The best case occurs when (a) no elementary region traversed by the relevant curve is invalid—that is, it never traverses the same elementary region more than once, and never traverses an elementary region that another curve traverses; and (b) for some region *R* that is checked for validity, *R* includes the relevant curve and nothing else. "Tracing" involves one parallel region-labeling operation applied to *R*, which involves $O(\log(diameter\text{-}of\text{-}R))$ steps. The worst case arises when every elementary region traversed by the relevant curve is invalid—in that case no valid curve chunks are available at all, and the entire curve must be traced at the pixel level. In this case, divide-and-conquer curve chunking provides no performance improvement. Worst case input configurations are unlikely to occur in practice. A more likely "bad case" for this technique is a situation in which most elementary regions traversed by the relevant curve are valid, but for which most attempts to *merge* regions fail. In this case, divide-and-conquer curve chunking may provide only a constant factor speed-up over pixel-at-a-time tracing. The speed-up fac-

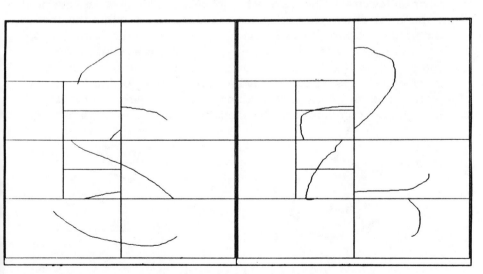

Figure 25. The result of two-label divide-and-conquer curve chunking. The array on the left represents label-0, the array on the right label-1. The tesselations in both arrays are the same.

tor is roughly the diameter of the elementary regions. Typical inputs fall somewhere between the best and "bad" cases, and *the number of steps required for tracing a curve is normally on the order of 10.*

Mahoney (1987) discusses a simple extension to the preceding divide-and-conquer curve chunking scheme that provides a substantial improvement in performance. The extended scheme suffers less fragmentation from intersection or close proximity between curves. Merging failures are averted by (a) distinctly labeling curve segments that are in close proximity and (b) applying the merging rules on a *per-label* basis. This *two-label divide-and-conquer curve chunking* scheme simultaneously generates (a) a tesselation of the input and (b) a two-valued distinguishing labeling within each region of the tesselation of the input and (b) a two-valued distinguishing labeling within each region of the tesselation (Figure 25). Given this representation, a tracing process may issue a label to the elements of *one of* the curve segments in each subregion region by specifying the relevant label to a *selective* region labeling operation.

Curve Chunks, Tracing, and The Establishment of Shape Properties and Spatial Relations

Many spatial judgments about curves involve the ordering of locations on the curve. The tasks of Figure 26 are examples. An important implication of the use of curve chunks for tracing is that the goal of extracting a curve as rapidly as possible may conflict with the goal of making judgements like these—the set of curve chunks representing a curve will only partially preserve ordering information. In the most extreme case, a curve may be represented by a single chunk, in which case no ordering information will be available at all. Therefore, it may sometimes take less time to establish curve connectivity

a

b

Figure 26. Judgements involving the ordering of locations on a curve. (a) Is the spiral curve a right-handed or left-handed? (b) Which is closer to a termination of the curve, the small circle or the dot.

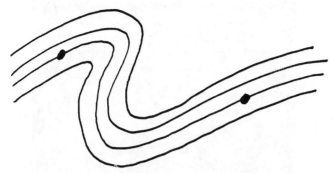

Figure 27.   Establishing that the dots lie on the same curve in this example may not require curve tracing.

(e.g., "same-curve") relations than relations involving curve ordering—the establishment of ordering relations may require a deliberately serial retracing of the curve, using chunks of artificially restricted extent. Thus, there is a possible distinction between the process by which a curvilinear entity is extracted and the process by which its properties and relations are computed. As such, in Figure 26 (a) it would take less time to decide that the two dots are on the same curve than that the curve in question is a right-handed spiral. Similarly, in Figure 26 (b) it would take less time to establish that the dot and the circle are on the same curve than to decide which is closer to an endpoint of the curve.

It is worth noting in this context that some properties and relations pertaining to curves may in certain circumstances be established without the use of curve tracing. The problem of evaluating tracing processes in the human visual system is complicated by this fact. For example, in Figure 27, it is possible to establish that the two dots lie on the same curve by counting curves and checking for breaks. Also, it may be possible to recognize certain curvilinear properties and configurations strictly in parallel. Perhaps the most striking demonstration of this is Fraser's illusion, Figure 28, in which nested concentric sets of disconnected spiral segments give rise to the irresistable impression of continuous nested spirals. This percept suggests that in this case local curve information is integrated in parallel in the human visual system, not by tracing.

CHUNKING FOR REGION COLORING

Like curve tracing, region coloring has a tradition of strictly serial pixel-at-a-time formulations which have time-performance linear in

Figure 28. The Fraser illusion (Fraser, 1908): perhaps the reason spirals are perceived where there are none is that descriptions of curves are being generated without the use of tracing.

the number of pixels in the region to be colored. Unlike tracing however, there is also a tradition of parallel coloring formulations; these are the cellular-array-based, parallel region growing methods. On the whole, however, these schemes have $O(region\text{-}diameter)$ run time, where the diameter of a region is defined to be the length of the longest path connecting two elements of the region.

Shafrir (1985) reports a thorough and very revealing study of the problem of fast region coloring. Shafrir's work established that high-performance region coloring may be achieved by describing the input in terms of extended *empty* regions. Region chunks make it possible to color regions in a relatively small number of steps. He investigated three high-performance schemes all of which involve (a) a preprocessing stage that detects regions of the input containing no boundary elements, and (b) a coloring process whose basic operation is the simultaneous activation of all the pixels in such a region. The schemes differ according to the shapes of the region chunks that the preprocessing phase can define, their degree of overlap, and the maximum communication degree of the hardware mechanism used to detect these regions. (Communication degree is defined to be the number of direct communication lines coming into a processor. All the schemes are based on a simple local parallel computation model.) The average case performance of the different coloring methods was established by simulation. Mean coloring times were recorded for five families of visually important input shapes. These shapes included circles, horizontal squares, horizontal rectangles, "stars" (randomly generated shapes whose convex kernel were large in comparison to their total area), and "snakes" (randomly generated shapes whose convex kernels were small in comparison to their total area). (See Figure 29.)

One of the schemes Shafrir investigated—the *quadtree model*—makes use of a quadtree-like input decomposition and processor network. The network's maximum communication degree is small (exactly 9). There is no overlap between regions. Region-vacancy is established by a hierarchical process. The basic computing step in this model is the operation of communicating a value between nodes that share a communication link. For compact shapes (circles, squares, stars) coloring time has only a small dependence on area of the shape. The position of the shape in the input and the position of the starting location within the shape did not significantly influence coloring time. For elongated shapes (rectangles and snakes), however-er, coloring time depended significantly on area.

Taking human performance as a guide, Shafrir found it un-satisfactory that the quadtree model's performance is significantly worse for simple elongated shapes (rectangles) than it is for compact ones. He developed another scheme—called the *rectangular model*—that is more satisfactory in that it colors simple elongated regions just as rapidly as compact ones. This scheme, illustrated in Figure 30, *directly* detects one-pixel-wide vertical and horizontal re-gion chunks at a series of lengths and positions governed by a scal-ing parameter S and an overlap parameter O. For S small enough and O large enough, all possible region chunks are detectable, but this requires $O(n^3)$ processors, where n is the diameter of the square input array in pixels. Shafrir tested the performance of the rec-

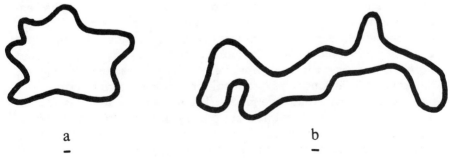

a   b

Figure 29. Two of Shafrir's test shapes: (a) A "star" and (b) A "snake."

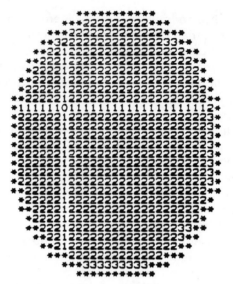

Figure 30.   Coloring of a circle by Shafrir's rectangular model. All locations labeled *n* were labeled at step *n*.

tangular model for virtually the full range of settings of S and O subject to an $O(n^2)$ limit on the number of processors and an $O(n)$ limit on communication degree. He found that the particular settings of S and O did not significantly affect performance—all choices with similar "density" (number of chunk detectors and communication degree) gave similar behavior. The rectangular model is insensitive to the area of the input in the case of circles, squares, rectangles, and stars. Coloring requires fewer than 10 steps for these simple inputs. For snakes, there appears to be a small dependence on the area of the input figure. Naturally, performance improves even further with higher density.

The third high-performance coloring scheme Shafrir investigated—the *square model*—was built along the same lines as the rectangular model, but makes use of square regions instead. Squares are defined at a series of diameters and positions governed by S and O. The square model proved more sensitive to area of the input figure, and substantially slower overall, than the rectangular model.

Mahoney (1987) makes a proposal for region chunking that is very similar to Shafrir's quadtree model. The main purpose of his discussion is to put representational support for high-performance region activation in the broader perspective of image chunking. Two main observations are made. The first is that region chunking is most effective when applied to the results of curve activation operations, rather than directly to the input boundary represenation.

The second observation is that it is possible for region and curve chunking processes to share their underlying machinery; minor modifications to the curve chunking processes lead to effective region chunking processes at no additional cost. Firstly, this is because the mechanism for generating a tesselation obviously does not depend on the application which the regions in the tesselation are to serve. Secondly, the validity check for curve chunking subsumes the one for region chunking—recall that a valid region was defined to be a vacant region or a region containing exactly one curve segment. This is not to say that curve and region chunking *must* use common machinery. Shared machinery connotes a savings in hardware cost, but it may also imply a compromise in the time performance of coloring, tracing, or both. The representational requirements of the best existing curve and region chunking schemes (the two-label divide-and-conquer scheme and Shafrir's "rectangular model," respectively) appear at the moment to be quite incompatible.

CONCLUSION

On the whole, low-level vision has been the study of extracting images from images. This chapter has argued that low-level processes can and must go much further than this—spatially uniform, parallel, bottom-up processes can build representations which make the inherently serial operations of scene analysis very time-efficient, even under the conservative computational restrictions of simplicity and locality. Image chunking is the "missing link" between traditional "very low-level" vision—the computation of pointwise properties—and the generation of detailed descriptions of complex spatial entities, properties, and relations.

Image chunking may be distinguished from the traditional notions of segmentation and grouping. Image chunks function much more as building blocks and "beacons" than as descriptors. Furthermore, there is a methodological distinction to be made, having to do with representation design—a more specific formulation of the application of a representation makes it possible to design a more effective representation for that application. The chunk representations are specifically designed to support particular basic operations of intermediate vision, whereas many proposals for "perceptual organization" have been designed to support *general goals* of vision, like recognition or reasoning, without detailed formulations of the processes that are to use the representations to achieve these goals.

Image chunking is not merely the exploitation of parallelism—it is the application of a crucial parallel-then-serial problem decomposition. The study of image chunking is as much concerned with what low-level processes *cannot* do effectively, as with what they can. The goals of image chunking are deliberately limited compared to those of many other scene analysis schemes that exploit parallelism, because parallelism is exploited most effectively if its limits are not exceeded.

In this paper we have emphasized the role of image chunks in providing time-efficiency in spatial analysis, but their importance goes beyond this—they also provide *expressive power*. Image chunks allow spatial analysis processes—visual routines—to be expressed in terms of entities that are more meaningful to the immediate goals of vision than are individual pixels. For example, the indexing operation—shifting of the processing focus to the location of a likely object—cannot be expressed on the basis of pixel-level representations of the scene, because the objects of interest in vision are typically extended. Similarly, it is more natural to express a "find-space" operation in terms of region chunks than pixels. We believe this notion of expressive power is crucial to the effective development of visual routines, both in man and machine.

REFERENCES

Ballard, D. H., & Brown, C. M. (1982). *Computer vision*. Englewood Cliffs NJ: Prentice-Hall.

Barrow, H. G., & Tenenbaum, J. M. (1978). Recovering intrinsic scene characteristics from images. In A. R. Hanson & E. M. Riseman (Eds.), *Computer vision systems* (pp. 3–26). New York: Academic Press.

Bolt Beranek and Newman Inc. (1985). Development of a butterfly multiprocessor test bed. Rep. 5872, Quarterly Technical Report, No. 1.

Cook, S. (1983). The classification of problems which have fast parallel algorithms. Proc. FCT-83, Springer-Verlag Lecture Notes in Computer Science.

Cornsweet, T. N. (1970). *Visual perception*. New York: Academic Press.

Duda, R. O., & Hart, P. E. (1973). *Pattern recognition and scene analysis*. New York: Wiley.

Edelman, E. (1985). Fast distributed boundary activation. M.Sc. Thesis, Department of Applied Mathematics, Feinberg Graduate School, Weizmann Institute of Science, Rehovot, Israel.

Fraser, J. (1908). A new visual illusion of direction. *British Journal of Psychology*, 2, 307–320.

Gottlieb, A., Grishman, R., Kruskal, C., McAuliffe, K., Rudolph, L., & Snir, M. (1983). The NYU Ultracomputer—Designing an MIMD Shared Memory Parallel Computer. *IEEE Transactions on Computers C-32* (2), 175–189.

Hillis, W. D. (1985). *The connection machine*. Cambridge, MA and London: MIT Press.

Hirschberg, D. S., Chandra, A. K., & Sarwate, D. V. (1979). Computing connected components on parallel computers. *Communications of the ACM, 2*(8).

Horn, B. K. P. (1986). *Robot vision*. Cambridge, MA and London: MIT Press.

Jastrow, J. (1900). *Fact and fable in psychology*. Boston: Houghton Mifflin.

Jolicoeur, P., Ullman, S., & Mackay, M. (1985). Boundary tracing: An elementary visual process. *Memory and Cognition, 4*, 219–227.

Julesz, B. (1981). Textons, the elements of texture perception, and their interactions. *Nature, 290*, 91–97.

Koch, C., & Ullman, S. (1984). Selecting one among the many: A simple network implementing shifts in selective visual attention. A. I. Memo 770. Cambridge, MA: MIT Technology Artificial Intelligence Laboratory.

Lim, W. (1986). Fast algorithms for labeling connected components in 2-D arrays. Thinking Machines Comporation Report NA86-1. Cambridge, MA: Thinking Machines Corporation.

Mahoney, J. V. (1987). Image chunking: Defining spatial building blocks for scene analysis. S.M. Thesis. Cambridge, MA: Dept. of Electrical Engineering and Computer Science, MIT.

Marr, D. (1982). *Vision*. San Francisco: W. H. Freeman.

Minsky, M., & Papert, S. (1969). *Perceptrons*. Cambridge, MA and London: The MIT Press.

Muller, M. J. (1986). Texture boundaries: Important cues for human texture discrimination. *IEEE Proceedings of Conference on Computer Vision and Pattern Recognition*, June 22–26, 1986, Miami Beach, Florida, pp. 464–468.

Pavlidis, T. (1982). *Algorithms for graphics and image processing*. Rockville, MD: Computer Science Press.

Rock, I. (1984). *The logic of perception*. Cambridge, MA and London: The MIT Press.

Rosenfeld, A. (1979). *Picture languages*. New York: Academic Press.

Rosenfeld, A., & Kak, A. C. (1976). *Digital picture processing*. New York: Academic Press.

Shafrir, A. (1985). Fast region coloring and the computation of inside/outside relations. M.Sc. Thesis. Rehevot, Israel: Department of Applied Mathematics, Feinberg Graduate School, Weizmann Institute of Science.

Shiloach, Y. & Vishkin, U. (1982). An O(log n) parallel connectivity algorithm. *Journal of Algorithms, 3*(1), March, 1982.

Treisman, A. & Gelade, G. (1980). A feature integration theory of attention." *Cognitive Psychology, 12*, 97–136.

Ullman, S. (1976). Relaxation and constrained optimization by local processes. *Computer Graphics and Image Processing, 10*, 115–125.

Ullman, S. (1984). Visual Routines. *Cognition, 18*, 97–159.

9
Here and *There* in the Visual Field*

Zenon Pylyshyn

Center for Cognitive Science
University of Western Ontario

INTRODUCTION

This paper develops and extends some ideas originally introduced in Pylyshyn, Elcock, Marmer & Sander (1978) in the light of more recent work we and others have been doing on low-level vision and its interface with cognition. Some of the design decisions we took in the early work have also been re-examined in the light of recent empirical findings.

One of the assumptions that was introduced in the Pylyshyn et al. work, and which forms the basis of some of our current investigations, concerns the pre-attentive, or pre-cognitive processing of *places* in the visual field. I use the terms pre-attentive or pre-cognitive to emphasize that what I am talking about is a very primitive and automatic process that is a precursor to such further processes as the encoding or recognition of the location of visual objects. There is something the visual system must do first, before it can even discern a spatial pattern or spatial relations among component features in a display: it must "pick out" or "individuate" the features among which it will recognize some spatial relations, such as "above," "part of," "inside" and so on. Before you can say that "this" is inside "that" you must have a way to, in effect, "point to" the two

---
* This research was supported in part by operating grant A2600 from the Natural Science and Engineering Research Council of Canada.

features to which the "inside" relation will apply—somewhat like the indexical words "here" and "there" are used in English to point to places.

In order to accomplish this "pointing" there is no need to have first recognized which feature-types are being pointed to. All you need is to *pick out* or in some way *index* the locations of the feature-tokens in question (whether this "picking out" entails encoding these locations, or indexing them in some more primitive manner, and whether the relevant "locations" are, at this level, characterized in retinal or spatial coordinates, are questions to which I will return presently). Recognizing the simple point that individuation must precede relational encoding leads to one of our basic postulates (the notion of FINSTs), which in turn gives us the tools we need to illuminate a number of other puzzles.

ASSIGNING FINSTs TO VISUAL FEATURES

In order to illustrate in a concrete (though somewhat fanciful) way what is intended by the FINST mechanism to be described presently, imagine the following. Suppose you place each of your fingers on a different object or feature in a scene. Now imagine that the objects are moving about or that you are changing your position while your fingers keep in contact with the objects (if you imagine the movements to be too large you may have to imagine that your fingers have the elastic powers of the cartoon character "plastic man"!). Even if you do not know anything at all about what is located at the places that your fingers are touching, you are still in a position to determine such things as whether the object that finger 1 is touching is to the left of or above the object that finger 2 is touching, or whether the object that finger 3 is touching is larger than the object that finger 4 is touching. Of course, you may not be able to determine this directly without further analysis, but your finger-contact gives you a way of referring to the objects so that some further processing of them can be undertaken. Moreover, the access that the finger-contact gives you makes it possible for you to keep referring to the *same* object-feature independent of its location in space, so you can track moving objects or otherwise *individuate* them and keep them distinct.

Such direct *mechanical* indexing provided by the sense of touch makes it possible to do something that cannot be done directly in vision. Touch provides a way of thinking "this" object or "that" object, for any object being touched, independent of its location in space.[1] The parallel case in vision appears to be different, since it would seem that the equivalent of pointing *directly* at a feature in a 3-D scene is not possible. That's because the only visual transducers we have are ones that respond to the 2-D retinal projection of the scene. Nonetheless—and this is a central assumption of our model—we can do something very analogous to pointing. Our theory posits a mechanism called a FINST, which allows one to accomplish, in a limited way, something that is functionally similar to indexing a feature in a 3-D scene (not just a distinct visual feature on the retina), much as a finger allowed us to index such a feature in the tactile example discussed above (which is why we originally called such an index an "instantiation finger," later abbreviated to "FINST").

A FINST is, in fact, a reference (or index) to a particular feature or feature cluster on the retina. However, a FINST has the following additional properties: (a) because of the way clusters are primitively computed, a FINST keeps pointing to the "same" feature cluster as the cluster moves across the retina (where by "same" feature I mean that the feature arises from the same distal object or property)[2] and (b) a FINST provides a cross-reference between the indexed feature and certain corresponding entities in an evolving internal description of the scene, as well as perhaps a representation of the scanning pattern by which the scene was examined. If the retinal feature cluster in question maintains a reliable correlation (over time) with some particular feature of the scene, then the pointer will succeed in pointing to the retinal projection of a particular scene location, independent of its location on the retina. Thus, for features that are currently on the retina, FINSTs allow the system to access their stable (2-D) spatial relations in a way that is *transparent to their retinal location*. In addition, they allow the system to relate parts of the representation of the scene being constructed to particular retinal

[1] Perhaps this has something to do with why Turvey (1977), quite erroneously, assumes that information processing psychologists believe "mechanical commerce with the environment is privileged in that it permits the direct detection of object properties." Of course nobody working on the information processing tradition believes that tactile perception is "direct" or unmediated by computational processes, any more than is the case in vision.

[2] Presumably clusters of microfeatures are aggregated by some local processes operating in parallel over a retinotopic representation. A common way of keeping track of such aggregates in computer vision systems is to maintain a "list of contributing points" which defines a putative edge or other feature. In that case keeping a FINST attached to a feature as it moves continuously about the retinal field amounts to something like the policy that if lists on successive frames of view share a significant subset of their elements they are identified as the same list (i.e. they are given the same name).

features. It does this by associating the FINST that is bound to a particular feature-token, with the corresponding part of the internal representation of the scene. Another way to put this is that the FINST mechanism allows symbols in the internal description to actually *refer* to individual entities in the scene.

The main purpose of FINSTs is to allow higher cognitive processes to refer to particular visual features prior to evaluating properties located at these indexed places. This is particularly important when relational properties involving several places are encoded. For example, we assume that in order for the cognitive system to encode a relational property holding among several places—such as COL-LINEAR(x,y,z) or INSIDE(u,v) or PARALLEL(m:line, n:line)—the arguments to these predicates must first be bound to features or places in the scene, that is these places must be assigned FINSTs. Once assigned, groups of features or "chunks" may also be formed and under certain conditions a FINST assigned to the entire chunk. A chunk which has a FINST bound to it may or may not also have FINSTs bound to its component parts. However, in order to evaluate a predicate such as PART-OF(x:element, C:chunk) both arguments have to be bound by separate FINSTs.

The question of how FINSTs are assigned in the first instance remains open, though it seems reasonable that a certain number of FINSTs are assigned in a stimulus-driven manner, perhaps by the activation of locally-distinct properties of the stimulus—particularly by new features entering the visual field.[3] In addition, there may be occasions when top-down processes might also cause a FINST to be assigned to a feature, or perhaps even an empty place. Many of these questions await further empirical exploration. Nonetheless, as I shall argue below, there are already numerous consequences of the assumptions that seem most secure. The assumptions discussed above are summarized in Table 1 for reference.

If we add to the above certain additional assumptions, which we will not discuss in this paper (but which are sketched in the appendix), the FINST binding mechanism also permits the motor system to make reference to visual locations (so motor commands can be issued in relation to visually perceived locations), as well as to a limited

---

[3] One of the ways in which a feature can stand out is in terms of its motion. For this reason the local processes that are responsible for "activating" the features must operate over time or over consecutive frames of view. It turns out that such processes are required in any case for independent reasons: they are required in order to solve the "correspondence problem" associated with maintaining the identity of clusters in successive presentation frames. The relaxation algorithm described elsewhere in this volume by Dawson & Pylyshyn can signal such local distinctiveness as a side-effect of solving the correspondence problem.

Table 1.   Summary of Some Assumptions of the FINST Model

1. Primitive stage 1 processes produce *feature-clusters* automatically and in parallel across the retina.

2. Certain of these clusters are selected or *activated* (also in parallel) based on their distinctiveness within a local neighbourhood (e.g., the so-called "popout" or *odd-man-out* features).

3. The activated clusters compete for a limited pool of internal referencing tokens called FINSTs. This also happens in parallel, and the initial assignment of FINSTs is stimulus-driven.

4. The stage 1 processes that create feature clusters also maintain their integrity: A FINST that is bound to a feature cluster keeps being bound to it as the cluster changes its location continuously on the retina, thus continuing to "point to" fixed places in a scene and serving like the indexical pronouns "here" or "there."

5. Some higher order patterns, consisting of aggregates of primitive feature clusters (e.g., contours) can also be assigned FINSTs under either top-down or bottom-up control.

6. Only FINSTed feature clusters can enter into higher cognitive processes. Relational properties like *Inside*(x,y), *Part-of*(x,y), *Above*(x,y), *Collinear*(x,y,z), . . . can only be encoded if features x, y, z, . . . are FINSTed. However, the visual system can encode relations among retinal features without in any way first identifying these features.

7. FINSTs can also be assigned in a top-down fashion. Higher level processes can, for example, direct a FINST to be placed at certain points defined in terms of already FINSTed locations. For example, they can allocate a FINST to places such as *MIDPOINT*(x,y) or *INTERSECTION*(u:line, v:line).

8. The model also allows a limited ability to bind FINSTs to non-retinal locations and to a few objects that can serve as both visual and proprioceptive references (See Appendix).

number of off-retinal places. The overall picture that is being presented may be illustrated roughly in the form of a diagram as in Figure 1.

This figure shows that FINSTs allow internal representations to refer to places in a visual scene that have not yet been assigned unique descriptions. They can, in other words, refer the way the English indexicals "here" or "there" (or "now") can refer. In addition they allow multiple references to be made simultaneously, and also allow the motor system to, in effect, issue commands to move a limb to certain visually perceived locations (see Appendix for more on the latter). The capacity to make such indexical references in vision has far-reaching implications. Below I discuss a few of the consequences of this primitive mechanism for explaining various empirical phenomena found in the experimental literature. In addition, there are some deeper implications of such a mechanism for understanding certain philosophical issues concerning the functioning of spatial indexicals in general (see, for example, the discussion of this issue by Peacock, 1983).

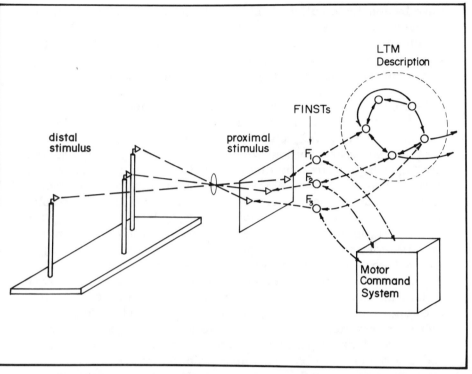

Figure 1. Schematic illustration showing how FINST indexes allow reference to be made to elements in the distal environment, as well as allowing
cross-modality binding of variables in semantic memory and in the motor
control system.

AN EMPIRICAL DEMONSTRATION OF THE FINST
MECHANISM: TRACKING MULTIPLE INDEPENDENT TARGETS

Perhaps the easiest way to illustrate the FINST hypothesis is to describe an experiment intended to be a fairly direct test of the basic
assumptions behind this notion.

Consider the following experiment (for more details, see Pylyshyn
& Storm, 1987). Suppose subjects are shown a field of identical randomly arranged points and are required to keep track of some subset
of them (called the "targets")—as they must if their task is to count
the targets, or to indicate when one of them flickers or moves. In such
a task, subjects might proceed by encoding the location of each of
the targets with respect to either a local or global frame of reference,
thus making it possible to distinguish and keep track of each target
by its coordinates. The encoding of relative positions might be facili-

tated by noticing a pattern formed by the points, thereby "chunking" the set in a single mnemonic pattern (see the discussion by Mahoney & Ullman in this volume). What clearly would not work in this situation is to remember visual characteristics of individual points in the target subset, since the targets and nontargets are visually identical.

Now suppose the points are set into random independent motion, and the subject is required to indicate (by pressing a button) whenever one of the target objects briefly changes its shape, or to indicate (by pressing another button) whenever a nontarget briefly changes its shape. In this case the distinctiveness of each point cannot be attributed to its location, since this is continually changing. Hence storing a code for the location of each point would not help to solve the problem, unless the location code is updated sufficiently frequently. The update frequency would have to be such that during the time between updates the target remained within a small region where it would not be confused with some nearby distractor. If location codes have to be assigned in series by moving attention to each in turn (as most people believe), this would entail sampling and encoding locations according to some sampling schedule in which points are scanned in sequence. If we had some idea of the maximum rate at which points could be visited and their locations encoded, we might be able to set up a situation that would cause this strategy to fail—say because the points would have moved far enough in the sample interval so there was a high probability that another point was now in the place occupied earlier by the point whose location code one was attempting to update. Under such conditions, subjects should no longer be able to do the multiple-tracking task described above.

We did carry out such an experiment—which I will describe below. The following, however, was the conclusion. Using some widely accepted assumptions concerning the location encoding process, we found that subjects could do very much better at this task than predicted by the sequential encoding procedure. What, then remains as a possible mechanism for carrying out this task? If our assumptions and our analysis of the situation are correct, it appears that subjects are able to simultaneously keep track of at least four distinct "places" in the visual field without encoding their location relative to a global frame of reference using some explicit symbolic location code. This is precisely what the FINST hypothesis claims: it says that there is a primitive referencing mechanism for *pointing to* certain kinds of features, thereby maintaining their distinctive identity without either *recognizing* them (in the sense of categorizing them), or explicitly encoding their locations.

Let us look now at the details of the experiment. Based on some preliminary studies we determined that subjects could track at least four randomly-moving points (in the shape of "+" signs) in a total field of eight such randomly-moving points. In order to design the task in such a way as to preclude its solution by a sequential-sampling procedure, we appealed to the generally held view that in order to encode the location of a point, a subject must attend to that point. As Ann Treisman and others have shown (see chapter 13), noticing that a stimulus contains a certain feature is not the same as noticing *where* that feature is: the two can be functionally dissociated. In order for the information about location to be available for such purposes as identifying where the point is in relation to some frame of reference or some other feature, it seems that the feature has to be attended to. Furthermore, it is widely believed (see the list of references in Table 2 below) that this sort of attention is unitary— that is, there is only one such locus which must be moved from place to place. *Attending*, according to this view, entails actually moving a locus of focused attention (independent of eye movements) to that location. A substantial number of studies now exist which conclude that a single locus of processing must be moved about in the visual field and that the movement is continuous. I am not endorsing this view—indeed the FINST hypothesis stands in opposition to a certain interpretation of the moving-attention view, but that is the main alternative we are addressing.

The velocity with which attention can move within the visual field has been estimated by various researchers to range from 30 to 250 degrees per second (i.e., from 33 to 4 msecs/degree). Some representative values are shown in Table 2. If we take 250 degrees/sec (or 4 msec/degree) as an estimated upper bound, then knowing the minimum path length to scan all four points being tracked, we can set the dispersion of the points and their velocity so as to maximize the chances that a subject using the scan-and-encode method will mis-

Table 2. Estimates of Velocity of Attention Scanning

speed (deg/sec)	Source
30.3	Eriksen & Schultz (1977)
38.5–41.7	Jolicouer, Ullman & Mackay (1987)
51.3	Finke & Pinker (1982)
52.6	Shulman, Remington & McLean (1979)
58.8	Kosslyn (1978)
53.8–76.9	Pinker (1980)
117.6	Tsal (1983)
250	Posner, Nissen & Ogden (1978)

take a distractor point for a target. This was done by a combination of making the mean speed of movement of the points sufficiently high (8 degrees/second) and arranging for each target to never be more than 1.5 degrees from a distractor. The determination of the probability of erroneously switching to tracking a distractor prior to the time of the probe (which consists of a square displayed for 83 msec over either a target or a distractor) was done by simulating a sequential scan of the stimuli used in the experiment and picking the point nearest the encoded location at each sampling cycle.

Time and distance parameters used in the simulation were based on measurements made on the actual displays used in the experiment, and on empirical observations. We examined the sequence of displays and measured the shortest path covering all four targets in each frame, then averaged these over all frames in each trial. This distance, together with estimates of the velocity of attention, was used to obtain the appropriate intersample time. Several different scanning strategies were simulated, including a complex strategy based on the assumption that subjects determined the speed and direction of the sampled point and used this to project where the point would be at the time of the next sample. In this strategy the

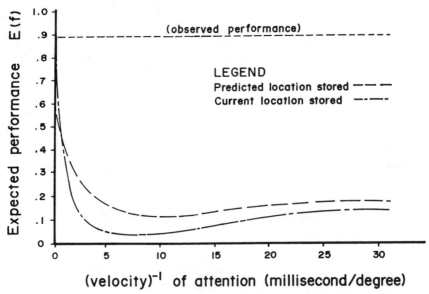

Figure 2.   Predicted performance in tracking four independent moving targets based on a serial tracking model, expressed as a function of the velocity with which attention can be scanned across the display. The superiority of observed performance suggests that parallel tracking is being used. From Pylyshyn & Storm (in press).

subjects were assumed to encode and store the projected location rather than the current location. The results of these simulations, together with the observed performance, are shown in Figure 2 where the predicted level of performance is plotted as a function of estimated velocity of attention. This shows clearly that subjects are not sampling the points in a sequence of move-encode operations.

Our conclusion, then, is that the four targets are being tracked in parallel, and that the tracking is not based on encoding the locations of points with respect to some frame of reference, but rather is based on a simple dynamically-maintained indexing scheme such as that proposed by the FINST hypothesis.[4]

SOME IMPLICATIONS OF THE FINST HYPOTHESES

Constructing a Spatially Stable Representation

Although the input to the visual system consists of continually moving images on the retina, we nonetheless perceive a world that remains stable with respect to a global frame of reference. This suggests that there is a stage (call it stage 1) at which the visual system operates upon representations fixed in a retinotopic frame of reference, and a later stage (call it stage 2) at which locations of perceived scene features are encoded in relation to a frame of reference or coordinate system that is fixed in space (or at least a 2-D projection of such a coordinate system).

Not a great deal is known about the different processes that operate at each of these stages. At the *first* stage, properties such as intensity gradients, are associated with specific locations on the retina. It is known that some visual aftereffects and certain types of masking occur at this stage. I will suggest later that there is some reason to expect that this may be a more important stage than generally believed, and that other perceptual effects may have their locus here, as opposed to the more commonly accepted stage 2 locus. In fact, the claim being made in this paper is that the FINST mechanism operates at this stage and provides a means of obtaining some of the stage 2 "geocentric" phenomena.

[4] Since the above experiment was reported, a number of other studies have been carried out using different equipment (i.e., an Amiga computer) allowing us to achieve smooth movement with faster speeds, to use up to 6 targets (whose shapes were allowed to vary dynamically), and to construct trajectories that avoided collisions by simulating an inverse-square law repulsion field about each object (instead of by making sudden direction changes just prior to a potential collision, as in the present study). We found that in this more complex setup, experienced subjects were able to perform even better than those in the original experiment.

The *second* stage is thought to produce a 2-D representation of the optical properties of the light emanating from a scene, though fixed in a coordinate system which is independent of eye movements. The output of this stage may correspond to what Marr (1982) has called the "primal sketch" (though the term "sketch" is misleading since it suggests some sort of picture, rather than the more neutral and more accurate notion of a datastructure). Once again what goes on at this stage is open to debate. Davidson, Fox & Dick (1973) have reported a nice demonstration of how stage 1 and stage 2 representations can be distinguished. They showed that whereas at least one form of visual masking takes place over a stage 1 representation (i.e., the location of the masking effect depends on the retinal locus of the mask), the perception of the relative location of two briefly displayed stimuli, under conditions in which an eye movement occurs between displays, is determined by the stage 2 representation (i.e., the perceived relative location of the two stimuli is a function of their spatial, not retinal, coordinates). There is also some evidence that apparent motion may be computed at stage 2 (Rock & Ebenholtz, 1962), and that the "correspondence problem" (Ullman, 1979) may be solved here as well, although in some of these cases it is controversial whether these processes operate over a 2-D or over a 3-D representation.

One fairly traditional view of how we construct and update a stage 2 geostable representation is that we "paint" the level 1 retinotopic representation onto an extended "image" of our environment, and that the reference point specifying where to transfer the current retinal image is moved in exact correspondence with the motor vector that drives the eye. This is called the "corollary discharge" view, because it claims that an "efferent copy" of the signal going to the eye muscles is also sent to the mechanism that superimposes the retinal image on the extended internal geostable image. While there is considerable doubt about the neurophysiological assumptions underlying the "corollary discharge" view, and even the critical role of eye movements is questionable (since Hochberg, 1968, has shown that passively presented glances can be integrated under certain circumstances), the idea that the visual system "points" a larger spatially-fixed image is frequently taken for granted.

However, the FINST model suggests that there is no need to posit an "extended image" in order to account for the invariance of perceived location with eye movements. Indeed, there is no reason to even assume that there is a stage at which the spatial location of features must be *explicitly* encoded—in the sense, for example, in which such locations might be encoded in terms of their Cartesian coordinates. To see why this is so, recall the "pointing fingers" analogy discussed earlier. The use of these mechanically-linked tactile sensors bypassed the need for an explicit encoding of locations in some global framework, because all the information that was needed could be read off the actual scene as required by using the indexes.

Similarly, FINSTs provide a way of indexing a number of places in a scene independent of their retinal locations. This, in turn, may provide the basis for achieving some of the same effects that can be derived from the postulated extended geostable image. For example, if FINSTs provide the reference points for determining the perceived locations of elements in a scene, then the fact that FINSTs remain bound to distal features as the eyes move about means that the perceived locations of such elements will remain invariant with eye movements. This is exactly what happened in the case of the tactile example discussed earlier, where fingers were used as indexes to distal features.

Although this is an intriguing possibility, there is more to spatial stability than maintaining such bindings to distal reference points. Below we discuss two considerations that are relevant to this issue. The first concerns the integration of off-retinal features. The "extended internal image" idea is intended, in part, to allow the perceptual integration of information on the retina with that which is no longer on the retina (but which was there as part of a previous glance). Thus we need to inquire about the nature of the stored off-retinal information. The second consideration is the question of precisely what we mean when we say that the stage 2 representation is geostable. I will briefly examine these two questions in turn.

Only a little is known concerning the first question. For example, it is known that off-retinal information is encoded in a sufficiently partial or abstract form that this information does not function exactly as it would if it were on the retina. Off-retinal information does not combine with retinal information to produce certain perceptual phenomena that occur when all the information is retinal. For example, impossible figures (such as the "devil's pitchfork") are not easily detected if the distances between inconsistent portions is large, or if the information is presented in "glances" in some arbitrary order. Similarly, the automatic interpretation of certain line drawings as depicting three dimensional objects does not occur as readily if the parts of the figure are not presented in an appropriate order or if the segmentation of the scene into glances fails to present entire critical features in individual glances (Hochberg, 1968). Such results suggest that the stage 2 representation may already have been processed or even interpreted to some extent, and therefore does not consist of an "image," insofar as this term implies something like a pattern laid out on an extended retina. In general, it is easy to overestimate the amount of information about off-retinal features that is available. It could well be that simple labels attached to FINSTs (perhaps including FINSTs which index off-retinal objects—see Appendix) might provide all the information needed to account for such things as the *anorthoscope* or *eye-of-the-needle* effect (Rock, 1981).

With respect to the second question, consider what properties a representation has to have in order to count as being geostable. One property that is surely implicated in the notion of geostability is that the world does not *appear* to move as our eyes move. Although reliance on such phenomenology is generally considered problematic (for example, there is evidence that we can respond to movements in the perceptual world of which we do not appear to have conscious awareness; see the chapter by Mel Goodale in the present volume), there is no doubt that the phenomenal experience of stability is an important aspect of what we mean by perceptual stability. A more objective reason for believing in the geostability of a percept in the face of eye and body movement, is that perceived space appears to connect with the motor system in a stable and consistent manner. For instance, if we were to point to some object or location in a scene we were viewing, the direction we would point would be independent of where the projection of that place fell on the retina. Similarly, pattern perception is coherent over time, so that when we recognize or identify objects we do so on the basis of properties that may have been picked up from many distinct glances. Integration of information from different glances, even when that information is overlaid on the same retinal coordinates, produces some of the most dramatic evidence for geostability (Hochberg, 1968).

In order to construct a representation that has the right spatial stability properties, the following, at least, should be true. *First,* under certain conditions of movement of features on the retina, it must be possible to identify these features as originating from the same place in the scene. *Second,* there must be some (perhaps partial or abstract) representation of the location of features that are not on the retina. Such a representation would allow the visual system, at some stage, to relate these off-retinal features to the location of retinal features. As noted earlier, however, one should not overestimate the amount of information actually stored at off-retinal locations. *Third,* we must have some way to coordinate movements (whether eye movements or pointing) with this integrated representation. What these three requirements amount to is the need for some sort of coordination between retinal features and off-retina features which allows one to identify sequences of proximal features as arising from the same *distal* feature, even if the sequence is discontinuous and interrupted (i.e., as the proximal feature moves off the retina and back again in the course of eye movements).

Although how this is carried out is far from being known, there is at least reason to doubt that the task requires the piecewise "painting" of an extended internal image. Indeed, the task does not even appear to require that locations of features be explicitly encoded— only that some means be available for indexing the features so that they can be addressed by primitive perceptual and motor operations. The model that Pylyshyn et. al. (1978) have sketched suggests one or two *minimal*[5] mechanisms which, along with the basic FINST mechanism discussed earlier, appear to be *sufficient* for tasks 1-3 above. Some of these additional mechanisms are sketched in the Appendix.

Implications for Visual Routines

I now turn to the relevance of the FINST hypothesis to the computation of spatial relations. Ullman (1984) has examined a number of relations that the visual system can compute, apparently with ease, and has asked how they might be computed. In the case of many of the relations (e.g., "inside") it is difficult to see how this relation could be computed by a parallel process since it requires checking on the relation between the location of a point and an arbitrary curve. All the possible algorithms that Ullman considers involve some serial process, such as "painting" a region beginning either at the point in question or at places along the curve, or extending radial lines from the point in question and noticing the parity of their crossings with the curve. In all these cases (as well as in the case of evaluating other visual predicates, such as whether two points are on the same contour) it appears that "the execution of visual routines requires a capacity to control locations at which elemental operations are applied (Ullman, 1984, p. 135)."

Although it is clear that a capacity to control locations at which to carry out processes is necessary, it is not clear that this sort of "con-

---

[5] I use the term "minimal" here in an informal sense. I simply mean that the mechanisms appear to embody the smallest set of assumptions necessary for accomplishing the task—though there is no proof that no "simpler" mechanism is possible, and indeed the very notion of simplicity used here is not made explicit. The mechanisms are minimal in the sense that a Turing Machine is a minimal mechanism for computing, namely it is very elementary, yet sufficient for the task.

trol" itself always involves moving a unitary locus of attention to that location, or encoding the location in some explicit way. Thus I do not agree with Ullman that "The marking of a location for later reference requires a coordinate system . . . with respect to which the location is defined."—at least it is not necessary that the location be "defined" by an explicit set of coordinate codes.

The FINST mechanism shows how one can mark a location in a manner that will subsequently allow attention to be directed to it if, for example, some serial process using a unitary focus of attention is needed for some computational task. The FINST mechanism also provides for the location to be referred to in certain primitive motor instructions. Yet FINSTs do not necessarily make the location in question *directly* available to any high-level process, such as one that generates a verbal location response, or even one that computes a spatial relation between the place in question and some other place. The latter requires, at the minimum, that *both* locations be FINSTed. Indeed, most predicates, and virtually all of Ullman's proposed visual routines, require that more than one place be indexed prior to the computation proceeding. Furthermore, many of the studies cited above which conclude that a single focus of attention is moved continuously in the visual field, used peripheral cues to induce the movement of attention. In these cases the cue itself would have to be located in order to serve as a directional indicator, and this would have to be done prior to attention being shifted to it. It is because of such considerations that indexing via FINSTs must be distinguished from "attending" in the usual sense where this term is understood to mean a single focus of processing.

Consider the following examples of relations requiring visual routines. Figure 3 shows some stimuli used to illustrate visual tasks (several of these are reported in Chapter 7 and in Ullman, 1984). In panel *a* the task is to decide whether point *x* (or *x'*) is inside the contour. In panel *b* the task is to say whether points *x* and *y* (or *x* and *y'*) are on the same contour (as in the studies reported by Jolicoeur in this volume). In panel *c* the task is to say whether there is a path from the centre of the circle to the circle itself. In panel *d* the task is to say how many points there are. In panel *e* the task is to say whether the three objects are collinear. There are, of course, an unlimited number of such tasks. Notice that in each case the task cannot be done without "marking" (to use Ullman's term) more than one visual object. In some cases all the objects in question are points. In others (such as *a* and *c*) they include contours.

It is not known whether entire contours can be FINSTed, though there are some reasons for thinking that at least simple ones can. For

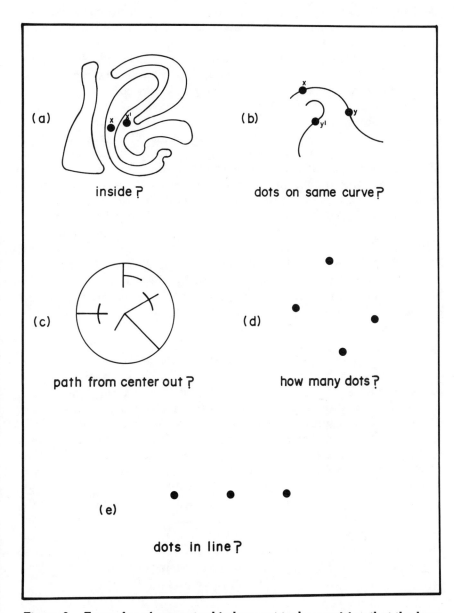

Figure 3.   Examples of perceptual judgement tasks requiring that the location of several features be indexed simultaneously, as hypothesized by the FINST model.

example, Rock and Gutman (1981) have shown that people can attend selectively to a contour of one color when it is intertwined with a similar contour of a different color, as shown by their inability to recognize the unattended contour as one they had seen before. Similarly, the phenomenon demonstrated by Treisman and Kahneman (1984), wherein a letter presented briefly in a particular moving box, primes recognition for that letter when it recurs in the same box, suggests that subjects can track the movements of contours such as boxes, and that they also use the identity of an object (such as a box in this case) to index other associated properties. On the other hand, the fact that the difficulty in evaluation the "inside" predicate depends to some extent on the size and shape of the bounding contour—at least when the contour becomes sufficiently complex—suggests that FINSTing contours may not be a simple primitive operation. Our proposal is that *some* larger aggregates can indeed by FINSTed. However, it may be that an entire contour such as that in Figure 3a requires several FINSTs to cover distinct segments of the curve, and that how accurately a FINST localizes features depends on such factors as their distinctiveness and on how many features compete for the pool of available FINSTS (recall that the FINST allocation process is resource-limited).

In any case, it is clear that some pre-attentive indexing must be going on. The assumption that there are limits on the number of such indexes that can be simultaneously maintained also seems plausible. Our tracking experiment suggests that at least four (and possibly as many as five or six) FINSTs are possible—and this number is a lower bound estimate obtained in a task designed to be particularly difficult. Subitizing (which requires that objects be marked rapidly as they are counted) also suggests about four or five FINSTs. I suspect that the amount of information that can be indexed in this way can be increased by "chunking" patterns, and then FINSTing the entire chunk, much as the amount of information held in short-term memory can be increased by chunking. Clearly there remain many unanswered empirical questions concerning exactly what kinds and how many features can be FINSTed, though the principle that a number of different features can be indexed in the way assumed by the FINST hypothesis seems reasonably secure.

Implications for Studies of Mental Imagery

One of the phenomena that inspired us to develop the FINST hypothesis in the first place was the widespread assumption that somewhere after stage 1 in visual processing, a representation is constructed that is like an "image." Indeed it is frequently referred to as an "image" or an "icon" or an "analogue representation." The nature of this representation is assumed to be similar whether constructed from stage 1 information in perception, or from long-term memory in the course of thinking using visual "images." Because we believe there are plenty of reasons to resist such a view (see Pylyshyn, 1981, or Chapter 9 of Pylyshyn, 1984), we decided to see how far we could get in accounting for some of the stage 2 phenomena without positing an internal, spatially extended image.

Among the phenomena that have led some people to assume the existence of a spatially extended object referred to as an "image" are such findings as the increased time taken to report properties of imagined objects when the image is small, the increased time taken to mentally scan longer distances in an image, as well as certain other parallels between imagery and perception (such as motor adaptation to imagined errors in pointing that parallel adaptation to observed errors in pointing induced by displacing prisms). I believe that we have managed to demonstrate clearly that at least the scanning phenomena, and arguably other similar phenomena as well, are due to the demands of the task, rather than to any intrinsic properties of "images" (see Pylyshyn, 1981, for the arguments). For example, in the image scanning case we showed that the increase in reaction time as a function of imagined distance is present only if subjects (a) interpret the task to be one of reproducing the effects (including temporal sequences and durations) they believe would have occurred in the corresponding real perceptual situation, and (2) believe that in the corresponding real situation there would be an increase in time taken as a function of the increasing distance. If either of these conditions is absent the scanning phenomenon does not occur with images, though it continues to occur in perception and also when the image is superimposed onto a perceived scene. The latter, I believe, can be accounted for by the FINST hypothesis, and I shall return to it later.

A particularly simple illustration of how the FINST hypothesis can deal with a phenomenon assumed to implicate imagery is a finding described by Shepard and Podgorny (1978). A subject inspects a grid on which a pattern, such as the capital letter "F," is shaded in. A colored spot appears briefly and the subject must press one of two buttons; one if the spot occurs in a grid square within the letter, the other if it occurs in one of the grid squares not inside the letter. Reaction time was found to vary systematically with the location of the spot on the display: it is generally shorter when the spot is inside the letter and is shortest when the square on which it occurs lies at the intersection of two or more letter-strokes (i.e., at a vertex). What was most interesting, however, is that exactly the same pattern of results is found when the subject is asked to *imagine* the letter on the grid, rather than presenting the actual letter pattern itself. This is taken as evidence that perception and imagery share common mechanisms, though in one case they are activated by a visual stimulus and in the other case by memory.

On one interpretation, this conclusion is innocent enough: there is bound to be much in common between perception and the rest of cognition. Perception is, after all, a cognitive process, involving (at some stage) interpretation, memory, and reasoning. But the result of the study I have described is usually interpreted in a much stronger way—as suggesting that there is a superposition of two "images" in the imaginal condition. This presumably occurs at stage 2 in our scheme. Since I have been arguing that there is no such entity as an extended geometric display at stage 2 (or anywhere past the retina), the stronger conclusion seems to me unwarranted. In fact, the mechanism for accounting for the result described is precisely the one we needed to explain the possibility of a stable stage 2 representation without appealing to any extended internal image.

If we assume that in *both* perceptual and imaginal conditions, the subject prepares for the task by placing FINSTs on the *actual display*—for example, on selected features or feature-aggregates consisting of groups of perceptually-integral grid squares such as letterstrokes—then we can explain the systematic pattern of reaction times in both cases in exactly the same way. Recall that our assumption has been that in order to compute a relation such as "part of" between two features, both have to be FINSTed. The grid squares that lie at the intersection of two line-features are part of two feature-clusters, and hence are indexed by two FINSTs. Thus when a point occurs at that location it has a good chance of being evaluated faster, since there are two paths to the correct answer. No special "image" is required, only the ability to index feature clusters that are on the retina—an ability that is required equally to explain both patterns of results.

A similar approach also explains a number of parallels between phenomena involving imagery and corresponding ones involving perception, providing that the imagery experiments are carried out by requiring subjects to "project" their images onto some scene they are currently viewing. In that case the metrical properties that such experiments report (e.g., the linear increase in scanning time with imagined distance) come from actual operations (e.g., attentional scanning) carried out among FINSTed features of a *real* (stage 1) image.

The difference between the current view and the "standard" imagery position may seem small to some, since in the above context both claim that there is something similar going on in perception and in imagery. The differences are, however, quite substantial, as I have argued at length in Pylyshyn (1981). The standard imagery view (e.g., as put forward by Kosslyn, Pinker, Smith, & Shwartz, 1979, and others) hypothesizes a stage 2 (or higher) representation that is embedded in a medium with certain intrinsic Euclidean properties. For example, in such a representation, if we represent 3 points, A, B, C, as being located on a horizontal line, then it *must* be the case that one of the points (say, point B) bears the relation *BETWEEN* to the other two points; two of the points (say, "B" and "C") must bear the relation *RIGHT-OF* to point A; placing point C on the line containing points A and B does not change the relation between points A and B; and so on.

We take such properties for granted because they are true of the physical world—that is, they are true of marks placed on a piece of paper, since the paper remains physically rigid in this application. But what makes these properties true of the *represented* world? There is nothing about the nature of representation *in general* which requires that from $COLLINEAR(x,y,z)$, $RIGHT\text{-}OF(x,y)$ and $RIGHT\text{-}OF(y,z)$ it follow automatically that $BETWEEN(y;x,z)$. To the extent that we wish this inference to be treated as valid in the represented domain we shall have to explicitly represent this fact. There are, of course, ways to build this inference in as part of the *implicit* structure of the representational scheme, but then we would no longer be able to represent a world where the inference did not hold. It is an *empirical* question whether this sort of representational constraint is part of our intrinsic functional architecture. I have argued (Pylyshyn, 1984) that the remarkable plasticity of human cognition precludes this sort of "built in" feature in a great many cases where that has been proposed.

Now this sort of problem does not arise in the case of perception—providing only that the perceptual process meets certain natural requirements—for example, that the geometry of the visual field is not distorted too radically by the visual system, that the spatial relations that are perceived remain reasonably constant when the stimulus itself does not change, and so on. For example, the question raised above concerning the constancy of the relation between A and B when C is added into the picture does not arise in perception so long as the object being observed has properties of rigidity and perception is reasonably veridical and time-invariant. That's simply a consequence of the fact that *in the world* adding C as a collinear point after A and B does not change the relation between A and B.

And that, in turn, is a consequence of *physical law*, not of perception.

Now in the hybrid case, where imaginal places or contours are projected onto a visual scene, we can have some of the Euclidean properties of real perception providing certain minimal requirements are met. Besides the requirement of time-invariance of perception that I already said was needed in the vision case, we also need some way to "pick out" or "individuate" certain places in a scene, in order that the cognitive system be able to refer to those very tokens. This would make it possible to assign property-labels to such places, and thereby to treat the stimulus (at least in certain respects) as though features of certain kinds actually were located at these places. Such a facility is precisely what FINSTs provide.

It is important to note, however, that this story does not require that we postulate an internal object with Euclidean properties: the Euclidean properties are inherited from properties of the real rigid stimulus being perceived at the time.[6]

If this account is correct, then it is easy to see why such "imaginal" phenomena as the increased time taken to examine more distant places in one's image (which people like Kosslyn, Ball, & Reiser, 1978, have taken to show that images "preserve metrical spatial information") should appear under "image projection" conditions. As it happens, we have found that these are the *only* conditions under which these sorts of phenomena appear reliably and independent of instructions. If subjects are merely asked to imagine some scene, phenomena such as the "scanning effect" (whereby greater imagined distances lead to increased scan times) appear *only* when subjects interpret their task as requiring them to imagine what would have happened had they actually viewed such an event taking place (these and other such experiments are discussed at some length in Pylyshyn, 1981).

SUMMARY AND CONCLUSION

To summarize, I have given a number of examples of how the simple idea of a primitive mechanism capable of individuating and index-

---

[6]Of course, it remains an interesting open empirical question just how perception-like the processing of information from such "bound" features can be. There have been claims that some illusions—such as the Muller-Lyer illusion—can be created by imagining arrowheads on lines (Bernbaum & Chung, 1981). Although the interpretation of such experiments is not unproblematic, they should not, in any case, be taken as supporting an "imagery" position: for one thing, there is much we don't know about the locus of such illusions in the visual case.

ing a small number of features (or feature-clusters, or "places") in a visual field can help elucidate a number of quite disparate empirical phenomena. These phenomena include ones that are involved in the use of "visual routines" to encode certain relational properties of stimuli, as well as a number of contentious findings involving experiments on mental images. I have argued that something very much like the FINST binding mechanism, is independently required for determining *where* visual operations (such as those in "visual routines") are to be applied. They also represent a resource-constraint assumption similar in spirit to the hypothesized limit on the number of chunks that may be held in short-term memory, or Newell's (1980) postulation of a cost associated with each variable that gets bound in the firing of a production. They also represent a mechanism whereby variables in predicates can be bound to particular places or elements in a stimulus so that these predicates may be evaluated in a concrete perceptual context. And finally, though I have not discussed this point here, there is a need for a mechanism such as FINSTs in order to deal with the generally difficult problem of assigning semantics to expressions containing indexicals.

In addition to exploring these entailments—and suggesting a number of others, such as those involving cross-modality binding of visual and motor spaces—I have presented some direct evidence bearing on one of the assumptions about properties of FINSTs. The assumption in question is that FINSTs can pre-attentively track a number of independently moving visually-identical objects under conditions where it is unlikely that the task is being done by serial time-sharing.

The wide range of phenomena addressed by this simple, independently motivated postulate, makes it a promising basis for investigating the interface at which attention and higher cognitive processes are brought to bear on the products of the earliest automatic and preattentive stages of vision.

APPENDIX: SUMMARY OF SOME ADDITIONAL
ASSUMPTIONS OF THE MODEL

This paper has been concerned primarily with the basic idea of FINSTs, since it has implication for the stage 4 processes that are my main focus. However, since I raised the question of the relevance of the FINST model to the construction of a stage 2 representation, I should perhaps take a few pages to sketch a few of the details of the model as they relate to this problem.

Constructing a Stage 2 Representation

The FINST idea, as presented so far, fails to address many aspects of the problem of constructing a geostable (stage 2) representation. For example, even though it does deal with some nontrivial problems that must be solved in order for a system to extract simple geometrical relations among features concurrently on the retina, it says nothing about what happens when features are no longer on the retina. In order to extend the usefulness of the FINST mechanism beyond the simple case, we have to address such additional questions as (a) How does the system maintain the identity of a feature cluster when the cluster disappears off the retina and later reappears, and (b) How, in general, does the system compute the spatial relation among features that are not on the retina concurrently? These are much more difficult problems because it is clear that their solution depends on proprioceptive as well as visual information, and because they involve memory. While we do not have the answer to the question of how the human visual system manages to achieve the skills referred to above, our approach has been to ask first for sufficient conditions for it to be possible.

Clearly we can represent proprioceptive information in such a way that it can be interpreted in terms of direction and distance traveled, and we can issue motor commands that result in our eyes or limbs moving to desired locations. Let us put aside, for the moment, the question of how this is done. Let us assume that the ability to issue a certain limited set of motor commands, which cause a limb or eye to move to selected sensed locations, is part of our primitive perceptual-motor capacity. Whatever the mechanisms by which these things are accomplished, they are sure to be quite different from those with which we are currently familiar—such as those being used in the design of industrial robot arms.

In developing a computational theory of aspects of perceptual-motor coordination we have adopted what we call a *minimal mechanism* strategy. In understanding how a cognitive or perceptual-motor function can be accomplished (how it is *possible*) it is inappropriate that we be constrained to work within well-known computational architectures—such as those of Von Neuman style computers. An alternative strategy, which we have adopted, is to take a small set of simple capacities that people appear to possess and see whether the assumption that these are primitive operations in the human organism allows us to develop a model that is *sufficient* for the task at hand, and which also accounts for certain otherwise puzzling phenomena. Of course, we expect that these primitive mechanisms will be simulated on a computer, but the *way* in which they are realized is not important to the theory—it is to be thought of as an implementation issue, rather than a theoretical hypothesis about the process. Though foreign to the usual practice is psychology, this minimalist top-down strategy has frequently been used to advantage in computational models. Good examples are Newell's (1973) production system architecture, and Marr and Nishihara's (1976) SPASAR mechanism for rotating 3D models into a cononical orientation in the process of recognition.

The simple operations that we have assumed to be primitive are those summarized in Table 2, together with the following additional ones that pertain specifically to the perceptual-motor coordination problem.

1. If X is a FINST bound to a certain retinal feature, then the operations **MOVE***(Fovea, X)* and **MOVE***(Pointer, X)* can be executed primitively.[7]

2. Only two moveable objects are assumed in the minimal version of the model: an eye and a limb (which we call the "pointer"). The positions of the pointer or fovea are also assumed to be available for use in motor commands; the pointer can primitively be moved into the fovea and the eye can primitively be moved to the location of the pointer. Hence **MOVE***(Fovea, Pointer)* and **MOVE** *(Pointer, Fovea)* are assumed to be primitive operations.

3. In addition to the position of the pointer and the fovea, a reference to one other object can occur as an argument in a motor command. This object is called the *Anchor*. We assume that it is possible to bind a FINST to the proprioceptively sensed *Anchor* and to store this binding so that when the FINSTed feature is no longer on the retina, the system can still issue a motor command to move the eye or the pointer to it. This simple cross-modality binding makes it possible, in effect, to coordinate visual and motor systems in a minimal way. The system can command either of the moveable objects to move to the location of the anchor by using the primitive operation **BIND-ANCHOR***(X)*, together with either **MOVE***(Fovea, Anchor)* or **MOVE***(Pointer, Anchor)*.

4. The objects *Anchor* and *Pointer* can serve in place of FINSTed

---

[7] Presumably the converse is also true, at least for ballistic movements—that is, they can be directed only at targets that are either FINSTed or are ANCHORed. Some support for this is found in single cell recording studies. Goldberg and Wurtz (1972), and Wurtz and Mohler (1976) have shown that at the level of the superior colliculus, the firing rate of cells whose receptive field coincides with the target of an eye movement increases, with the increase occurring well before the eye movement itself begins. This suggests that the activation of such cells may correspond to the assignment of FINSTs to those locations.

features when evaluating perceptual predicates (such as *Above*, *Inside*, and so on) even if they are not on the retina. In other words these two objects provide a limited means for evaluating spatial relations among pairs of places when both places are not visible concurrently.

These ideas are presented primarily to make the methodological point that neither a spatially extended iconic store nor an explicit encoding of the locations of features is logically demanded by the requirement of having a geostable representation. This follows from the fact that the minimal assumptions sketched above allow one to build up routines that can give access to the location of all pairs of features that have been FINSTed, whether or not they are currently on the retina, and also make it possible to provide some coordination of visual and proprioceptive locations. It must be emphasized that this is a minimal set of operations: much more powerful operations are almost certain to be available in the human perceptual-motor system. Nonetheless, these operations appear to be sufficient for such tasks as drawing 2D figures. If the hypothesis that they form a universal set holds up it would at the very least show that it is possible to achieve some of the stage 2 phenomena without the need to store an iconic (or other analogical) image, and without having to explicitly encode the locations of features according to a global coordinate system (such as a Cartesian or spherical coordinate system).

REFERENCES

Bernbaum, K., & Chung, C. S. (1981). Muller-Lyer illusion induced by imagination. *Journal of Mental Imagery, 5,* 125–128.

Davidson, M. L., Fox, M. J., & Dick, A. O. (1973). Effects of eye movements on backward masking and perceived location. *Perception and Psychophysics, 14,* 110–116.

Eriksen, C. W., & Schultz, D. W. (1977). Retinal locus and acuity in visual information processing. *Bulletin of the Psychonomic Society, 9,* 81–84.

Finke, R. A., & Pinker, S. (1982). Spontaneous imagery scanning in mental extrapolation. *Journal of Experimental Psychology: Learning, Memory, and Cognition, 8*(2), 142–147.

Fodor, J. A. (1983). *The modularity of mind: An essay on faculty psychology.* Cambridge, MA: MIT Press.

Goldberg, M., & Wurtz, R. (1972). Activity of superior colliculus in behaving monkeys. I. Visual Receptive Fields in Single Neurons. *Journal of Neurophysiology, 35,* 542–559.

Hochberg, J. (1968). In the mind's eye. In R. N. Haber (Ed.), *Contemporary theory and research in visual perception* (pp. 309–331). New York: Holt, Rinehart & Winston.

Jolicoeur, P., Ullman, S., and MacKay, M. F. (1988). Curve tracing: A possible basic operation in the perception of spatial relations. *Memory and Cognition.*

Kosslyn, S. M. (1978). Measuring the visual angle of the mind's eye. *Cognitive Psychology, 10,* 356–389.

Kosslyn, S. M., Ball, T. M., & Reiser, B. J. (1978). Visual images preserve metrical spatial information: Evidence from studies of image scanning. *Journal of Experimental Psychology: Human Perception and Performance, 4,* 46–60.

Kosslyn, S. M., Pinker, S., Smith, G., & Shwartz, S. P. (1979). On the demystification of mental imagery. *Behavioral and Brain Science, 2,* 535–548.

Marr, D., & Nishihara, H. K. (1976). Representation and recognition of spatial organization of three-dimensional shapes. *MIT A.I. Memo* 377:1–57.

Marr, D. (1982). *Vision.* San Francisco: Freeman.

Newell, A. (1980). HARPY, Production Systems, and Human Cognition. In R. A. Cole (Ed.), *Perception and production of fluent speech.* Hillsdale, NJ: Erlbaum.

Newell, A. (1973). Production systems: Models of control structures. In W. Chase, (Ed.), *Visual information processing.* New York: Academic Press.

Peacock, C. (1983). *Sense and content.* Oxford: Clarendon Press.

Pinker, S. (1980). Explanations in theories of language and of imagery. *Behavioral and Brain Sciences, 3*(1), 147–148.

Posner, M. I., Nissen, M. J., & Ogden, W. C. (1978). Attended and unattended processing modes: The role of set for spatial location. In H. L. Pick & I. J. Saltzman (Eds.), *Modes of perceiving and processing information.* Hillsdale, NJ: Erlbaum.

Pylyshyn, Z. W. (1984). *Computation and cognition: Toward a foundation for cognitive science.* Cambridge, MA: MIT Press.

Pylyshyn, Z. W. (1981). The imagery debate: Analogue media versus tacit knowledge. *Psychological Review, 88,* 16–45.

Pylyshyn, Z. W., Elcock, E. W., Marmor, M., & Sander, P. (1978). Explorations in visual-motor spaces. *Proceedings of the Second International Conference of the Canadian Society for Computational Studies of Intelligence,* University of Toronto.

Pylyshyn, Z. W. , & Storm, R. W. (*in press*). Tracking of multiple independent targets: evidence for a parallel tracking mechanism. *Spatial Vision.*

Rock, I. (1983). *The logic of perception.* Cambridge, MA: MIT Press.

Rock, I. (1981). Anorthoscopic perception. *Scientific American, 244,* 145–153.

Rock, I., & Ebenholtz, S. (1962). Stroboscopic movement based on change of phenomenal rather than retinal location. *American Journal of Psychology, 75,* 193–207.

Rock, I., & Gutman, D. (1981). The effect of inattention on form perception. *Journal of Experimental Psychology: Human Perception and Performance, 7,* 275–285.

Shepard, R. N., & Podgorny, P. (1978). Cognitive processes that resemble perceptual processes. In W. K. Estes (Ed.), *Handbook of learning and cognitive processes* (Vol. 5). Hillsdale, NJ: Erlbaum.

Shulman, G. L., Remington, R. W., & McLean, J. P. (1979). Moving attention through visual space. *Journal of Experimental Psychology: Human Perception and Performance, 15,* 522–526.

Treisman, A., & Kahneman, D. (1984). Unpublished manuscript.

Tsal, Y. (1983). Movements of attention across the visual field. *Journal of Experimental Psychology: Human Perception and Performance, 9,* 523–530.

Turvey, M. T. (1977). Contrasting orientations to the theory of visual information processing. *Psychology Review, 84,* 67–88.

Ullman, S. (1984). Visual routines. *Cognition, 18,* 97–159.

Ullman, S. (1979). *The interpretation of visual motion.* Cambridge, MA: MIT Press.

Wurtz, R. W., & Mohler, C. W. (1976). Organization of monkey superior colliculus: enhanced visual response of superficial layer cells. *Journal of Neurophysiology, 39,* 745–765.

III
Architectural Issues in Early Vision

10
Pathways in Early Vision*

Patrick Cavanagh

Département de Psychologie
Université de Montréal

Between the retinal ganglia and the higher centres of vision, visual information divides into several independent pathways each passing through physiological structures specialized for the analysis of different visual attributes. This paper first discusses the relative inputs of color and luminance information to these visual pathways and then examines the coding primitives available in each of the pathways. Finally, the capacities of the individual pathways are evaluated by examining which perceptual tasks—for example, shape from shading, relative depth from occlusion—can be performed on representations in each pathway.

There has been a great deal of interest in multiple visual representations recently, in both biological and artificial vision: the specialized regions in monkey prestriate cortex identified by Zeki and others (see Zeki, 1978, van Essen, 1985), the independent feature maps proposed in cognitive psychology (Treisman & Gelade, 1980; Treisman & Souther, 1985), and the intrinsic images used in computer vision (Barrow & Tenenbaum, 1978). The representations described in each of these fields varies but there is little disagreement concerning the potential importance of multiple representations to visual function. How can we examine these multiple representations in humans? Psychophysical techniques do not permit us to access individual visual areas directly but we may be able to restrict information to particular pathways through the visual areas by using stimuli defined by a single attribute. The stimulus information of a figure de-

* This research was supported by Grant A8606 from the Natural Sciences and Engineering Research Council of Canada.

fined by relative motion, for example, may pass preferentially through the prestriate region MT (van Essen, 1985) and so can probe its characteristics while a stimulus defined only by color may probe a different area (visual area V4, Zeki, 1978).

Physiological studies have produced a sketch of the pathways followed by various types of information in the visual system (see Maunsell & Newsome, 1987). At the level of the retina, only luminance and color appear to be coded and they follow separate neural channels through the lateral geniculate bodies and on to the cortex. The opponent-color information is transmitted by the parvocellular layers of the LGN and the luminance information by the magnocellular layers. More recent studies show that the actual situation is somewhat more complex with the red/green opponent-color information and the high spatial resolution luminance information being combined in the activity of R/G cells of the LGN and being decoded by cortical areas (Ingling & Martinez-Uriegas, 1985). In the first visual cortex V1, orientation, spatial frequency, motion, and binocular disparity emerge as new properties derived from the luminance and opponent-color information. The cells responsible for this derivation are often multidimensional, responding to preferred values of several attributes. There does appear to be some spatial clustering, however, of non-oriented, color-selective cells, surrounded by oriented luminance cells (Livingstone & Hubel, 1983). This initial, local specialization becomes progressively accentuated as color cells project to specific regions of V2 and then to V4, while oriented cells project to other areas. There is a similar segregation of directionally selective cells projecting from the lower layers of V1 to MT, an area specialized for motion, and from V1 to specific areas of V2 and then to MT. There is, in addition to this motion-specific pathway, a subcortical pathway through the superior colliculus and then to MT. Many aspects of these pathways remain to be determined and some may be organized according to response factors rather than stimulus factors. Several areas of the parietal cortex, for example, are known to be involved in eye movements (Andersen, Essick & Siegel, 1985). A simplified version of the various visual pathways is presented in Figure 1. (Of the attributes that we will examine here, only two, color and motion, have clearly defined physiological pathways. The others, luminance, binocular disparity, and texture may be involved in several pathways or none at all. Until further evidence is available, they can only be considered as reasonable candidates for pathways.) Following the multiple representations of the prestriate region in Figure 1, a combined representation of stimulus shape is proposed that then serves as input to high-level inference processes.

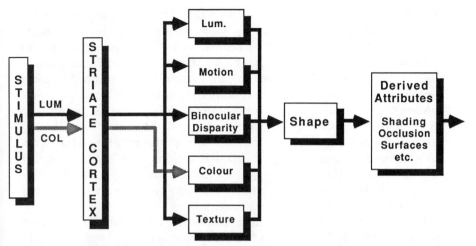

Figure 1. Pathways in the visual system. Luminance (non-opponent) and color (opponent) pathways bring information from the retinal ganglia to the striate cortex where multifunction cells begin the analysis of orientation, motion and binocular disparity. Following the striate cortex, information is routed to areas performing specialized analyses of various attributes: color (area V4, Zeki, 1978) and motion (area MT, van Essen, 1985), for example. Luminance, binocular disparity, and texture are other stimulus attributes that may receive specialized analyses in separate areas of prestriate cortex. Each of these specialized areas generates a two-dimensional representation of the attribute being analyzed, contributing to an overall representation of stimulus shape—external and internal contours—from which higher level attributes such as shading, occlusion and surfaces can be derived.

The first issue that we have addressed is the relative inputs of the color and luminance pathways coming from the lateral geniculate to the various pathways that start out in area V1 and then diverge. Since there is an area specialized for the analysis of color (V4, Zeki, 1978), it is often thought that color information only contributes to the pathway leading to this area and not at all to other pathways that are involved in the analysis of motion or binocular disparity. However, we will see that color does contribute to motion (Cavanagh, Boeglin, & Favreau, 1985; Cavanagh, Tyler, & Favreau, 1984, Cavanagh & Favreau, 1985; Cavanagh & Anstis, 1986) and binocular disparity (de Weert & Sadza, 1983; Grinberg & Williams, 1985), although only weakly.

The next question we have examined is the coding primitives available in the different pathways. There have been many papers that have demonstrated orientation and size preferences for cells in area V1 and related these preferences to the spatial organization of the cells' receptive fields. Receptive field shapes can therefore suggest likely dimensions of encoding for stimulus shape. We can examine coding dimensions in other pathways by determining receptive field shapes for cells that respond to stimuli defined by attributes such as relative motion or texture, for example. In addition to using receptive field shape as a physiological index of coding primitives, we can also use aftereffects to psychophysically probe the underlying dimensions of encoding. Tilt and size aftereffects have been used to support the notion of orientation and size coding and these tests can be extended to other stimulus representations such as random dot stereograms (Tyler, 1975), equiluminous colors (Favreau & Cavanagh, 1981) and kinematograms to evaluate whether size and orientation are encoded for these representations.

Finally, we will examine the abilities of each pathway to support various perceptual tasks. The first research on the perceptual capacities of what I would call individual pathways was by Julesz (1971) on images defined only by binocular disparity. (Prestriate areas V3 and V3A may be involved in stereopsis, according to Zeki, 1979, but this is not well established.) Julesz not only asked what were the necessary conditions for depth to be visible in a random dot stereogram when no monocular cues were present, but more important, he asked what sorts of things could be seen with images defined this way. That is, the analysis of binocular disparity does not lead simply to the extraction of depth but also to the representation of the shapes of regions defined by their different depths. Julesz examined whether such shapes could produce classical visual illusions, identifiable letters and various other perceptual phenomena. We have extended Julesz's approach to an additional set of pathways and made comparisons across these pathways (Cavanagh & Leclerc, 1985; Cavanagh, 1985a, 1985b, 1987). Other laboratories have followed related programs examining vision for shapes defined solely by color (Gregory, 1977; as had Gestalt psychologists Lehmann, 1904, and Liebmann, 1927), texture (Nothdurft, 1985; Prazdny, 1986), and motion (Regan & Beverley, 1984; Prazdny, 1985, 1986).

COLOR INPUT TO VISUAL PATHWAYS

Motion

If motion and color are analyzed by different areas in the prestriate cortex, the perception of the motion of stimuli defined only by color should pose problems for the visual system. We found, first, that motion could be seen, although somewhat degraded, for equiluminous colored stimuli. We then asked whether this motion perception was mediated by a separate pathway specialized for colored stimuli or by a single motion pathway responding to both color and luminance information.

Ramachandran and Gregory (1978) had initially reported that motion could not be seen for equiluminous stimuli. Their stimuli were produced by alternating two fields of random dots, both containing a square region of identically organized dots whose position differed slightly in the two fields (Figure 2). The background regions were identical in the two fields. Rapid alternation between the two fields gives the impression of an oscillating square, floating above the background if black and white dots are used. Ramachandran and Gregory (1978) reported that when the black and white dots were replaced with red and green and adjusted to be of equal luminance, the oscillating central square was no longer visible. The reduced acuity of the color pathway did not seem to be a factor as, in all cases, the individual red and green dots were clearly seen. They suggested that there was a functional independence of color and motion analyses in the visual system and that the motion pathway responded only to luminance information.

Figure 2.   A kinematogram is generated by alternating the left and right random dot fields shown here. They are superimposed spatially and exchanged at a rate of about 3 Hz. There is a background area of dots that remains fixed in the two fields and a central square area of dots that is displaced slightly from one field to the next. The kinematogram produces the impression of a sharply defined square that floats above the background.

It is not clear why the visual system would want to analyze motion only for luminance information and, certainly, physiological recordings do suggest some response of color-selective cells to motion (Michael, 1978, reports color-opponent, complex cells in V1)—although these findings are controversial (Lennie, Sclar & Krauskopf, 1985, report only nonoriented, color-opponent cells in V1). We therefore tried to replicate Ramachandran and Gregory's (1978) findings and found that motion was visible for equiluminous kinematograms although over a more restricted range of displacements and alternation rates (Cavanagh, Boeglin & Favreau, 1985). In fact, Ramachandran and Gregory had used a blank interval of 50 msec between the two random dot fields. The display during the interstimulus interval was black and the flicker that this produced when alternating with the random dot fields had masked the weaker motion signal from the colored stimuli. We were able to reproduce their findings for their condition and show that when the dark ISI was removed the perception of motion returned.

We also studied the perception of motion in simple stimuli: drifting equiluminous sinewave gratings (Cavanagh, Tyler, & Favreau, 1984). These appeared to be significantly slowed and occasionally stopped when compared to stimuli having luminance contrast. Relative speed judgements as a function of luminance contrast of a drifting, red/green or blue/yellow sinewave grating are shown in Figure 3. Observers were presented with a drifting color grating in the top half of the field and a comparison luminance grating in the bottom half. They adjusted the speed of the luminance grating until it appeared to match that of the color grating. When the relative luminance of the two colors approached equality, the apparent speed decreased to 40 to 60% of the actual speed. If we used stimuli that were moving quite slowly (less than 0.5° of visual angle per second), they could even appear to stop moving at equiluminance. The colored bars could be seen clearly and although it was apparent that the bars occasionally changed position this produced no subjective impression of motion. For any stimulus that could produce this stopped-motion phenomenon, the perception of motion could be reinstated by increasing the speed of the stimulus. We concluded that the input of color to motion was simply weaker, so that a particular stimulus could be above its pattern threshold and be seen clearly as colored bars, but below its motion threshold so that the bars did not appear to move (Figure 4).

Evidently, motion could be seen for colored stimuli, but it was not clear which pathway was involved in the response. Could there be cortical areas specialized for the motion of equiluminous, colored

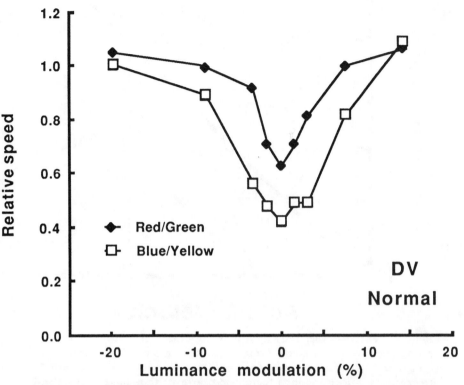

Figure 3.   The apparent speed of a drifting, colored grating decreases at equiluminance (luminance contrast 0%). The observer adjusted the speed of a 15% contrast, 0.5 cpd luminance grating to match the apparent speed of a 0.5 cpd, colored grating (red/green or blue/yellow) for several luminance contrasts between the two colors. This setting is then divided by the actual speed (1° of visual angle per second) to determine relative speed. The display subtended 8° of visual angle and the central 2° was masked in the case of the blue/yellow grating.

stimuli (Figure 5)? Using a motion aftereffect paradigm, we (Cavanagh & Favreau, 1985) showed that this was not the case. Following adaptation to drifting luminance gratings, motion after-effects could be seen on equiluminous colored tests, and vice versa, implying a single, common site for motion adaptation. In addition, following adaptation to a luminance grating, the motion aftereffect seen on a colored test could be nulled by moving the colored test in the opposite direction, an interaction that required a motion pathway accessed by both color and luminance information.

We next measured the strength of input of colored stimuli to this

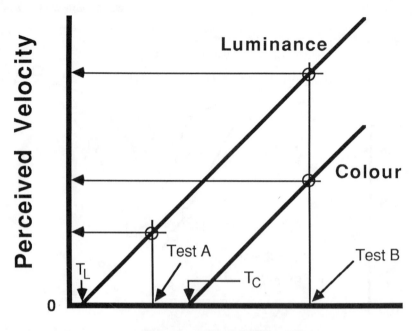

Figure 4. Perceived velocity as a function of actual velocity. If a colored grating is above its threshold for form perception but below that for motion perception (T_C), its bars can be seen but its motion is not visible (*Test A*). If the grating's velocity is then increased above its threshold, its motion becomes visible but it appears to move more slowly than a luminance grating drifting at the same speed (*Test B*). This independence between form (defined by color) and motion perception implies separate analyses of form and motion. T_L and T_C are the threshold velocities for luminance and color stimuli, respectively.

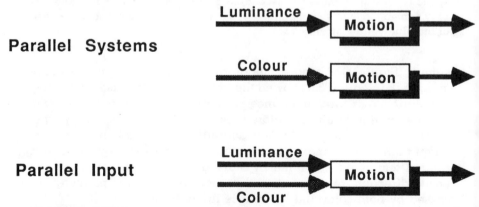

Figure 5. Parallel systems: one specialized for the motion of luminance-defined stimuli and the other for color-defined stimuli. Parallel input: a single motion system responding to both luminance and color.

common motion pathway using a motion nulling paradigm (Cavanagh & Anstis, 1986). We can measure the contrast of an unknown grating by varying the contrast of an otherwise identical grating moving in the opposite direction (Figure 6). The direction of perceived motion of the combined gratings is determined by the grating with the higher contrast. When the two gratings have equal contrast, a motion null—counterphase flicker—is obtained. We asked observers to adjust the contrast of a luminance grating until it just nulled the motion of a color grating moving in the opposite direction. This contrast setting was taken as the "equivalent luminance contrast" of the color grating. The equivalent contrast for color stimuli was, at best, about 10%, one tenth the maximum contrast possible for luminance stimuli. When these strengths were expressed as multiples of the respective color and luminance thresholds, however, the two types of stimuli had approximately equal influence on the perception of motion.

Although we had shown that color stimuli influenced the motion system, we had not proven that they did so through the opponent-color pathways. The luminance pathway responds to the relative luminance difference between two colors so it should be possible to adjust the luminance of one of the colors relative to the other until there is no response in the luminance pathway. However, if the cells in the luminance pathway have a variety of equiluminance points (Derrington, Krauskopf & Lennie, 1984) then no color pair can be simultaneously equiluminous for all cells. The motion seen for an equiluminous stimulus may therefore be due either to the contribution of the opponent-color input to a motion pathway or to this residual noise response in the luminance pathway. To test the possibility of opponent-color input, we presented a green/purple stimulus that fell along the tritan confusion line, differentially stimulating only the blue-sensitive cones (these cones have no input to the luminance pathway, Eisner & MacLeod, 1980; Cavanagh, Anstis & McLeod, 1987). We found an equivalent luminance contrast of about 4%. Moreover, following bleaching of the blue-sensitive cones, this stimulus had zero equivalent contrast. Since bleaching the cones eliminates their response to the stimulus, and eliminated the motion as well, we concluded that the input had passed through the blue-sensitive cones and, therefore, through the opponent-color pathway.

Overall, these studies indicate that there is a common motion pathway for opponent-color and luminance information and that the color input to the motion system is as effective, when considered in terms of threshold contrast multiples, as luminance input. On the other hand, the speed of equiluminous color stimuli is seriously misjudged. An understanding of this misperception of color velocity may

Luminance Contrast
10 %

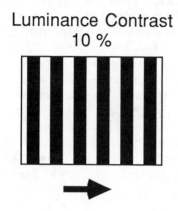

Luminance Contrast
X %

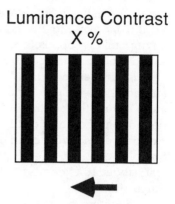

Chrominance Contrast
70 %

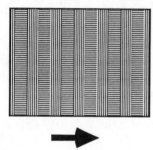

Luminance Contrast
X %

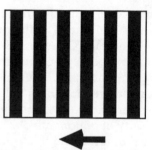

Figure 6. Opposing motion paradigm. On top: if two luminance gratings with the same spatial and temporal frequencies are superimposed and drifted in opposite directions, motion is seen in the direction of the grating having the higher contrast. If both gratings have the same contrast, no motion is seen, only counterphase flicker. The contrast of the rightward moving grating can therefore be measured by adjusting the contrast of the leftward moving grating until no motion is seen: the contrast setting to obtain the motion null would be 10%. Similarly, on the bottom, a chromatic grating drifting to the right is superimposed on a luminance grating drifting to the left and the contrast of the luminance grating is adjusted until a motion null is obtained. The resulting contrast is the "equivalent luminance contrast" of the chromatic grating.

provide an important test of the several models of velocity judgement (Watson & Ahumada, 1985; Adelson & Bergen, 1985; van Santen & Sperling, 1985).

Stereopsis

Initial studies by Lu and Fender (1972), Gregory (1977), and others demonstrated that depth in random stereograms was lost at equiluminance even though depth was not lost for figural stereograms. However, de Weert and Sadza (1983) showed that observers actually could judge depth in equiluminous random dot stereograms even though the subjective impression was very weak. De Weert and I have made some preliminary measurements showing, in addition, that the difference between random dot stereograms and figural stereograms may be due only to a difference in spatial frequency content and not to any qualitative difference. Our measurements showed that the depth perceived in a stereogram decreased as the colors approached equiluminance in much the same way that the apparent speed of drifting colored gratings decreased at equiluminance. Grinberg and Williams (1985) showed that depth could be seen in random dot stereograms that only stimulated the blue-sensitive cones. Since these stimuli are restricted to the opponent-color pathway, they concluded that color information did contribute to stereopsis.

Overall, evidence points to a color contribution to at least color, motion, and binocular disparity pathways. Whether or not color influences all pathways, it is certain that its influence on motion, stereo, and texture is less than that of luminance information. This may be simply a function of the higher thresholds for color. As was seen for motion, the contribution of color was equivalent to that of luminance in terms of threshold multiples but since the maximum obtainable color saturation was only about 40 times the threshold value while the maximum luminance contrast was approximately 400 times its threshold, the effect of color is typically much less than that of luminance.

CODING PRIMITIVES

How is image information coded in each representation? The visual areas that we have discussed, V1, V2, V3, and V4 and MT, are retinotopically organized, preserving adjacency. Image information could be coded simply in terms of local value, such as a point by point color value (perhaps both hue and saturation). This type of coding requires sampling the retinal array for only small, local areas and suggests that physiological studies would find only circular receptive fields of a fixed size. Studies have shown circular receptive fields for retinal and LGN cells but most cortical cells show quite a different structure. Hubel and Wiesel (1968) described oriented fields (Figure 7) that prefer a particular orientation and width of a line as the optimal stimulus. Moreover, many cortical cells respond to stimuli within the same retinal region, with different cells responding to different orientations and sizes. Thus, at each retinal location, the stimulus information is represented by the pattern of activity across a set of cells that code, among other things, orientation and size.

Others have tried to identify the receptive field structure, and thus coding primitives, for stimuli defined by color and by binocular disparity. The evidence concerning color is contradictory. Michael (1978) has reported that many cortical cells with oriented receptive fields that respond best to color contrast in the absence of luminance contrast. Lennie, Sclar, and Krauskopf (1985), however, have not been able to find any oriented cells responding principally to color and not luminance. Poggio, Motter, Squatrito & Trotter (1985) have studied neurons that respond to bars defined by random-dot ster-

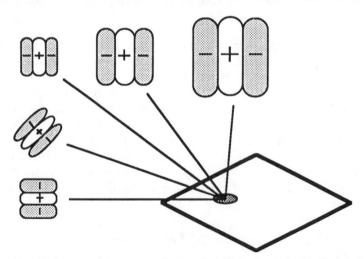

Figure 7. Coding primitives can be established physiologically by demonstrating that there are many cells responding along different stimulus dimensions for each retinal area. Recordings in the striate cortex show that for each retinal area there are cells tuned to several different sizes and orientations.

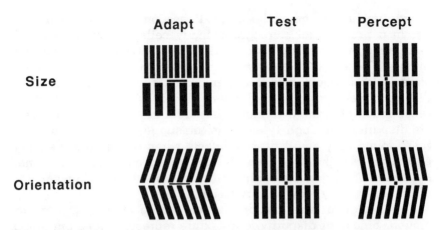

Figure 8. Coding primitives can be established by psychophysical techniques demonstrating size- and orientation-specific aftereffects.

eograms and claim much reduced or absent orientation sensitivity for these cells.

Psychophysical techniques have also been used to identify the underlying coding primitives. Blakemore and Sutton (1969) and Campbell and Maffei (1971) as well as many others have demonstrated size and orientation specific aftereffects that demonstrate encoding of size and orientation for stimuli defined by luminance. In general, if an observer is exposed to a stimulus having a specific size or orientation for several minutes, subsequently viewed stimuli of slightly different values will change their perceived size and orientation (Figure 8).

Olga Favreau and I (Favreau & Cavanagh, 1981) were able to show that luminance and color stimuli could induce simultaneous size aftereffects in opposite directions. This demonstrates that there must be parallel encoding of size information for both color and luminance. Eisner (1978) also showed that a tilt aftereffect could be induced for equiluminous stimuli, implying orientation coding for color. Since the evidence concerning oriented color-selective cells in V1 is contradictory, the site of the adaptation producing these size and orientation-specific aftereffects may be further along the color pathway. Zeki (1978) has shown that many cells in area V4 that are selective for color also show orientation preferences.

Tyler (1975) has tested the possibility that binocular disparity may be coded according to size and orientation. He claimed to find tilt and spatial frequency aftereffects for gratings defined by random-dot stereograms. These effects are not dependent on the size and

orientation tuning of the binocular cells responding to retinal disparity. In a random dot stereogram, many cells will respond to the
texture elements having various disparities between the two retinal
images. These texture elements may have random orientations and
sizes. Tyler's effects, however, were specific to the bars defined by
disparity over a large spatial extent. Any cells responding to these
areas must integrate over many first-order cells responding to retinal disparity. Although Tyler's data are suggestive, Wolfe and Held
(1982) found only small tilt aftereffects when attempting the same
experiment. No published reports have appeared for stimuli defined
by texture or motion but our initial tests indicate that, if these aftereffects exist, they may be very weak.

To summarize, size and orientation coding may be present for
motion, binocular disparity, and texture representations but this is
not as firmly established as it is for color and luminance.

IMAGING CAPABILITIES

In order to study the capacities of each pathway for shape analysis,
our computer graphics system constructs images defined by a single
attribute: color, relative motion, binocular disparity, texture, or luminance. Starting from a video image of a black and white stimulus,
the black areas are replaced with, for example, a random texture
moving in one direction, and the white areas with a similar texture
moving in the opposite direction. Figure 9 demonstrates this attribute replacement for texture and binocular disparity.

The areas of the prestriate cortex may restrict their input information in order to perform a highly specialized analysis of one particular attribute, such as, for example, color constancy in the case of
area V4 (Zeki, 1978). However, these areas, in addition to performing
specialized analyses, are also capable of representing shape: a two-
dimensional map of regions differentiated by the attribute in ques-

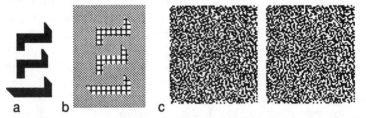

Figure 9.   Attribute replacement generates figures defined by texture (b) or
binocular disparity (c) from an original figure (a) defined by luminance.

tion. Gregory (1977, 1979) has suggested that each attribute may provide a rough map of the visual stimulus and that these maps may be aligned to the luminance representation which he considers the master map.

We were interested in examining what perceptual abilities were supported by the shape information in a single representation in the absence of any luminance "master map." In particular we found that information signalled by stimuli having explicit contours, for example, T-junctions indicating occlusion (Figure 10), was effective no

Figure 10.   Stimuli with explicit contours. These stimuli are defined by a difference in texture and should be invisible if you squint your eyes. The stimuli: a Necker cube seen as a wire frame—the background can be seen through the areas between the cube contours; an occlusion figure with one square, opaque sheet covering one corner of a similar sheet—the areas between the contours hide the background; a smiling face; the letter E. In general, the same interpretations are reported for these stimuli whether defined by luminance, texture, color, relative motion or binocular disparity. However, large individual differences are found in the latter two cases when there are conflicts between depth implied by the picture and depth indicated by relative motion or binocular disparity.

matter which visual pathway was used (Cavanagh, 1985a), Simple, two-dimensional letter shapes could be easily identified. Three-dimensional objects defined by complete contours in line drawings involving occlusion and perspective were interpreted in the same fashion whether represented by luminance, color, or texture. When relative motion or random dot stereograms were used to present these same drawings, the depth suggested by stimulus shape (due to occlusion and perspective, for example) sometimes conflicted with the depth indicated by the relative motion or binocular disparity used to present the figure. Many observers could see the depth implied by the drawing despite the conflict, although others could not. In cases where there was no conflict between the depth inferences in the picture and the depth used to present the picture (based on either relative motion or binocular disparity), the pictures were interpreted in the same manner as for luminance, color, or texture presentations. Shape information involving explicit object contours therefore appears to be represented equally well in any of the pathways. The depth and surface inferences based on these shape representations probably occur after this level of separate pathways and accept shape descriptions from any pathway. There was no indication that luminance information had any priviledged role to play in these images.

The results for stimuli involving implicit contours were strikingly different. We studied two stimuli of this type: shadows (Cavanagh & Leclerc, 1985) and subjective contours (Cavanagh, 1985b). In both cases, a luminance difference was necessary between the two image areas (e.g., the open grid and the fine dot areas of the E in Fig. 9b). If the parts of the stimuli were presented without a luminance difference, they were interpreted as separate, unconnected islands of color or texture (see Figure 11). If a luminance difference was then introduced, the overall global organization of the stimulus would become visible. Moreover, it was not sufficient for the mean luminance of the two image areas to be different; this luminance difference also had to be in the same direction all along the edge between the regions. Studies of filtered images showed that any frequency band signaling inappropriate or inconsistent edge polarity could veto the edge as a potential shadow border or suppress a subjective contour. Thus the luminance pathway is essential for shadows and subjective contours but it appears that it is the edges that are signaled at this low level and not the entire shadow region or subjective surface. We verified that the reduced resolution and contrast inherent in texture or color representations were not the cause of the failure to see shadows and subjective contours.

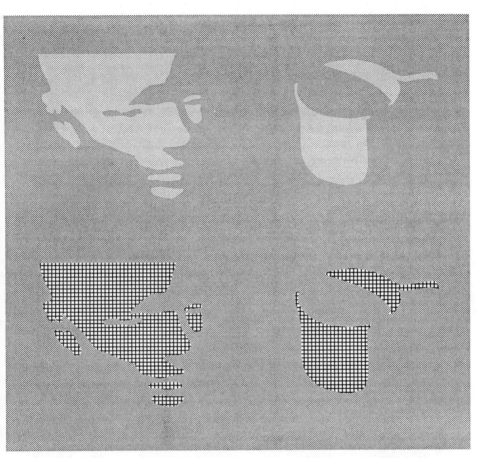

Figure 11. Stimuli with implicit contours: shadows. Many of the contours of these figures are shadow contours, not object contours and many of the object contours, both external and internal self-occlusions, are implicit. The interpretation of these figures changes when an appropriate luminance difference is present between the shadowed and non-shadowed areas (top row) compared to when no luminance difference is present (bottom row). The interpretation of surface relief due to shadows occurs for stimuli defined by luminance and but not for stimuli defined by texture (shown here), color, relative motion or binocular disparity.

It might seem self-evident that shadows would require luminance information to be properly interpreted: a real shadow is always darker than the adjacent nonshaded region. Shadow analysis may therefore be part of the specialized luminance analysis just as seeing colors is part of the specialized colour analysis. This is not a convincing argument, however, since the inference of depth from shadows must be based on their shape (necessarily in our stimuli which did not have any luminance gradients), not on their darkness. Rainbows can be identified in black and white images because of their shape but depth from shadows is not perceived in images defined only by color even though all the essential shape information is there. Moreover, it seems unlikely that an early level of the visual system such as the luminance pathway would be independently capable of the depth and surface inferences involved in interpreting shadows. It is reasonable to assume that higher level analyses are participating in these inferences but perhaps these analyses access only luminance information and, in particular, the location of appropriate luminance borders having consistent polarity. By ignoring shape information in other pathways, the visual system would give up opportunities to reject areas as shadows because of impossible colors or inappropriate depths, motions or textures. This is what our data showed as observers saw depth in shadow images having appropriate luminance patterns even when they violated the color, depth, motion, and texture contraints of natural shadows (Cavanagh & Leclerc, 1985).

Luminance may be a natural aspect of shadows, but this is not the case with figures producing subjective contours. Theories of subjective contours are generally based on high-level inferences of occluding surfaces. Gregory (1972), Rock and Anson (1979) and Kanizsa (1979) all suggest that the occluding surface is hypothesized to simplify the interpretation of the image. Thus, in Figure 12 (top left), it is easier to see eight circular disks covered by two square, opaque sheets (one disk completely covered) than seven, irregular, three-quarter pie shapes. These cognitive explanations are based entirely on stimulus shape and therefore should be unaffected by the manner in which the shapes are presented. On the contrary, subjective contours were only visible when there was a luminance difference between the regions defining the shapes (Cavanagh, 1985b; Pradzny, 1985; Brussell, Stober, & Bodinger, 1977). In this case, the lack of equivalence between the different pathways when presenting the same stimulus shapes contrasts with the equivalence of the pathways for explicit shapes presented as line drawings (Figure 10). It is evident that the perception of subjective contours cannot be a shape-based phenomenon. This conclusion is also supported by the physiological studies of von der Heydt, Peterhans and Baumgartner (1984) who report cells in area V2 that respond to subjective contours. It may be that low level analyses (based on these area V2 cells that, according to our findings, should respond only to luminance) signal contours in areas of the figures where there are no physical contours.

Figure 12.   Stimuli with implicit contours: subjective contours. When there is a luminance difference between the inducing elements and the background, subjective contours can be seen in these figures (top row). When these same figures are presented without a luminance difference (bottom row), subjective contours are weak or absent.

Higher level analyses may then use these contours to assert appropriate subjective occluding surfaces.

CONCLUSIONS

The results of the work that I have described here are summarized in Figure 13. Color information appears to contribute to motion and binocular disparity analyses. This can only improve the adap-

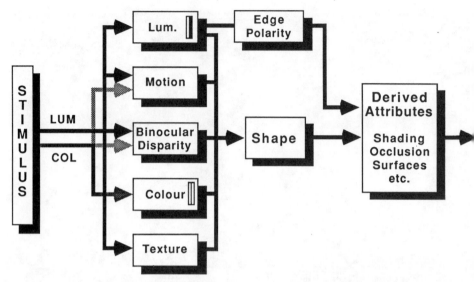

Figure 13.   Pathways in early vision. Color information contributes to motion and binocular disparity analyses in addition to color analysis. Both color and luminance analyses make use of size and orientation coding primitives. Shape defined by explicit contours is signaled by all pathways but shape defined by implicit contours is only signaled by the luminance pathway, most likely in terms of assertions of constant polarity edges in the stimulus.

tiveness of the visual system by allowing some motion and depth responses to stimuli defined only by color or only by binocular disparity. The responses are, however, rather weak so that reducing luminance differences between a stimulus and its background remains a very effective camouflage technique. Coding primitives involving size and orientation are used in the color and luminance pathways and perhaps in the other pathways as well, although this remains to be demonstrated convincingly.

Shape information was equally well signaled by all the pathways but shadows and subjective contours appeared to depend on a special purpose process in the luminance pathway. It is possible that this process involves signaling appropriate luminance borders that maintain a consistent contrast polarity all along their length.

The ability to probe individual pathways in the visual system opens many new possibilities for understanding vision. Each pathway may be thought of as a pared down visual system, less complex than the whole and easier to understand. On one level, much of the visual research done using luminance-defined stimuli needs to be repeated on the remaining pathways to determine their capabilities and this work is underway in several laboratories. On another level, new experiments testing how information is integrated across the individual representations will identify important principles in image understanding.

In the research that I have reported, I have assumed that the stimuli used were able to isolate visual pathways. A basic goal of this research is to examine to what extent this isolation is actually achieved. Possible techniques include identifying cortical magnification factors for these stimuli and comparing them to physiological data (Gattass, Sousa & Covey, 1985) for the visual areas involved.

REFERENCES

Adelson, E. H., & Bergen, J. R. (1985). Spatiotemporal energy models for the perception of motion. *Journal of the Optical Society of America A, 2,* 284–299.

Andersen, R. A., Essick, G. K., & Siegel, R. M. (1985). Encoding of spatial location by posterior parietal neurons. *Science, 230,* 456–458.

Barrow, H. G., & Tenenbaum, J. M. (1978). Recovering intrinsic scene characteristics from images. In A. Hanson and E. Riseman (Eds.) *Computer vision systems,* 3–26. New York: Academic Press.

Blakemore, C., & Sutton, P. (1969). Size adaptation: A new aftereffect. *Science, 166,* 245–247.

Brussell, E. M., Stober, S. R., & Bodinger, D. M. (1977). Sensory information and subjective contour. *American Journal of Psychology, 90,* 145–156.

Campbell, F. W., & Maffei, L. (1971). The tilt after-effect: A fresh look. *Vision Research, 11,* 833–840.

Cavanagh, P. (1985a). Depth and surface inferences in line drawings. *Journal of the Optical Society of America A, 2,* P51.

Cavanagh, P. (1985b). Subjective contours signalled by luminance, vetoed by motion or depth. *Bulletin of the Psychonomics Society, 23,* 273.

Cavanagh, P. (1987). Reconstructing the third dimension: Interactions between color, texture, motion, binocular disparity and shape. *Computer Vision, Graphics and Image Processing, 37,* 171–195.

Cavanagh, P., & Anstis, S. M. (1986). Do opponent-color channels contribute to motion? *Investigative Ophthalmology & Visual Sciences Supplement, 27,* 291.

Cavanagh, P., MacLeod, D. I. A., & Anstis, S. M. (1987). Equiluminance: spatial and temporal factors and the contribution of blue-sensitive cones. *Journal of the Optical Society of America A, 4,* 1428–1438.

Cavanagh, P., Boeglin, J., & Favreau, O. E. (1985). Perception of motion in equiluminous kinematograms. *Perception, 14,* 151–162.

Cavanagh, P., & Favreau, O. E. (1985). Color and luminance share a common motion pathway. *Vision Research, 25*, 1595–1601.

Cavanagh, P., & Leclerc, Y. (1985). Shadow constraints. *Investigative Ophthalmology and Visual Sciences Supplement, 26*, 282.

Cavanagh, P., Tyler, C. W., & Favreau, O. E. (1984). Perceived velocity of moving chromatic gratings. *Journal of the Optical Society of America, A 1*, 893–899.

de Weert, C. M. M., & Sadza, K. J. (1983). New data concerning the contribution of colour differences to stereopsis. In J. D. Mollon & L. T. Sharpe (Eds.), *Colour Vision: Physiology and Psychophysics.* (553–562). London: Academic Press.

Derrington, A. M., Krauskopf, J., & Lennie, P. (1984). Chromatic mechanisms in lateral geniculate nucleus of macaque. *Journal of Physiology, 357*, 241–265.

Eisner, A., & MacLeod, D. I. A. (1980). Blue sensitive cones do not contribute to luminance. *Journal of the Optical Society of America, 70*, 121–123.

Eisner, A. (1978). Hue difference contours can be used in processing orientation information. *Perception & Psychophysics, 24*, 451–456.

Favreau, O. E., & Cavanagh, P. (1981). Color and luminance: Independent frequency shifts. *Science, 212*, 831–832.

Gattass, R., Sousa, A. P. B., & Covey, E. (1985). Cortical visual areas of the macaque: Possible substrates for pattern recognition mechanisms. In C. Chagas, R. Gattass, & C. Gross (Eds.), *Pattern recognition mechanisms* (1–21). Berlin: Springer.

Gregory, R. L. (1972). Cognitive contours. *Nature, 238*, 1972, 51–52.

Gregory, R. L. (1977). Vision with isoluminant colour contrast: 1. A projection technique and observations. *Perception, 6*, 113–119.

Gregory, R. L. (1979). Stereo vision and isoluminance. *Proceedings of the Royal Society of London, B 204*, 467–476.

Grinberg, D. L., & Williams, D. R. (1985). Stereopsis with chromatic signals from the blue-sensitive mechanism. *Vision Research, 25*, 531–537.

Hubel, D. H., & Wiesel, T. N. (1968). Receptive fields and functional architecture of monkey striate cortex. *Journal of Physiology, 195*, 215–243.

Ingling, C. R., & Martinez-Uriegas, E. (1985). The spatio-temporal properties of the r-g X-cell channel. *Vision Research, 25*, 33–38.

Julesz, B. (1971). *Foundations of Cyclopean perception.* Chicago: University of Chicago Press.

Kanizsa, G. (1979). *Organization in vision.* New York: Praeger.

Lehmann, A. (1904). Die irradiation als Ursache geometrisch-optischer Täuschungen. *Archiv für die gesamte Physiologie (Pflügers) CIII.*

Lennie, P., Sclar, G., & Krauskopf, J. (1985). Chromatic sensitivities of neurons in striate cortex of macaque. *Investigative Ophthalmology and Visual Sciences Supplement, 26*, 8.

Liebmann, S. (1927). Über das Verhalten farbiger Formen bei Helligkeitsgleichheit von Figur und Grund. *Psychologische Forschung, 9*, 300–353.

Livingstone, M. S., & Hubel, D. H. (1983). Specificity of cortico-cortical connections in monkey visual system. *Nature, 304*, 531–534.

Lu, C., & Fender, D. H. (1972). The interaction of colour and luminance in stereoscopic vision. *Investigative Ophthalmology, 11,* 482–489.

Maunsell, J. H. R., & Newsome, W. T. (1987). Visual processing in monkey extrastriate cortex. *Annual Review of Neuroscience, 10,* 363–401.

Michael, C. R. (1978). Colour sensitive complex cells in monkey striate cortex. *Journal of Neurophysiology, 41,* 1250–1266.

Nothdurft, H. C. (1985). Orientation sensitivity and texture segmentation in patterns with different line orientation. *Vision Research, 25,* 551–560.

Poggio, G. F., Motter, B. C., Squatrito, S., & Trotter, Y. (1985). Responses of neurons in visual cortex (V1 and V2) of the alert macaque to dynamic random-dot stereograms. *Vision Research, 25,* 397–406.

Prazdny, K. (1985). On the nature of inducing forms generating perceptions of illusory contours. *Perception & Psychophysics, 37,* 237–242.

Prazdny, K. (1986). Psychophysical and computational studies of random-dot Moire patterns. *Spatial Vision, 1,* 231–242.

Ramachandran, V. S., & Gregory, R. (1978). Does colour provide an input to human motion perception? *Nature, 275,* 55–56.

Regan, D., & Beverley, K. I. (1984). Figure-ground segregation by motion contrast and by luminance contrast. *Journal of the Optical Society of America, A 1,* 433–442.

Rock, I., & Anson, R. (1979). Illusory contours as the solution to a problem. *Perception, 8,* 655–681.

Treisman, A., & Gelade, G. (1980). A feature-integration theory of attention. *Cognitive Psychology, 12,* 97–136.

Treisman, A., & Souther, J. (1985). Search asymmetry: A diagnostic for preattentive processing of separable features. *Journal of Experimental Psychology: General, 114,* 285–310.

Tyler, C. W. (1975). Stereoscopic tilt and size aftereffects. *Perception, 4,* 187–192.

von der Heydt, R., Peterhans, E., & Baumgartner, G. (1984). Illusory contours and cortical neuron responses. *Science, 224,* 1260–1262.

van Essen, D. C. (1985). Functional organization of primate visual cortex. In A. Peters & E. G. Jones (Eds.), *Cerebral cortex* (259–329). New York: Plenum Publishing.

van Santen, J. P. H., & Sperling, G. (1985). Elaborated Reichardt detectors. *Journal of the Optical Society of America, A 2,* 300–321.

Watson, A. B., & Ahumada Jr., A. J. (1985). Model of human visual-motion sensing. *Journal of the Optical Society of America, A 2,* 322–342.

Wolfe, J. M., & Held, R. (1982). Binocular adaptation that cannot be measured monocularly. *Perception, 11,* 287–295.

Zeki, S. M. (1978). Functional specialization in the visual cortex of the rhesus monkey. *Nature, 274,* 423–428.

Zeki, S. M. (1979). Functional specialization and binocular interaction in the visual areas of rhesus monkey prestriate cortex. *Proceedings of the Royal Society London, B 204,* 379–397.

11
Modularity in Visuomotor Control: From Input to Output

Melvyn A. Goodale

Department of Psychology
University of Western Ontario

INTRODUCTION

In the last 10 years, a number of theoretical approaches to perception have begun to challenge the once-influential "New Look" school with its strong emphasis on top-down control of stimulus encoding. Indeed, a central feature of these recent assaults on the New Look has been an explicit denial of the idea that an individual's beliefs and expectations can influence the way in which most incoming sensory information is processed. There has instead been a return to an earlier, almost nativistic, account of perception. In his influential book, *The Modularity of Mind*, Jerry Fodor (1983) has argued persuasively that input systems to central cognitive mechanisms are not only fast, domain-specific, and obligatory, but they are also quite impervious to other information and beliefs that the subject may have about the world. Moreover, these perceptual modules appear to function as independent and parallel "smart systems" whose internal operations are "cognitively impenetrable" (Pylyshyn, 1984) and refractory to analysis by central cognitive processes. Thus, according to this point of view, mental architecture is characterized by a vertical and modular organization of input systems. In addition, the arrangement of the modular systems is thought to correspond to a neural architecture characterized by hardwired parallel inputs to localized and distinct targets in the central nervous system.

This kind of modular approach to perception characterizes much of the field of computational vision. David Marr (1982), a theoretical pioneer in this recent tradition, postulated that the so-called "primal sketch" provided by early vision consists of a number of independent visual primitives, each of which is committed to the analysis of a particular stimulus feature. Support for a neural architecture consistent with this conception can be found in the multitude of neurophysiological studies carried out on the visual system, beginning with the pioneering work of Hubel and Wiesel (1961, 1962), in which single units differentially sensitive to particular stimulus features have been repeatedly demonstrated. Only recently, however, has empirical work begun to provide some insight into the organization of early vision at the psychological level. Using simple but elegant studies of texture segregation and grouping, Anne Treisman (1985) has shown that differences along single stimulus dimensions such as color and shape are very easy for subjects to see whereas differences that depend on a conjunction of properties such as color and shape are much harder to see. Indeed, if a visual target is defined by a simple feature such as color, it will often "pop out" of the display almost immediately, regardless of how many distractors are present in the stimulus array. Treisman has interpreted this kind of result as suggesting that the coding of simple features such as color and shape takes place in parallel by means of independent and primitive input modules whereas any combination of the different dimensional values provided by these modules takes place much later in visual processing. While this approach to visual perception has proved to be empirically powerful and theoretically rich, it has concentrated almost entirely on the input side of visual processing and has virtually ignored one of the most important functions of vision and the visual system, that of controlling motor output.

MODULARITY IN VISUOMOTOR CONTROL

For most workers in cognitive science, vision has become identified with visual perception, and the function of the visual system has been described in terms of providing an integrated representation of the external world. This is an old tradition in psychology (and in physiology, for that matter) and the recent modular approach to vision is, in this respect, no different from the New Look school. What has been forgotten in both cases is that vision evolved in vertebrates and other organisms to control the movements that these organisms make in living their often precarious lives (Goodale, 1983a). Vision helps them make the movements involved in foraging for food and in avoiding predators. It helps them move toward others of their kind, around barriers, across chasms, and away from danger. In "designing" these control systems, natural selection appears to have favored the development of a visual system that consists of a network of relatively independent sensorimotor channels, each of which supports a particular kind of visually guided behavior. Modularity is indeed the rule, but it is a modularity that extends far beyond the input systems themselves. In a very real sense that I will try to describe below, the modules are often visuomotor rather than simply visual in nature. They may be fast, domain-specific, obligatory, and cognitively inpenetrable but they extend from input to output, from the retina to the motor units producing the visually guided behavior in question. The vertical architecture of such a system, I contend, can be fully understood only by studying the organization of its motor outputs as well as its sensory inputs.

Although a sensorimotor approach to the study of the visual system is an unusual one in psychology, it is not an unfamiliar one in ethological studies of animal behavior. Investigators studying vision in frogs and other "lower" vertebrates have shied away from the traditional psychophysical paradigms employed in psychological laboratories and have instead looked directly at the visually guided movements made by these creatures as they catch their prey, avoid obstacles, and escape from predators (see Ingle, Goodale, and Mansfield, 1982; and Ewert, Capranica, and Ingle, 1983 for reviews of work in this area). Many of these sequences of behavior would be called fixed action patterns by ethologists such as Lorenz (1958) and Tinbergen (1951). Not only are such patterns of behavior species-specific and highly stereotyped, but many of them also have the added advantage that they can be readily elicited in the laboratory by relatively simple visual stimuli.

In a now classic series of studies of vision in the frog, *Rana pipiens*, David Ingle (1973, 1982) was able to show that visually elicited feeding, visually elicited escape, and visually guided avoidance of barriers are mediated by entirely separate pathways from the retina to the motor nuclei. The most recent account of their organization (Ingle and Hartline, in press) suggests that turning briskly toward small prey-like stimuli is a modular system which consists on the input side of projections from Class II retinal ganglion cells to superficial laminae in the optic tectum. Cells in these laminae give rise in turn to projections, largely crossed, to nuclei in the pons, medulla, and spinal cord that control turning. On the other hand, the "visual escape module," which mediates rapid jumping away from large looming stimuli, consists at its forward end of projections from a quite different set of retinal ganglion cells including Class IV and perhaps some Class III cells. These projections end in tectal laminae much deeper than those receiving input from the Class II cells. Moreover, the tectofugal projections that constitute the downstream side of the visual escape module appear to be largely uncrossed and project to nuclei in the rostral medulla. The visuomotor module that mediates the avoidance of barriers does not involve the optic tectum at all but is instead comprised of retinal projections to pretectal nuclei in the posterior thalamus which send efferent projections to premotor nuclei in the brainstem. Notice that the modular organization of these visuomotor systems is not limited to separate input channels but instead extends right through to the motor nuclei that encode the movements produced by the animal. These are truly *visuomotor*, not visual modules.

Thus, instead of a monolithic visual system dedicated to the construction of a representation of the external world, the frog appears to possess a number of separate visuomotor systems or modules, each of which is dedicated to the sensory control of a different pattern of behavior. It seems that the visual system of the frog did not evolve to provide an integrated representation of the world in which it lives, but rather to control the different kinds of movements that it needs to make to survive and reproduce in that world. To understand how these visuomotor systems are organized, it is necessary to study both the selectivity of their sensory inputs and the characteristics of the different motor outputs they produce.

VISUOMOTOR MODULES IN MAMMALS

The vertical architecture of visuomotor control is not limited to amphibia. As we shall see, mammals too show evidence of modularity in the organization of their visuomotor pathways. The evidence is limited, however, not because investigators have looked for it and failed to find it but because almost all studies of vision in mammals (including humans) have approached the problem in perceptual and cognitive terms and have largely ignored the visual control of motor output. Indeed, it is the theoretical commitment to vision *qua* perception that has shaped the methodology employed in most laboratories that study the functional architecture of the visual system. Instead of studying the different kinds of motor behaviors that are normally controlled by visual inputs (in a manner similar to that employed in the study of vision in the frog by Ingle and others), investigators working with mammals have typically looked at the performance of their subjects on some form of visual discrimination task.

The logic of the visual discrimination paradigm rests on the assumption (usually implicit rather than stated) that the function of vision is to provide an integrated representation of the external world upon which the animal can act. Thus, the argument goes, it makes good sense to concentrate on the way in which normal animals and perhaps animals with specific brain lesions process and store information that might be used in the construction of that representation. By varying the stimulus parameters and training conditions of a visual discrimination task, one can gain important insights into how animals extract information from the stimulus array and how they code and store that information. It makes little sense, therefore, to look at the large range of different visually guided movements or responses that animals make, since, except for a few visual reflexes, the behavior the animal generates is always in relation to the perceptual representation or model the visual system provides. According to this view, any modularity in the visual system is limited to the input side and the different input lines eventually converge on a higher-order network which controls the animals behavior. As such, the organization of the visual system can be studied quite independently of motor output. Indeed, it does not matter what the actual motor behavior is that the animal produces in a visual discrimination task. The animal could be pressing a lever, jumping from one platform to another, running down an alleyway, pulling a string, knocking aside the cover of a food-well, or picking up an object. What matters is that the animal discriminates. It is the decision, not the motor act, that interests most investigators. Modularity on the output side, when it is considered at all, is seen as largely independent of input modularity and the province of motor physiology or the psychology of motor skills.

One important exception to this research tradition has been the study of eye movements. Here, perhaps because the movements are so striking and so clearly related to the spatial and temporal characteristics of the visual stimulus, investigators have routinely recorded the movements directly while at the same time systematically varying the stimuli controlling those movements. This has made it possible to relate the activity of single cells in this visuomotor system not only to the characteristics of the visual stimulus, but also to the characteristics of the motor output. Moreover, it has been possible to relate systematic changes in the latency, amplitude, velocity, and acceleration of different kinds of eye movements to lesions in specific areas of the central nervous system. While the interpretation of data from these experiments has often been difficult, real progress has been made in describing the neural networks underlying a number of different oculomotor movements and in developing control-system models to explain the relationship between oculomotor performance and stimulus characteristics (see Carpenter, 1977 for review). In fact, eye movements are undoubtedly the best understood example of visually guided behavior in mammals.

Unfortunately, the study of other kinds of visuomotor behavior in mammals has lagged far behind the study of eye movements. While this neglect is due, in part, to the fact that measuring multijoint movements such as locomotion and grasping is technically much more difficult than measuring eye movements, it is also a result of the belief that vision and the visual system can be studied quite independently from motor output. But as we have seen in the frog, the visual system of vertebrates can perhaps best be studied by looking at both sensory input and motor output. Of course, this is easy to say, but often very hard to do. It means that one must be prepared to record the different patterns of visuomotor behavior in as much detail as possible. In short, we must do for other kinds of visuomotor behavior in mammals what we have done for eye movements—provide an accurate and continuous record of the animal's behavior so that the spatiotemporal organization of the movements it produces can be related to the spatial and temporal characteristics of the stimuli eliciting and/or controlling those movements.

Over the past few years in our laboratory, we have been taking this kind of approach to the study of visually guided movement in a small diurnal rodent, the Mongolian gerbil (*Meriones unguiculatus*). In all this work, we have used quantitative analysis of film and video records of the behavior of both normal gerbils and gerbils with lesions in specific retinofugal targets. While the picture that is emerging from these experiments is much more complex than that seen in the frog work, there are important similarities in the neural substrates of visuomotor behavior in these two vertebrates.

In the gerbil, for example, projections from the retina to the optic tectum and from there to the contralateral brainstem and cord appear to play an important role in mediating brisk head and body turns toward novel and baited targets initially presented in the visual periphery (Ellard, 1986; Ellard & Goodale, 1986; Goodale, 1983b; Goodale & Milner, 1982; Mlinar & Goodale, 1984). This projection system is remarkably similar to the module controlling visually guided feeding in the frog. Unlike the frog system, however, the orientation system in the gerbil can be influenced by input from visual areas in neocortex. Thus, animals with lesions of area 17, unlike normal gerbils, fail to make adjustments in the amplitude of their orienting head movements to compensate for the movement of visual targets (Ingle, 1977). Instead of making "anticipatory" overshoots, they initiate brisk and accurate head turns to the point in space where the moving target first appeared. These results suggest that although the optic tectum is essential for initiating brisk head turns toward moving targets, the output of this system is modulated in some way by the geniculostriate network (possibly via cortical projections to the optic tectum itself, since electrophysiological studies in hamster [Rhoades & Chalupa, 1978a, 1978b] and cat [Wickelgren & Sterling, 1969] have revealed that the directionality of single units in the optic tectum is dependent on direct projections from area 17). Frogs, of course, show no compensation for the movement of the prey when they turn and snap, and appear to depend instead on rapidity of movement and the fact that they have a wide and sticky tongue.

A second projection system in the gerbil (retina—optic tectum—ipsilateral brainstem) appears to mediate visually elicited escape reactions in a manner remarkably similar to that already described in the frog (Ellard, 1986). In both species, lesions of the optic tectum or interruptions of the ipsilateral tectal efferents abolish visually elicited escape reactions.

Finally, a third projection system in the gerbil (retina—pretectum—brainstem) is involved in the control of locomotion around barriers. Gerbils with lesions of the pretectal nuclei, like frogs with similar lesions, have difficulty getting around large barriers placed in front of them (Goodale, 1983b; Goodale & Milner, 1982).

In summary then, the modular structure of visuomotor behavior in the gerbil appears to depend on a neural architecture that is quite similar to that of a much simpler vertebrate, the frog. In both species, visually guided orientation movements toward small visual targets, visually elicited escape, and visually guided locomotion around barriers are mediated by quite separate pathways from the retina right through to the motor nuclei in the brainstem and spinal cord. None of these parallels in the vertical architecture of the visuomotor system (or the architecture itself) would have been discovered, however, had we not been prepared to record the visually guided behavior of the gerbil in great detail. No amount of visual discrimination training would have revealed either the functional organization of visuomotor behavior or the architecture of its neural substrate.

ARE THERE VISUOMOTOR MODULES IN HUMAN BEINGS?

Although I have chosen thus far to emphasize the similarities in the modular organization of visuomotor behavior in two different vertebrates—similarities, by the way, which suggest a common modular structure throughout the vertebrate line—it is clear that in the higher mammals at least, there is considerable elaboration of both visually guided behavior and visual processing within neocortex, an elaboration that does not exist in other vertebrates such as amphibia. The visually guided behavior of a primate, for example, is incredibly more rich and varied than that of a frog. Different visual inputs have access to a multitude of different output channels in the primate, and these visual inputs contribute to the formation of goals and to the control of goal-directed actions that are quite independent of particular motor outputs. Moreover, the animal has access to stored information derived from previous visual input and can use this information to direct its ongoing behavior. Thus, vision in the primate (and other mammals as well) plays an important role in cognitive processes that are clearly independent of fixed motor programs. Much of this visually controlled activity is mediated at least in part by neural circuits in visual areas of neocortex beyond area 17. It is the unraveling of this circuitry that has provided the focus of a good deal of the research effort of visual neuroscientists over the last 100 years.

Nevertheless, the presence of highly plastic visuocognitive mechanisms in primates does not mean that much of their visuomotor behavior does not continue to depend on relatively independent input-output modules whose functions remain largely outside the influence of perceptual and cognitive processes. The independence of some visuomotor modules and certain perceptual processes can be readily demonstrated in normal human subjects. In a recent series of investigations, for example, we were able to show that the mechanisms that preserve the perceptual identity of a visual target as its position is shifted on the retina during an eye movement are dissociable from those that mediate reaching movements directed at that target (Goodale, Pelisson, & Prablanc, 1986; Pelisson, Prablanc, Goodale, and Jeannerod, 1986).

The experiments were originally designed to investigate how sudden displacements of a visual target would affect the trajectories of a moving limb aimed at that target. On the basis of previous work (Prablanc, Pelisson, & Goodale, 1986), we expected to find evidence for a dynamic control system in which extravisual information about limb position is compared with visual information about the position of the target. In carrying out these experiments, however, we also found evidence that the adjustments in the trajectory of the moving limb took place without the subjects being aware that the target had been displaced.

To explain our experimental procedure, it is necessary to describe how visually guided pointing movements typically unfold. When one reaches toward a target that suddenly appears in the peripheral visual field, not only does the arm extend toward the object, but the eyes, head, and body also move in such a way that the image of the object falls on the fovea. The onset of electromyographic activity in the ocular and brachial musculature is very nearly simultaneous (Biguer, Jeannerod, & Prablanc, 1982). Nevertheless, because the eye is influenced much less by inertial and gravitational forces than the limb, the saccadic eye movements directed at the target are typically completed while the hand is still moving. Indeed, the first and largest of the saccades made toward the target is often completed before (or shortly after) the hand has begun to move. (A second saccade—the so-called "correction saccade"—puts the image of the target right on the fovea.) This means that during a pointing movement, the target will be located on or near the fovea as the hand reaches toward it. Theoretically then, detailed information about the position of the target provided by central vision could be used to correct the trajectory of the hand as it moves toward the target. Popular models of how reaching movements are programmed have argued just this (Keele & Posner, 1968; Paillard, 1982). According to these models, the initial movement of the limb is ballistic and is programmed on the basis of visual information about the position of the target derived from the peripheral retina. As soon as the target is foveated, however, amendments to the motor program (and thus corrections in the trajectory) are made on the basis of the dynamic feedback provided by central vision. This feedback depends upon comparing the position of a seen hand with a seen target after the eye has foveated the target.

Although this account of reaching sounds quite reasonable, some basic observations about reaching movements cast doubt on its validity. Subjects who cannot see their hand, for example, are more accurate when the target remains visible throughout the reach than when it disappears after the reach has begun (Prablanc, Goodale, Pelisson, Biguer, & Jeannerod, 1984; Prablanc et al., 1986). It is this kind of result that suggests that visual feedback about the relative positions of the hand and target is not the only source of information about the accuracy of the reaching movement. Information derived from proprioception or even "efference-copy" of the motor commands must also play a role in these situations. Unfortunately, little attention has been paid to the way in which such information can influence an ongoing reaching movement. Instead, it seems to have been assumed that any adjustments that are made to the trajectory of a reaching movement are dependent upon seeing the hand moving with respect to the target.

It was for this reason that we decided to examine whether subjects could modify an ongoing reaching movement on the basis of new visual information about the position of the target when vision of the moving hand was not available. If they could, then this would provide evidence that modification of a reaching movement need not depend on seeing where the hand is with respect to the intended target. Any adjustments to the trajectory of the moving limb would instead have to depend on information derived from proprioception or efference-copy of the motor commands sent to the limb. In carrying out this work, we took advantage of the fact that during a reaching movement the eyes begin to move much sooner than the hand. It was possible therefore to change the position of the target during the point of highest velocity of the first saccade, just as the hand had begun to move. Since subjects typically undershot targets with their first saccade, they had an opportunity to make a larger or smaller correction saccade and to modify the trajectory of their limb movement to accommodate the change in the position of the target. To our surprise, we found that not only did the subjects make the necessary adjustments in their eye and arm movements but they did so without any awareness that the target had moved from its original position. As will soon become apparent, it appeared as if the visuomotor network mediating visually guided reaching movements was quite independent from the mechanisms preserving target stability during a shift in fixation.

Our experiments were carried out in the following way. Subjects were seated in turn in a darkened room with the index finger of their right hand resting on a small visual target located in a central position on a platform directly in front of them (Figure 1a). They were instructed to move their finger as quickly and as accurately as they could to the new location of the target as it jumped from its central position to one more peripheral. At the beginning of a trial, the subjects could see their hand and arm through a large half-reflecting mirror. The target consisted of a small light-emitting diode (LED) positioned above the mirror in such a way that its virtual image appeared below the mirror on the same surface as the subject's hand. The subject's head was fixed with a bite bar so that horizontal eye movements could be monitored by electro-oculography. The position of the subject's index finger at the beginning and end of a pointing movement was derived from resistive paper that covered the surface on which the virtual image of the target was located. The trajectory of the pointing movement was reconstructed from a photograph of a flashing infrared LED (2 ms at 10 Hz) attached to the end of the subject's finger.

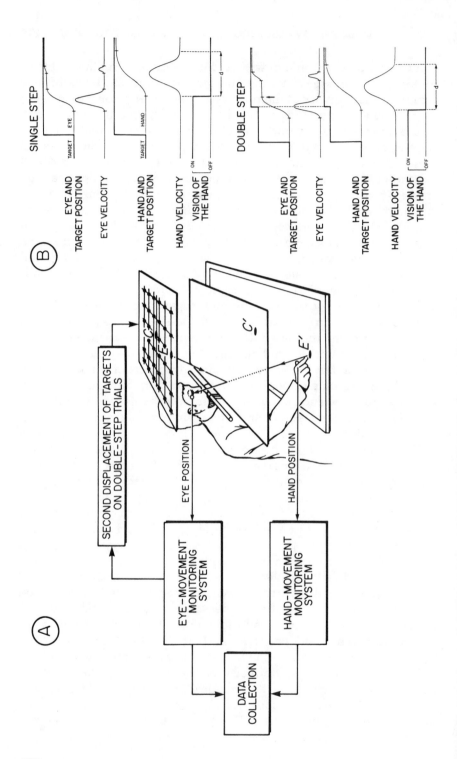

SINGLE STEP

EYE AND
TARGET POSITION

EYE VELOCITY

HAND AND
TARGET POSITION

HAND VELOCITY

VISION OF
THE HAND

DOUBLE STEP

EYE AND
TARGET POSITION

EYE VELOCITY

HAND AND
TARGET POSITION

HAND VELOCITY

VISION OF
THE HAND

SECOND DISPLACEMENT OF TARGETS
ON DOUBLE-STEP TRIALS

EYE POSITION

HAND POSITION

EYE-MOVEMENT
MONITORING
SYSTEM

HAND-MOVEMENT
MONITORING
SYSTEM

DATA
COLLECTION

As soon as the subject's index finger left the surface on which the central target had been located, the subject's view of his hand and arm disappeared and reappeared only after the target moved back to the central position and the hand began to move back in preparation for a new trial. This meant that as the subject moved his hand toward the peripherally located target he had no visual feedback about his movements.

In our first experiment (Figure 1b), on half the trials (which we called single-step trials), the target jumped from its central position to a randomly determined position 30, 40, or 50 cm to the right and stayed there until the subject's finger re-contacted the surface. On the remainder of the trials (double-step trials), the target jumped to one of these positions, followed by a second jump to a new position 10% further away from the center. The second jump occurred just after the first saccade reached its peak velocity. The single- and double-step trials were presented in random order. (In fact, the randomization of these two kinds of trials was critical to the experimental procedure. Changing retinal feedback from a saccade in a systematic way results in a progressive and adaptive change in the gain of the saccades (Henson, 1978). By presenting the subject with single- and double-step targets in random order, we were able to keep the gain of the oculomotor response constant.) Subjects were not told, of course, that the target would sometimes change position during the first eye movement.

The results of this first experiment were clear and unambiguous.

Figure 1. Apparatus and paradigm used in reaching experiments. A. This schematic diagram illustrates the apparatus used to present targets and to record hand and eye movements. The upper surface consisted of a matrix of LEDs each of which could be independently illuminated. C represents the LED that would be illuminated to present the central target; E represents the LED that would be illuminated to present a target in the peripheral visual field. When the subject looked through the semi-reflecting mirror, he would see the virtual image of one of these LEDs (C' or E') on the pointing surface below the mirror. Vision of the hand could be prevented by turning off the illumination below the semi-reflecting mirror. By recording the eye movements on-line, it was possible to produce a second displacement of the target (E/E') just as the eye movement reached peak velocity. B. Schematic diagram illustrating the spatiotemporal organization of the movements of the target, eye, and hand on single- and double-step trials. The duration of the hand movement is indicated by d. The second displacement on double-step trials was triggered when the first saccade reached peak velocity. Vision of the hand was prevented as soon as the hand began to move.

As Figure 2 illustrates, although the subjects consistently undershot the position of the target, the distributions of their final positions in the single- and double-step conditions were separated by a distance that was equivalent to the size of the second displacement of the target. In other words, the subjects were correcting the trajectory of their limb movement to accommodate this displacement. Moreover, the duration of a limb movement made to a displaced target corresponded to the duration of movement that would have been made to a nondisplaced target located at the same final position (Figure 3). Thus, no additional processing time was required during the double-step trials. The velocity and acceleration profiles that were derived from the movement trajectories obtained on these trials also showed no "breakpoint" to indicate a reprogramming of the manual response.

In a second experiment, in addition to the double-step trials described above, we included double-step trials in which the second displacement of the target sometimes moved it *back* toward the central position (from 46 cm to 44 cm, for example). These backward

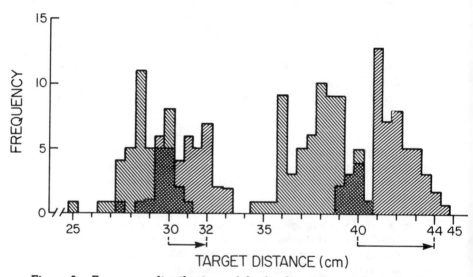

Figure 2. Frequency distributions of the final positions of the index finger for the four subjects in Experiment 1 on single- and double-step trials. Only the 30- and 32-cm and the 40- and 44-cm targets are shown. The distributions on single-step trials are indicated by the oblique lines running up to the left. The distributions on double-step trials are indicated by the oblique lines running up to the right. Overlap in the two distributions is indicated by cross-hatching. The size and direction of the second displacement on double-step trials is indicated by the small arrows on the abscissa.

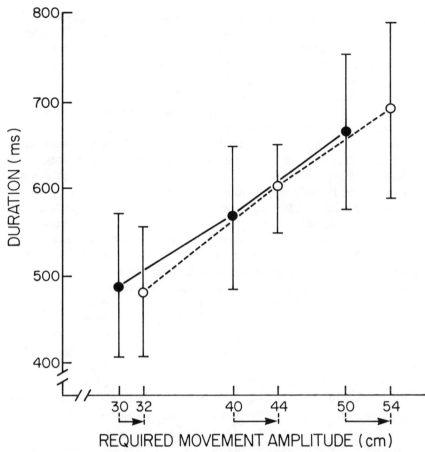

Figure 3.   Mean duration of hand movements plotted as a function of their amplitude. The amplitude-duration curve for single-step trials is indicated by a solid line and filled circles. The amplitude-duration curve for double-step trials is indicated by a dotted line and open circles. Vertical lines indicate standard deviations. As the graph plainly shows, the two lines fall on one another.

movements of the target were much smaller than the average correction saccade for targets at that eccentricity and consequently the final position of the target still placed it beyond the terminal position of the first saccade. In other words, the required correction saccade was in the same direction as the first, but was smaller than similar correction saccades on single-step trials (and was certainly much smaller than the correction saccades observed on those double-step trials where the target had moved further away from the central position). Just as they had in the first experiment, subjects in this second study compensated for the second displacement of the target on double-step trials. This time, however, they made larger limb movements when the target was jumped away from the central position on the second displacement and smaller limb movements when it was jumped toward the central position. Thus, it appears that modifications to the trajectory of the moving limb can accommodate displacements in the position of the target in either direction.

The results of both these experiments strongly suggest that when rapid eye and hand movements are directed at a visual target, an initial set of signals is sent to the muscles controlling both the eye and the hand based on information about the position of the target that is available on the peripheral retina. After the first saccade, however, the combined input from retinal and extraretinal signals updates the information about target position, which is then immediately used to "fine-tune" the trajectory of the hand. If this hypothesis is correct, then it becomes much clearer why the duration of a movement is the same for a given position of the target, independent of whether that position was achieved by a single or a double displacement of the target. In either case, the same post-saccadic information would be used to update the motor signals controlling the trajectory of the hand (provided the amplitude of the second displacement in the double-step condition is not too large). In other words, the apparent correction in the trajectory that occurred on double-step trials was nothing more than the normal updating of the motor programming that occurs at the end of the first saccade on an ordinary trial.

The modifications we observed in the trajectory of the pointing movement on double-step trials were made in a situation where visual feedback about the position of the limb was unavailable. Thus, the new information about the location of the target in visual space that was obtained at the end of the first saccade must have been combined with information about the current position of the hand that was derived from nonvisual sources such as proprioceptive feedback and/or efference copy of the signals sent to the upper limb musculature. Just how these different sources of information were compared is not clear. Whatever the mechanisms might be, these experiments have shown that efficient modulation of pointing movements can occur in the absence of visual feedback about limb position.

What was most important from the point of view of modular organization of visuomotor control, however, was the finding that at no time in either experiment did the subjects realize that the target had jumped to a new location while they were reaching towards it. Nor did the subjects detect anything different about their reaches on double-step as opposed to single-step trials. In short, they failed to perceive the change in target position even though their visuomotor output was modified to accommodate that change. Indeed, had the subjects perceived the change in target position that occurred on double-step trials and, on the basis of this perception, quite consciously altered the trajectory of their reach, then the duration of the movements on these trials would have fallen well outside the amplitude-duration curve obtained on single-step trials. The fact that this did not occur (Figure 3) is additional evidence that subjects were treating single- and double-step trials in an identical fashion and, as I have already argued, suggests that the "correction" on double-step trials was nothing more than the normal fine-tuning of the trajectory that occurs at the end of the first saccade on single-step trials.

If one is to argue that a certain class of visuomotor behavior can be affected by changes in the visual stimulus even though those changes are not reported, then it becomes important to determine whether or not subjects can ever perceive the second displacement of the target in this kind of situation. It was for this reason, that two additional experiments were carried out recently by Denis Pelisson and Claude Prablanc. In the first of these experiments, two subjects were tested in the same apparatus described above but instead of pointing were simply required to look at the target when it appeared in their peripheral visual field. They were told that on some occasions a second displacement of the target would occur and that if they saw this second jump they should press the button provided. In fact, two-thirds of the 150 trials were single-step trials and only one third were double-step trials. The subjects reported seeing the target jump on only 16.5% of the trials and, even then, their hit rate did not differ significantly from chance, with 64.5% of their button presses occurring on single-step trials and 35.5% on double-step trials. Moreover, the subjects complained of never being confident that the target had jumped as they looked towards it.

In the second experiment, every trial was a double-step trial. On half of the 120 trials, the target jumped away from the central fixation point on its second displacement while on the other half of the trials it jumped in the opposite direction (but only 2 cm). But as before, since the first saccade typically undershot the original position of the target by 10% or more, correction saccades in the same direction as the first saccade were required for even those targets that jumped back towards the central fixation point. The six subjects in this experiment were told that the target would jump randomly either "inwards" or "outwards" as they looked toward it and that they should press the button in their left hand if they thought it had jumped inwards and the button in their right if they thought it had jumped outwards. They were asked to respond on every trial and had three seconds to decide before the target returned to the central position. Despite the fact that the subjects (four of whom had participated in the original pointing experiments) were encouraged to use any visual cue that was available, their responses were correct on only 54.5% of the trials. In addition, they showed no response bias and pressed the left-hand button as often as they did the right (50.4% vs. 48.6%). Again, they maintained that they were never able to see the target jump but were just "guessing."

The failure to perceive the displacement may reflect the broad tuning of perceptual constancy mechanisms that preserve the apparent stability of a target in space as its position is shifted on the retina during an eye movement. The results of our experiment certainly suggest that artificially-induced displacements in real space of up to 10% of the initial displacement will not disrupt this constancy (provided, of course, the target changes its position during a saccade). The visuomotor module controlling reaching, however, is clearly much more sensitive to the exact position of the target image on the retina and compensates almost perfectly for changes in real position to which the perceptual systems seem quite refractory. In short, the visual mechanisms that construct the percept of the target appear to be quite independent of those that mediate visuomotor output directed at that target.

Using rather different paradigms, Bridgeman and his colleagues (Bridgeman, Lewis, Heit, & Nagle, 1979; Bridgeman, Kirch, & Sperling, 1981) have reported similar dissociations between what they call the "cognitive and motor-oriented systems of visual position perception." Like us, they found that some position information seems to be available for controlling pointing movements that is unavailable to more cognitive systems.

Other visuomotor control networks in human beings also seem to work quite independently of normal perceptual processing. For instance, David Lee and his students (Lee & Aronson, 1974; Lee & Lishman, 1975) have shown that movement of the entire visual field can be a powerful stimulus for the control of upright posture—despite the fact that subjects report having no perception of the retinal error signals driving the postural adjustments! In these experiments, subjects stood within a "room" which was, in fact, suspended from the ceiling of a much larger room. Because it was suspended just above the stationary floor on which the subjects were standing, the smaller room could be gently oscillated back and forth. When this happened, subjects would typically sway forward and backward approximately in phase with the oscillations of the room, even when those oscillations were as small as 6 mm. As Lee has remarked, "the subjects were like puppets visually hooked to the swinging room" (Lee & Thomson, 1982). Nevertheless, the subjects remained quite unaware that the room was oscillating or that they were swaying rhythmically back and forth.

NEUROLOGICAL STUDIES OF VISUOMOTOR FUNCTION

These dissociations between perception and visuomotor control (particularly those obtained in our reaching experiments) are reminiscent of some of the dissociations that have been reported in the neurological literature. Although it has been known for some time, for example, that damage to striate cortex in human beings produces a severe loss of visual capacity in correlated regions of the visual fields, in the last 15 years a surprising number of visual abilities have been shown to survive such damage (Perenin & Jeannerod, 1975; Weiskrantz, Warrington, Sanders, & Marshall, 1974). Thus, patients with striate lesions are often able to point accurately to visual targets located within their scotoma or "blind" hemifield, even though they do not report seeing the target under ordinary perimetry testing. The paradoxical ability of these individuals to reach out and touch stimuli that they deny perceiving has been given the rather whimsical name "blind-sight" (Weiskrantz et al., 1974).

One can also find in the neurological literature evidence to suggest that brain damage in human beings can result in the obverse of blind-sight, producing a profound disturbance in certain visuomotor abilities with a relative sparing of visual perception. As long ago as 1918, the British neurologist Gordon Holmes described the behavior of a young soldier with a gunshot wound in the occipital lobe that spared primary visual cortex. Despite the fact that the young man had no trouble recognizing and distinguishing visible objects and could even read individual words with little difficulty, he had enormous problems making even very simple visually guided movements. Although, for example, he had no trouble recognizing a pencil held up in front of him, when asked to reach towards it he would often reach in entirely the wrong direction or grope for it like a man reaching in the dark. At the same time, he had no difficulty moving his hand toward any object that touched his face or body. Thus, the deficit could not be characterized as a simple problem in reaching. Nor was it a sensory deficit, either, (at least in the usual sense of the term) since he appeared to have fair acuity, good color vision, and reasonably good stereopsis. Similar observations have been made more recently by Perenin and Vighetto (1983). Although the patients in this study were not so seriously disturbed as the young soldier studied by Holmes, they nevertheless had great difficulty reaching toward objects in the field contralateral to their lesion even though they could verbally describe the spatial position and relative location of objects in that field. Like the evidence from the blind-sight work, observations of this sort suggest that the neural circuitry underlying the visual control of reaching movements is relatively independent of that underlying perception and other more "visuocognitive" abilities.

The modular organization of visuomotor control in human beings has been further elaborated by the work of Brouchon (1984). She described the behavior of a patient with biparietal cortex lesions who was unable to reach accurately toward visual targets in what Brouchon calls immediate "prehensive" space even though the same patient could indicate with reasonable accuracy the position of visual targets located some distance away. This finding suggests that not only can visuomotor modules in human beings work independently of higher-order perceptual/cognitive mechanisms, they can also function to a certain extent independently of one another, just as they do in other vertebrates.

Although these neurological studies provide support for the suggestion that visuomotor pathways in human beings are organized as relatively independent and parallel networks, they do not give us much insight into the detailed neural architecture of such a system. At present, it is possible to do little more than speculate about the various neural substrates mediating the "residual" visual abilities of different groups of patients. Part of the reason for this is the failure of neurologists and neuropsychologists to study visually guided behavior (or any complex sensorimotor output other than eye movements) in any detail. The kind of study described above, in which some effort was made to look at visuomotor control, is quite uncommon in neurology and neuropsychology. Indeed, there has been a strong tendency in these disciplines to interpret a visual deficit almost entirely in terms of a disturbance in sensation, perception, cognition, or attention. In some ways, neuropsychological investigations of human vision resemble the traditional visual discrimination approach to the study of visual function in other mammals. While they have produced an enormous body of data about perceptual disorders in brain-damaged humans and have devised a number of different classification schemes for these disorders (the "agnosias"), much less attention has been paid to the role of vision in the control of reaching, locomotion, posture, and other spatially organized behavior. Until this happens, however, our understanding of the modular organization of the visuomotor system and its underlying neural architecture will remain quite incomplete.

SOME SPECULATIONS AND A CAVEAT

Throughout this chapter, I have argued that the modular approach to the visual system that characterizes much of contemporary cognitive science should be applied not simply to visual input channels but should instead be extended right through to motor output. I have suggested that the many different movements a vertebrate makes in space are mediated by a number of relatively independent visuomotor channels, each of which is sensitive to a particular constellation of input parameters and each of which utilizes a particular set of effector organs. The apparently vertical architecture of such a system, I suggested, corresponds to a neural architecture of independent and parallel pathways from receptors through to the motor nuclei. Visuomotor, not visual, modularity is the rule.

One difficulty with the concept of independent visuomotor modules is the problem of integrating and coordinating behavior. Too strong a commitment to independent sensorimotor channels can result in a theoretical stance that differs little from simple-minded reflexology. Visuomotor modules there may be, but they must somehow be orchestrated in order to permit the animal to operate in a stimulus-rich world in which a number of visually guided behaviors might be required at the same time. Moreover, as I intimated earlier, postulating a set of independent visuomotor modules will not account for the subtle and varied behavior of higher vertebrates. While inhibitory interactions between modules could account for some of the observed coordination in visually guided behavior, much of the necessary orchestration (in mammals at least) might be provided by neocortical elaborations of more basic modular functions within the visuomotor system. In other words, the more recently evolved geniculostriate system (and other neocortical systems) might influence the production of visuomotor behavior by modifying the input/output functions of other, more ancient, retinofugal systems or by gating and/or delaying the input to or output from those systems. There is recent evidence from our laboratory, for example, to suggest that the temporal coordination of eye and limb movements during visually guided reaching may depend on timing mechanisms located within the left hemisphere, even though the actual movements themselves are clearly programmed elsewhere in the brain (Fisk & Goodale, 1985; Goodale, in press). Whatever the particular nature of the control might be, it is clear that some sort of hierarchical organization must be present within the vertical architecture of visuomotor networks in higher mammals.

This kind of approach to the visual control of behavior has some parallels in the robotics literature. The usual approach to the problem of building an intelligent and autonomous mobile robot is to decompose the control system into a series of functional processes from perception, through modeling, planning, and task execution, to motor control. In some ways, this approach resembles the kind of thinking that one can see in much of the traditional perception and cognition literature where a similar horizontal organization of processing is assumed to operate between perception and motor output. Recently, however, Brooks and his colleagues have taken a radically different approach to the problem (Brooks, 1986). Instead of opting for a horizontally organized system, they have chosen to build a mobile robot with several different layers of control in which each layer corresponds to a particular "level of competence." The system is designed so that lower levels of control can work quite independently of higher levels. In some ways then, the vertical organization of this kind of control network resembles the kind of modular organization I have outlined for the vertebrate visual system. Where it differs, however, is in the clearly hierarchical arrangement of the different layers of control. While we have established the existence of parallel visuomotor modules in the vertebrate brain, we have yet to work out the way in which these modules are integrated in the behaving animal.

REFERENCES

Biguer, B., Jeannerod, M., & Prablanc, C. (1982). The coordination of eye, head, and arm movements. *Experimental Brain Research, 46*, 301–304.

Bridgeman, B., Kirch, M., & Sperling, A. (1981). Segregation of cognitive and motor aspects of visual function using induced motion. *Perception and Psychophysics, 29*, 336–342.

Bridgeman, B., Lewis, S., Heit, G., & Nagle, M. (1979). Relation between cognitive and motor-oriented systems of visual position perception. *Journal of Experimental Psychology: Human Perception and Performance, 5*, 692–700.

Brooks, R. A. (1986). Achieving artificial intelligence through building robots. A.I. Memo 899, Artificial Intelligence Laboratory, M.I.T.

Brouchon, M. (1984). A possible dissociation between locomotory and prehensive space organization in a patient with biparietal cortex lesions. Paper presented at NATO workshop on Optic Tectum, St. Andrews, Scotland, September 1984.

Carpenter, R. (1977). *Movements of the eyes.* London: Pion Ltd.

Ellard, C. G. (1986). The role of the descending tectal efferents in the visuomotor behavior of the Mongolian gerbil: Evidence for functional specificity of crossed and uncrossed pathways. Unpublished doctoral thesis, University of Western Ontario.

Ellard, C. G., & Goodale, M. A. (1986). The role of the predorsal bundle in head and body movements elicited by electrical stimulation of the superior colliculus in the Mongolian gerbil. *Experimental Brain Research, 64,* 421–433.

Ewert, J. P., Capranica, R. R., & Ingle, D. J. (Eds.) (1983). *Advances in vertebrate neuroethology.* New York: Plenum Press.

Fisk, J. D., & Goodale, M. A. (1985). The organization of eye and limb movements during unrestricted reaching to targets in contralateral and ipsilateral visual space. *Experimental Brain Research, 60,* 159–178.

Fodor, J. (1983). *The modularity of mind.* Cambridge, MA: MIT Press.

Goodale, M. A. (1983a). Vision as a sensorimotor system. In T. E. Robinson (Ed.), *Behavioral approaches to brain research* (pp. 41–61). New York: Oxford University Press.

Goodale, M. A. (1983b). Neural mechanisms of visual orientation in rodents: targets versus places. In A. Hein & M. Jeannerod (Eds.), *Spatially oriented behavior,* New York: Springer.

Goodale, M. A. (In press). Hemispheric differences in motor control. *Behavioural Brain Research.*

Goodale, M. A., & Milner, A. D. (1982). Fractionating orientation behavior in rodents. In D. J. Ingle, M. A. Goodale, & R. J. W. Mansfield (Eds.), *Analysis of visual behavior,* Cambridge, MA: MIT Press.

Goodale, M. A. Pelisson, D., & Prablanc, C. (1986). Large adjustments in visually guided reaching do not depend on vision of the hand or perception of target displacement. *Nature, 320,* 748–750.

Henson, D. B. (1978). Corrective saccades: Effects of altering visual feedback. *Vision Research, 18,* 63–67.

Holmes, G. (1918). Disturbances in visual orientation. *British Journal of Ophthalmology, 2,* 449–506.

Hubel, D. H., & Wiesel, T. N. (1961). Integrative action in the cat's lateral geniculate body. *Journal of Physiology, 155,* 385–398.

Hubel, D. H., & Wiesel, T. N. (1962). Receptive fields, binocular interaction, and functional architecture in the cat's visual cortex. *Journal of Physiology, 160,* 106–154.

Ingle, D. J. (1973). Two visual systems in the frog. *Science, 181,* 1053–1055.

Ingle, D. J. (1977). Role of visual cortex in anticipatory orientation towards moving targets by the gerbil. *Neuroscience Abstracts, 3,* 68.

Ingle, D. J. (1982). Organization of visuomotor behaviors in vertebrates. In D. J. Ingle, M. A. Goodale, R. J. W. Mansfield (Eds.), *Analysis of visual behavior*, Cambridge, MA: MIT Press.

Ingle, D. J., Goodale, M. A., & Mansfield, R. J. W. (1982). *Analysis of visual behavior*, Cambridge, MA: MIT Press.

Ingle, D. J., & Hartline, P. (In press). Towards a unified view of the optic tectum.

Keele, S. W., & Posner, M. I. (1968). Processing of visual feedback in rapid movements. *Journal of Experimental Psychology, 77,* 155–158.

Lee, D. N., & Aronson, E. (1974). Visual proprioceptive control of standing in human infants. *Perception and Psychophysics, 15,* 529–532.

Lee, D. N., & Lishman, J. R. (1975). Visual proprioceptive control of stance. *Journal of Human Movement Studies, 1,* 87–95.

Lee, D. N., & Thomson, J. A. (1982). Vision in action: The control of locomotion. In D. J. Ingle, M. A. Goodale, & R. J. W. Mansfield (Eds.), *Analysis of Visual Behavior.* Cambridge, MA: MIT Press.

Lorenz, K. Z. (1958). The evolution of behavior. *Scientific American, 199,* 67–78.

Marr, D. (1982). *Vision.* San Francisco: W. H. Freeman.

Mlinar, E. J., & Goodale, M. A. (1984). Cortical and tectal control of visual orientation in the gerbil: Evidence for parallel channels. *Experimental Brain Research, 55,* 33–48.

Paillard, J. (1982). The contribution of peripheral and central vision to visually guided reaching. In D. J. Ingle, M. A. Goodale, & R. J. W. Mansfield (Eds.), *Analysis of visual behavior.* Cambridge, MA: MIT Press.

Pelisson, D., Prablanc, C., Goodale, M. A., & Jeannerod, M. (1986). Visual control of reaching movements without vision of the limb. II. Evidence of fast unconscious processes correcting the trajectory of the hand to the final position of a double-step stimulus. *Experimental Brain Research, 62,* 303–311.

Perenin, M. T., & Jeannerod, M. (1975). Residual vision in cortically blind hemifields. *Neuropsychologia, 13,* 1–7.

Perenin, M. T., & Vighetto, A. (1983). Optic ataxia: A specific disorder in visuomotor coordination. In A. Hein & M. Jeannerod (Eds.), *Spatially oriented behavior.* New York: Springer-Verlag.

Prablanc, C., Goodale, M. A., Pelisson, D., Biguer, B., & Jeannerod, M. (1984). Visual control of reaching movements without vision of the limbs. *Investigative Ophthalmology and Visual Science, 25* (Suppl.), 96.

Prablanc, C. Pelisson, D., & Goodale, M. A. (1986). Visual control of reaching movements without vision of the limb: I. Role of retinal feedback of target position in guiding the hand. *Experimental Brain Research, 62,* 293–302.

Pylyshyn, Z. W. (1984). *Cognition and computation: Toward a foundation for cognitive science.* Cambridge, MA: MIT Press.

Rhoades, R. W., & Chalupa, L. M. (1978a). Functional and anatomical consequences of neonatal visual cortical damage in the superior colliculus of the golden hamster. *Journal of Neurophysiology, 41,* 1466–1494.

Rhoades, R. W., & Chalupa, L. M. (1978b). Functional properties of the corticotectal projections in the golden hamster. *Journal of Comparative Neurology, 180,* 617–634.

Tinbergen, N. (1951). *The study of instinct.* London: Oxford University Press.

Treisman, A. (1985). Preattentive processing in vision. *Computer vision, graphics, and image processing, 31,* 156–177.

Weiskrantz, L., Warrington, E., Sanders, M. D., & Marshall, J. (1974). Visual capacity in the hemianopic field following restricted occipital ablation. *Brain, 97,* 709–729.

Wickelgren, B., & Sterling, P. (1969). Influence of visual cortex on receptive fields in the superior colliculus of the cat. *Journal of Neurophysiology, 32,* 16–23.

12
How Does Human Vision Beat the Computational Complexity of Visual Perception?*

John K. Tsotsos

Department of Computer Science,
University of Toronto

This paper demonstrates how serious consideration of the deep complexity issues inherent in the design of a visual system, can constrain the development of a theory of vision. We first show how the seemingly intractable problem of visual perception can be converted into a much simpler problem by the application of several physical and biological constraints. For this transformation, two guiding principles are used that are claimed to be critical in the development of any theory of perception. The first is that analysis at the "complexity level" is necessary to ensure that the basic space and performance constraints observed in human vision are satisfied by a proposed system architecture. Second, the "maximum power / minimum cost principle" ranks the many architectures that satisfy the complexity level and allows the choice of the best one. The best architecture chosen using this principle is completely compatible with the known architecture of the human visual system, and in addition, leads to several predictions. The analysis provides an argument for the computational necessity of attentive visual processes by exposing the computational limits of bottom-up early vision schemes. Further, this argues strongly for the validity of the computational approach to modelling the human visual system. Finally, a new explanation for the pop-out phenomenon so readily observed in visual search experiments, is proposed.

---

* Many thanks are due to Allan Jepson and Steve Zucker for several productive discussions and for their useful suggestions. The author is a Fellow of the Canadian Institute for Advanced Research. This research was conducted with the generous assistance of the Natural Sciences and Engineering Research Council of Canada.

286

INTRODUCTION

The task of visual perception can be shown to be intractable for brute-force architectures in a straightforward manner, and in fact, sub-problems, such as polyhedral scene labeling, have been shown to be inherently NP-Complete (Kirousis & Papadimitriou, 1985). Yet, human vision is an effortless and exquisitely precise sense. How can this be? In the past, researchers have resorted to processing limits and attention in order to cope with this dilemma. Neisser, for example, first claimed that any model of vision that was based on spatial parallelism alone was doomed to failure, simply because the brain was not large enough (Neisser, 1967). This led him to his two-stage process of perception: a pre-attentive phase followed by an attentive phase. However, it is difficult to couch such a model in computational terms, there are so many missing details. Moreover, the reason for the need for attention is less than satisfactory. Stating that the brain is simply not large enough does not yield any useful constraints on the architecture of the visual system. Yet, Neisser's claim hints at the difficult issues of computational complexity that must be addressed. More recently, Feldman and Ballard concluded that time complexity considerations lead to massively parallel models being the only biologically plausible ones, since only they satisfy the 100 step rule (Feldman & Ballard, 1982). That is, since neurons compute at a rate of about 1000 Hz, and since simple perceptual phenomena do indeed occur in about 100 milliseconds, then biologically plausible algorithms can require no more than 100 steps. They did not, however, explain exactly how "massive" these networks must be (also, see Zucker, 1985). Feldman and Ballard also stress the importance of conservation of connections. Although the emphasis is correct, their application of this constraint leaves many questions unaswered and, in particular, they did not demonstrate that their set of conserving techniques is sufficient. Rumelhart and McClelland claim that the time and space requirements of a theory of cognitive function are important determinants of the theory's biological plausibility (Rumelhart & McClelland, 1986a). However, they do not provide any details on how such constraints may be satisfied. There have been several attempts at producing a unified "grand theory" of visual perception integrating many results from psychology, neurophysiology, neuroanatomy and computer science. Specific proposals are found in Treisman, 1986; Barrow & Tenenbaum, 1978; Feldman, 1985a; Ballard, 1986; and Marr, 1982, while philosophical positions are presented in Crick, Marr & Poggio, 1980; Poggio, 1984; Hildreth & Hollerbach, 1985; and Dobson & Rose, 1985a. None of these theories have been successful, although the exercise has been extremely valuable to progress in the field.

The problem that researchers face is that results from these disparate disciplines are not immediately compatible, and in fact, are often contradictory. Biological as well as computational scientists propose explanations and algorithms for various individual phenomena based on experimental results, or attribute functional significance to particular neural structures. Even though we recognize that certain neural structures are directly connected or that certain phenomena must be somehow related, the explanations are not immediately compatible with one another. That is, there has been very little work on "the big picture" within which the individual results may fit (see Dobson & Rose, 1985b and Maxwell, 1985 for excellent treatments of the problems with the research methodologies both in the neurosciences as well as in artificial intelligence). Thus, grand theories are easy prey for criticism, criticism that in one important sense is unfair at this point in the development of our discipline. (See for example the commentaries on the theories of Feldman, 1985a and Ballard, 1986.) There is no test that can be applied to a grand theory in order to determine whether or not fundamental considerations are satisfied. In this paper, a simple demonstration leads to the conclusion that parallelism, on its own, of biologically plausible degree is insufficient to satisfy the time complexity constraints for vision, and it is reasonable to speculate that it is insufficient, on its own, for any cognitive task. In addition, this paper proposes that satisfaction of the space and time complexity constraints be one of the elements of the test that new theories of visual perception must pass.

Computational complexity issues are broad and pervasive in the development of a theory of perception. The key philosophy underlying the research to be described in this paper is that the complexity considerations of the nature of the perceptual task are critical, and lead directly to "hard" constraints on the architecture of visual systems, both biological and computational. It is surprising that Marr did not even mention computational complexity issues, even as part of the computational level of his theory (Marr, 1982). According to Marr, the computational level of a theory addresses the questions: What is the goal of the computation? Why is it appropriate? and, What is the logic of the strategy by which it can be carried out? The representational and algorithmic level of the theory asks: How can this computational theory be implemented? What is the representation for the input and output? What is the algorithm for the transformation? And, finally, the implementational level of the theory asks: How can the representation and algorithm be realized physically?

Complexity issues span these three levels. Much past work in computer vision, motivated by Marr's philosophy, has tacitly assumed that the language of continuous mathematics is equivalent to the language of computation. Mathematical modeling is *not* equivalent to computational modeling. In proposing a mathematical solution for a problem, say that of solving optic flow equations, one has not also solved the problem computationally, even if simulated on a computer. There are still issues of representation, discretization, sampling, numerical stability, and computational complexity (at least) to contend with. The key first component of the computational level, in my mind, is the consideration of complexity issues, and Marr did not explicitly include this in his definition. Thus, I claim that there is another level of analysis required for any theory of perception—the *complexity level*.

This paper is concerned *only* with the complexity level. In particular, strategies for how a tractable solution can be achieved are discussed involving both time and space considerations. I will not attempt to ascribe functional significance to specific brain areas, I will not claim particular neural models, I will not propose representation schemes, and I will not choose a set of specific visual entities. There will be no algorithm proposed for a computer vision system. On the other hand, this exercise is claimed to be a critical one in the computational modeling of perception, and indeed, in the computational modeling of any aspect of intelligence. One of the key problems with AI solutions for tasks involving intelligence is that the solutions are so fragile with respect to "scaling up" with problem size. That is, theoretical solutions are derived, usually without theoretical regard to the amount of computation required, and then if an implementation is produced, it is tried out on a few small examples only. The standard claim then is that if faster or parallel hardware were available, a real-time solution would be obtained. There is something very unsatisfying about this type of research. In particular, parallel solutions, such as those proposed by the connectionist community, although motivated by complexity considerations, seem to neglect detailed considerations of computational complexity altogether (see the collections of papers on the subject in Rumelhart & McClelland, 1986b and Feldman, 1985b). For example, few if any deal with the time and space requirements of the relaxation procedures that they use, particularly in the context of time-varying input (but see Tsotsos, 1987a for empirical results on this). The issues raised by time and timing are in general not well handled (Tsotsos, 1986). If one is in the business of realizing systems, and proving that they behave in the required manner, the first requirement of a realizable system is that the task attempted and/or the proposed solution be computationally tractable.

Overview of Results

We show that in addition to spatial parallelism, the other characteristics of a sufficient visual architecture are:

- hierarchical organization through abstraction of prototypical visual knowledge, in order to reduce search time at least logarithmically;
- localization of receptive fields, noting that the physical world is spatiotemporally localized and that objects and events, and their physical characteristics, are not arbitrarily spread over time and space;
- maps are summarized via a pooled response, using the observation that not all visual stimuli require all possible parameter types for interpretation, and thus leading to separable, logical maps; and,
- hierarchical abstraction of the input arrays, in such a way as to maintain semantic content yet reducing the number of retinotopic elements.

These optimizations may be considered as sufficient, but not necessary, conditions to satisfy the time complexity constraint for the architecture of a visual system with performance comparable to human pre-attentive vision.

Applying connectivity constraints to this architecture, that is, determining how all the elements are to be connected and the resulting cost of connection, many further characteristics of primate visual systems are derived and others predicted:

- processor columnar organization;
- inverse magnification within the processor layer with respect to the retinotopic array;
- tuning of abstract computations, rather than direct access to more detailed maps;
- token coarse coding;
- physical separation of some maps;
- predictions for the best architecture for immediate perception;
- predictions for the overall configuration of the visual system in terms of the size and number of maps; and,
- pre-attentive vision is shown to be simply a special case of the

visual process, and not a component separable from attentive vision.

Background

Since this paper draws from several different disciplines, it is useful to briefly overview the relevant literature on the elements of computational complexity, computer vision, neurophysiology, neuroanatomy, and psychology that are relevant for the remainder of this paper.

A standard text on complexity is (Garey & Johnson, 1979). The time requirements of an algorithm are conveniently expressed in terms of a single variable, reflecting the amount of input data needed to describe a problem instance. A "time complexity function" for an algorithm expresses its time requirements by giving, for each possible input length, an upper bound on the time needed. The emphasis is on worst case measures—at least one instance out of all possible instances has this complexity. An intractable problem is one which no polynomial time algorithm can possibly solve for all instances. If a problem is in the class NP, then there exists a polynomial p(n) such that the problem can be solved by a deterministic algorithm having time complexity $O(2^{p(n)})$. Conversely, all decision problems that can be solved by a nondeterministic polynomial time algorithm are in the class NP. A problem is NP-Complete if it is in the class NP, and it polynomially reduces to an already proven NP-Complete problem. The first such problem is that of "satisfiability" (Cook's Theorem) (Cook, 1971). Problem complexity refers to the lower bound on the average or worst case of a problem over all possible algorithms. On the other hand, algorithm complexity refers to the complexity of a particular algorithm. The work described in the paper is not in either of these classes. The research addresses a sufficient solution and proposes lower or upper bounds on some key parameters, and thus presents an argument rather different than that found in other papers on complexity.

There have been a number of specific concepts proposed by the computer vision community that have proved very useful, although it is not the case that the issue of system architecture has been solved. For a survey of the area of image understanding systems, see Tsotsos, 1987b. Briefly, these key concepts include:

1) the cycle of perception (Mackworth, 1978);
2) processing cones (Uhr, 1972);
3) intrinsic images (Barrow & Tenenbaum, 1978);

4) 2½ D sketch (Marr, 1982);
5) cooperative computation (Hummel & Zucker, 1980); and,
6) model-directed, goal-directed, data-directed control (Tsotsos, 1980; Brooks, 1981; Marr, 1982).

No successful attempts have been made to integrate these concepts into a single framework.

The "biological hardware" that must be used to realize the architectures that are addressed in this paper has specific characteristics. A standard source for basic neuroanatomy and neurophysiology is (Kandel & Schwartz, 1981). The locale of most visual sensory processing is the cerebral cortex. The cortex is composed of two flat sheets of neurons, each about 1000 sq. cm. in area, and is remarkably homogeneous. While the brain has about 10^{12} neurons, there are about 10^{10} neurons in the cortex. Cortical neurons are organized in at least two spatial dimensions: there are six major layers of neurons within the cortex; and, collections of neurons also display a columnar organization orthogonal to the layers. The parameters, or functional significance, of the columns are not well understood. There are about 80,000 neurons per square millimeter of cortex, except in primates where area 17 (V1) has about 200,000 neurons per square millimeter. About 20% of the cortex is devoted to vision; but, many visual neurons are also innervated by other sensory (non-visual) pathways, and this complicates the division between brain areas. The speedup in processing due to parallelism is at least one, but also is surely less than the number of neurons, that is, 10^9. Each neuron can receive input from about 1000 other neurons and can provide output for about 1000 other neurons, on average. The number of fan-out synapses ranges from a few to several thousand, while for fan-in, the range is from a few hundred to a few tens of thousands. Since the system is remarkably fault tolerant, one may hypothesize that no single neuron performs a critical function, but rather, that assemblies of neurons constitute the basic processing units.

What are the characteristics of the visual processing mechanism implemented on this "hardware"? There are a variety of good sources with detailed discussions on the brain and the visual cortex, and the reader is referred to them for a more in depth treatment (Crick & Asunama, 1986; Churchland, 1986; Mansfield, 1982; Cowey, 1979; Barlow, 1981; Stone, Dreher, & Leventhal, 1979; Allman, Miezin, & McGuinnis, 1985; Desimone, Chein, Moran, & Ungerleider, 1985) and the collection of papers in (Rose & Dobson, 1985). It is quite well accepted now that there are several maps in the visual cortex. Each map is a complete representation of visible space, and has its own specific mapping characteristics, seemingly different from the other maps. Cowey proposes a number of reasons for the development of retinotopic maps in the cortex (Cowey, 1979). Firstly, a retinotopic representation in the cortex is needed in order to minimize connectivity lengths between neurons performing processes on the same part of visual space. If they are arbitrarily positioned, connectivity lengths would be much longer, and in addition, the task of wiring itself would be very difficult. Secondly, the reason for requiring more than one map is the same. Not only should neurons performing processes on the same region of space be close to one another, but also, neurons performing processes that deal with the same qualities of visual space should be close to one another. This implies that the segregation of maps is also a functional one. Finally, there are areas that are not discernably retinotopic. Cowey proposes that this must be the case due to the need to communicate with other neural tasks (such as motor control). Perception may be thought of as a combined activity across many neural areas, but then communication to other brain centers would require tremendous bandwidth between areas if the visual area responsible for visual output were retinotopic. If there are single neurons or neural assemblies (i.e., grandmother units) that may be thought of as world-centered "labels" for specific concepts, then communication could proceed in a more economical fashion. That is, the labels form the primitives of the language of communication between neural areas.

There may be 15 to 20 physical visual areas, but only some are organized retinotopically. Since many areas have more than one population of neurons, there are more *logical* maps than physical ones. The boundary between retinotopic and nonretinotopic maps is rather fuzzy, and some maps may be more appropriately described as exhibiting fuzzy retinotopy. The areas commonly accepted as being retinotopic include V1, V2, V3, MT, V4, while those that are more nonretinotopic include IT, posterior parietal cortex, and the frontal eye fields. MT and V4 seem to be the most abstract retinotopic areas. Maps seem to be organized hierarchically, as a partial ordering, such that, generally, the greater the distance from the retina, the physically smaller in spatial extent the maps are, and the larger the receptive fields. Further, there is more than one pathway from the retina to higher levels of processing (see Ungerleider & Mishkin, 1982; Stone, Dreher & Leventhal, 1979).

Hubel and Wiesel are responsible for the discovery that in V1, there is a distinct columnar architecture that seems to have some functional significance, namely that of the hypercolumn (Hubel & Wiesel, 1977). Hubel and Wiesel proposed that in V1 the basic processor unit was the hypercolumn, each containing a complete collection of neurons sensitive and selective for all the basic visual entities. The receptive fields within a hypercolumn all were overlapping and were specific for a given region of visual space. Crossing into a neighboring hypercolumn reveals the same collection of neural sensitivities, but for an adjacent region of visual space. A layer of such hypercolumns may be thought of as representing visual space with a resolution equivalent to that of an image where each hypercolumn is represented by a pixel. It is known that the area of each hemisphere of V1 in humans is in the range 1500–3700 sq. mm., with the average being approximately 2100 (Stensaas, Eddington, & Dobelle, 1974), and that each hypercolumn is approximately 1 sq. mm. in area, Therefore, there are 1500–3700 hypercolumns in V1, or on average, 2100. The other areas have fewer.

According to van Essen and Zeki, (1978), V4 contains at least three separate representations of the visual field, and the columns are about one-quarter the size of the columns in V1. V4 seems about one-quarter the size of V1 (visually inspecting van Essen & Maunsell's [1983] diagrams for monkey). If these ratios translate over to the human visual system as well, then one could estimate that each representation in V4 contains 500–1230 or so columns. MT seems to be less than one-tenth the size of V1, and its columns are also about one-quarter the size of those in V1. Thus, one could estimate that MT contains about 600–1480 columns, if it contains only one visual field representation. A key fact one must keep in mind is the tremendous variability between humans for each parameter (van Essen, Newsome & Maunsell, 1984).

There is much evidence supporting inseparability of early visual operations, thus individual data elements are computed over multiple dimensions. Evidence for retinal measurement inseparability is summarized in (Fleet, Hallett, & Jepson, 1985). A summary of examples of inseparability in other areas is provided in Cowey, 1979. For instance, Zeki (1978) describes the characteristics of neurons in V4 and MT. Zeki found that in area V4, 42.2% of neurons are orientation selective, 4.6% are direction selective, 1.8% are directionally biased, 55.4% are opponent color units, and 12.4% are color biased. By contrast, area MT has 44.8% orientation selective, 88.8% are directionally selective, 11.2% are directionally biased, and effectively no units are color selective or biased. The conclusion is that there are significant numbers of neurons that are both orientation and color selective in V4, while in MT there are many neurons that are both orientation and directionally selective. Similar results exist for many of the other representations. Thus, although a single neuron outputs a single value as a firing rate, that response depends on more than one stimulus quality.

Moran and Desimone (Moran & Desimone, 1985) have discovered that single neurons in trained monkeys, as early as V4 (but not in V1), can be tuned so that separate stimuli within the same receptive field can be individually attended, via topdown control, depending on spatial location and/or stimulus quality. They were motivated by the apparent contradiction that as receptive fields increase in size moving up the hierarchy of processing, they "see" more of visual space, yet information is still filtered out. They first determined which were the effective and ineffective stimuli for a given neuron using bars of various orientations, sizes, and colors. The training required from eight to 16 trials, and the task used was a "match-to-sample" one. Effective stimuli were presented in one part of the receptive field with ineffective stimuli simultaneously at another part. Attention to the effective stimulus lead to strong response, but attention to the ineffective one lead to a significant attenuation of response even though the effective stimulus was still in the receptive field. Attention directed outside the receptive field has no effect on the response of the neuron. The stimuli could be switched from effective to ineffective and back again with correct responses. They claim that unwanted information is filtered from the receptive fields of neurons in extrastriate cortex as a result of selective attention, on either stimulus location and/or stimulus quality. The attenuation was quite pronounced in V4, somewhat smaller in IT, and not found in V1. (We may speculate that the attenuation in IT was smaller because receptive fields in IT are larger and seem to be specific to more varied and complex stimuli.) This is a key experimental result. It points out that the Marr view, that of an early visual system that is purely bottom-up, has no direct biological counterpart, since top-down effects appear so early in the visual system. Also, there is serious impact on the validity of the visual search paradigm for visual primitive determination. This issue will be addressed in more detail below.

There is more evidence for a neurophysiological counterpart to attention. Neural responses in area 7 of the posterior parietal cortex were found to be enhanced if attended, regardless of response behavior, that is, how the stimulus is attended (Bushnell, Goldberg, & Robinson, 1981). Enhancement was spatially selective. Lynch and colleagues (Lynch, Mountcastle, Talbot, & Yin, 1977) claim that there is matching process between neural signals and the nature of objects and the internal goals of the organism, and this matcher determines and sends appropriate trigger signals that direct visual attention. Further, they gathered experimental evidence characterizing the neural mechanism for directed visual attention. They found that it is composed of neurons in the parietal lobe that are preferentially active before and during steady fixations, visually evoked saccades and slow pursuit movements.

The psychology community has developed many theories on attention as well. The timing constraints used in the complexity arguments in this paper are applicable only to so-called pre-attentive vision, a term coined by Neisser (1967). Neisser claimed that perception was a two-stage process: pre-attentive processing, followed by attentive processing. Pre-attentive vision is a process that is claimed to:

1) handle a certain class of visual data completely;
2) be involved in perceptual grouping and texture segregation;
3) yield little precision in feature localization;
4) extract separate feature maps;
5) not handle disambiguation of feature conjunctions; and,
6) be a spatially parallel process.

On the other hand, attentive vision:

1) initiates visual search;
2) can disambiguate feature conjunctions;
3) yields precise descriptions;
4) is directed by pre-attentive grouping;
5) is needed to bring feature maps into register with each other;
6) uses a spotlight of attention that is indivisible, constrained in shape, but variable in size;
7) determines external goal satisfaction; and,
8) is a serial process.

There has been much activity in the psychology community attempting to discover the exact stimulus characteristics that lead to pop-out (Beck, 1982; Julesz, 1978; Nakayama & Silverman, 1986), as well as the stimuli that cannot be recognized completely preattentively (Ullman, 1983).

Treisman and colleagues have enhanced the theory of pre-attentive/attentive vision and have produced a framework they term "feature integration theory", summarized in Treisman (1985). Two main categories of stimuli are used: disjunctive and conjunctive displays. In a disjunctive display, the target is identified by only one feature type, such as color, while in a conjunctive display, the target is defined by more than one, such as color and orientation. A typical disjunctive display could be a field of blue vertical lines, with an imbedded target being a blue horizontal line. A conjunctive display example would be a field of randomly selected colored letters, where the target is, say, a red letter "A". They claim that attention must be directed serially to each stimulus in the display whenever conjunctions of more than one separable feature are needed to correctly characterize or distinguish the possible objects presented. They also address the issue of visual primitives. Treisman claims that color, brightness, terminators, blobs, closure, tilt, and curvature are good candidates, while intersection, juncture, number, and connectedness are improbable ones. Finally, Nakayama and Silverman claim that stereoscopic depth "pops out" when conjoined with color or form or motion (Nakayama & Silverman, 1986), but significantly, the resulting experimental values appear as if two pop-outs occur, in sequence, rather than simultaneously. This leads to a potentially different explanation, namely, that specialized abilities perhaps allow eye movements to focus in at the correct depth plane, and then popout proceeds for the remaining stimulus quality.

THE MODEL OF COMPUTATION

A very simple abstract model will be proposed for the task of visual perception, involving four main elements, keeping in mind that the interest is primarily in the complexity of the task:

- A stimulus array with P elements. This is a retinotopic representation, that is, one whose physically adjacent elements represent spatially adjacent regions in the visual scene.
- At each array element, one or more tokens representing physical parameters of the scene may be computed. These tokens are of a given type, and for each type there are many possible token instances. Types are not necessarily independent. Tokens are distinct from measurements (usually taken to mean the output of some convolution operation), features (usually imply some level of interpretation), and primitives (nondecomposable elements), but are intended to be the elements that comprise the output of early vision. It is thus assumed that the output of early vision is retinotopic. A map is defined as a retinotopic representation of only one type of visual parameter. Maps are logical abstractions, and not necessarily physically separable entities. There are M

maps in the system and the types will be left unspecified and abstract.

- A knowledge base of visual prototypes, each one representing a particular visual object, event, scene, or episode. There are VP of these prototypes. Each prototype may be considered as an invariant description of a visual entity (invariant for size, location, rotation, and other parameters as appropriate).

- A large pool of identical processors, each having the capability of choosing a subset of the stimulus array locations, fetching a subset of the tokens representing physical characteristics at each location, accessing one visual prototype, and then matching the token set to the prototype. Collections of location/token elements are termed receptive fields, and thus, a receptive field is defined as the area of the visual scene in which a change in the visual stimulus causes a change in the output of the processor to which it is connected. The match process is the basic operation of the model. Matching here means that the processor determines whether or not the collection of tokens over the selection of locations optimally represents an image-specific projection of the prototype. This is clearly not a simple task. The output of a processor is match success or failure, with an associated goodness of fit measure. It is assumed that the entire sequence of processing steps, regardless of what they may be, are collapsed into a single processing layer. Each processor completes this operation in PS seconds. The final output of the system is also available in PS seconds, and thus the actual time required for this process does not matter.

The specific representations do not matter for this discussion. The costs associated with this model of computation are in the number of retinotopic elements (P), tokens (M × P), visual prototypes (VP), and processors (PP) and the connectivity costs both in number of connections and their total length.

A time complexity function will be formulated in such a way as to address the number of comparisons of location/token sets to prototypes that need be performed within a single bottom-up processing pass, in the worst case, with no prior information about the scene. It is claimed that the output of a single bottom-up pass through the entire visual system corresponds to pre-attentive vision, and leads to the pop-out phenomenon observed in perceptual experiments for certain stimuli (Treisman, 1985). If a percept "pops out", the perception is immediate and effortless. Note that in these experiments, if a target is not provided, it is not the case that nothing is recognized—access to the subject's entire knowledge base is still required. Recognition in this case may not be complete nor unambiguous pre-attentively, however. Thus, the entire knowledge base of visual prototypes must be included in the analysis that is to be presented. Stimuli are not matched only to targets; targets may be the first candidates for matching, but if they all fail, then other elements of stored visual knowledge are considered. This definition of pre-attentive vision may be expanded by noting that further processing beyond the first bottom-up pass will not yield a different interpretation. A minimal set of optimizations are then introduced to change the architecture of the system so that the timing constraint is satisfied. The implications of the resulting architecture and complexity function are then examined, and lead to many characteristics of primate visual systems as well as to several predictions.

THE NATURE OF THE COMPUTATIONAL TASK

Neisser, among others, claimed that a spatially parallel model of perception is inadequate quantitatively (Neisser, 1967). Neisser was motivated by the fundamental dilemma faced by all theories based on spatial parallel processing: If more than one item of the same kind is present in the visual field, how are they distinguished? In order to deal with the entire visual field at once as well as all the possible interpretations, one requires a much larger brain and too much experience. Neisser's claim is easy to demonstrate. Given VP visual prototypes, P elements of a retinotopic array, and M types representing visual parameters at each array element, then:

$$VP \times 2^{P \times M} \tag{1}$$

operations are required in the worst case. The number of possible subsets of location/type pairs is the powerset of all locations times parameter types. (The null set is included here, but has little effect at this stage of the discussion. It will be deleted later when it will make a difference.) Another possible complexity function would include M as a multiplier of the powerset of locations, rather than in the exponent of the powerset. However, this implicitly makes the assumption that only one type of parameter is required to define a visual entity, and this is true only in very special circumstances. The expression in equation (1) allows an arbitrary subset of parameters to be required for any visual entity. Figure 1 illustrates this configuration. The expression in equation (1) does not enumerate the number of images,

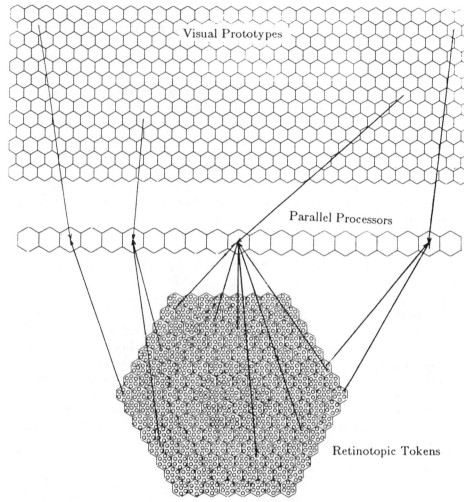

Figure 1. The basic architecture. The retinotopic array is at the bottom, the layer of parallel processors in the middle, and the collection of visual prototypes at the top.

rather it enumerates the number of data items that must be considered and comparisons that must be performed with those data items in the worst case. This is clearly combinatorially explosive. Interestingly, there has been a recent proof that the common "blocks world" problem, addressed by so many researchers in the late 1960s and the early 1970s is inherently NP-Complete in the number of lines (Kirousis & Papadimitriou, 1985). The specific theorems that they proved are: (a) It is NP-complete, given an image, to tell whether it has a legal labeling and, (b) It is NP-Complete, given an image, to decide whether it is realizable as the projection of a scene.

We can demonstrate the implications of this complexity measure by using a few relevant estimates for human vision for the amount of input data and the number of visual prototypes in memory. In the "Visual Dictionary" (Corbeil, 1986), 25,000 items are included pictorially. The world categorized is one of black and white outline diagrams, with little shading, no color, no motion, and no specializations or brand names for common objects. Thus a conservative lower estimate for the number of prototypes is VP = 100,000. A reasonable but arbitrary upper estimate would be VP = 10,000,000, for demonstration purposes. M is surely 1 at the photoreceptors. An upper-bound is rather difficult to estimate; one must answer the question: how many independent parameters are required to describe each point in visual space? Intuitively, there seem to be many: location in three dimensions; wavelength; energy; surface orientation; surface roughness; and a temporal derivative on at least some of these quantities. At the photoreceptors, all of these types are rolled up into a single continuous signal. An upper estimate on M of 12 will be used for demonstration purposes. P is the number of locations in the retinotopic representation. For illustrative purposes, three values will be used, an upper, a middle, and a lower value. The number of receptors in the retina (130,000,000) is the upper value, the number of retinal ganglion cells (approximately 1,000,000 and roughly the same as the number of pixels in a 1K × 1K image) is the middle value, and the size of a 256 × 256 image is the lower value (65536 pixels). It will become apparent that the particular choices for these parameters have no effect on the general conclusions, and it must be emphasized that the numerical choices for these parameters are for demonstration purposes only.

A time complexity function for a task expresses an upper bound on its time requirements by giving for each possible input, the largest amount of time needed, in terms of the input length. A priori, there is no way to predict which portions of the visual field will represent an image-specific projection of a given prototype, and thus in the most simple brute-force algorithm, a single processor in the worst case must consider each receptive field against each stored prototype, and in the average case, half that number of comparisons. It should be obvious that a parallel scheme requires much serious consideration of the problems of communication, shared resources, synchronization, task scheduling, etc. It is assumed that they can be resolved—there would be no impact on the results claimed in this paper. However, the effective speed-up due to parallelism is clearly smaller than the number of available processors.

Given PP as the degree of effective speed-up due to parallelism, then the amount of time taken to perform the worst case number of operations as presented in equation (1) is given by:

$$PS = \frac{2^{P \times M} \times VP \times PS}{PP} \tag{2}$$

where each processor requires PS seconds to complete one operation and the output of the system is also available in PS seconds. Table 1 gives values for PP for the estimates on P, M, and VP described above. The inescapable conclusion is that with this simplified architecture, the task is intractable: *parallelism alone is not the answer.* Although the problem cannot be solved without parallelism, it is interesting to note that Feldman and Ballard claim that massive parallelism is sufficient to satisfy the timing constraints. Moreover, it is easy to show that this problem is in the class NP. A nondeterministic scheme would simply guess which of the possible input data subsets would lead to a non-null search problem solution.

If a task is computationally intractable, then the only realizable solutions are approximating ones. Kirousis and Papadimitriou—even though they are not vision researchers—recognized the apparent contradiction contained in their theorems. Biological vision is an existence proof and thus, they claim, one of two possibilities arise: (a) vision is easier since other cues such as color can be used or (b) the probabilistic distribution of real scenes is biased for the development of ingenious fast algorithms. The first speculation of Kirousis and Papadimitriou is easily dismissed. If a richer world is considered, then not only are more cues available, but the space of possibilities is also dramatically increased, and thus the addition of other cues can only worsen the tractability of the task. This, of course, assumes that no information is available a priori. A drastic improvement may be possible for specific situations when the view-

Table 1. **Values of PP for Varying Values of P, M, and VP for the Basic Architecture.**

PP	VP = 10⁵		VP = 10⁷	
P	M = 1	M = 12	M = 1	M = 12
13,000,000	$10^{39,133,905}$	$10^{469,608,805}$	$10^{39,133,907}$	$10^{469,608,807}$
1,000,000	$10^{301,035}$	$10^{3,612,365}$	$10^{301,037}$	$10^{3,612,367}$
65,536	$10^{19,733}$	$10^{236,744}$	$10^{19,735}$	$10^{236,746}$

er has some knowledge of what to expect. The second claim is more believable, yet seems to be very difficult to prove. There are two more views possible on the nature of approximating solutions. One approach is to search for polynomial time algorithms for various specific visual computations (see, for example, Poggio, 1982 and Mackworth & Freuder, 1985). Although progress has been made in this direction, we are far from the development of such an algorithm for the entire problem of vision. Finally, remember that complexity measures reflect worst case situations. Suppose the brain is large enough to handle the sizes of problems that normally occur in the real world, and is designed such that performance degrades gracefully for the more complex situations. Then, one may ask the question "How large a problem can the brain handle?" In part, this question motivated the approach in this research. Put differently, "What are the limits of a bottom-up single pass early vision process?" Only by first answering this question can the computational need for attention be justified and a strategy for attentive processing be developed.

In formulating an answer for this question, one must employ some criterion for deciding between competing configurations. In computer science, "computing power" is commonly used to rate various computer systems. This is defined as the number of operations performed per second. A similar decision principle can be stated for choosing "best" configurations for vision systems:

The Maximum Power/Minimum Cost Principle

The power of a pre-attentive vision system is defined as the amount of data that may be processed per degree of parallelism (or per processor for simplicity), within a single bottom-up pass. Power is increased by increasing VP, M, or P or by decreasing the required degree of parallelism PP. The cost of a system is a function of the number of units allocated for the maps, the number of processors, the required fan-in and fan-out, the number of total connections, and the total connection length. Preferred configurations are those that maximize power while minimizing cost. The goal is to maximize the richness of the visual world that is immediately accessible to the system, within the hardware constraints.

DEMONSTRATING COMPLEXITY SUFFICIENCY

A time complexity function has been formulated for a brute force architecture attacking the first, bottom-up pass of visual information processing. The goal now is to discover a sufficient set of global optimizations so that a biologically plausible architecture is obtained, that yields a sufficient, but not necessary, solution to the timing constraints. Then, space complexity considerations, using connectivity, will be presented that lead to specific predictions on the size of problem that human vision solves. The side effects of the complexity considerations will also be presented. The result will be an argument supporting computational modeling of human vision, and an argument for a sufficient architecture for biologically motivated designs of vision systems.

Biologically plausible values for PP, P, and M are:

- For PP: The speed-up due to parallelism is clearly at least one, but it surely cannot be as large as the number of neurons in the brain, 10^{10}. Realizable parallel processing systems require considerations of local memory, synchronization, communication, and so on, and presumably, a collection of neurons is required to accomplish this for each degree of speed-up. Since about 20% of the cerebral cortex is devoted to visual processing, the value of PP that is biologically plausible is significantly less than 10^9.
- For P: Stensaas and colleagues measured the size of striate cortex (V1) in humans, and found that its extent averaged approximately 2100 sq. mm., and ranged from 1500–3700 sq. mm. for each hemisphere (Stensaas et al. 1974). Hubel and Wiesel claimed that the basic processing unit within the visual cortex is a hypercolumn, a localized collection of neurons, organized in columns, providing a complete set of processors for orientation, color, motion, etc. (Hubel & Wiesel 1977). The receptive fields within each hypercolumn are all localized to the same region of visual space, thus the representation is retinotopic. Since a hypercolumn is approximately 1 sq. mm. in extent, then V1 contains on average 2100 hypercolumns. Extrastriate areas are smaller, and have smaller hypercolumns (where hypercolumns have been found). It is assumed that the output of early vision corresponds to the output of the most abstract, retinotopic extrastriate areas. If, in an abstract sense, the elements of the retinotopic representations in this paper are equated with hypercolumns with respect to map resolution, then acceptable values for P are those less than 2100.
- For M: According to van Essen and Maunsell, the division between retinotopic and non-retinotopic areas, although fuzzy in general, may be placed after areas MT and V4 and before IT, area 7, and the frontal eye fields (van Essen & Maunsell, 1983). Thus, at the assumed output of early vision, V4 seems to have three sepa-

rate representations of visual space, and MT has one (van Essen & Zeki, 1978). Since the total number of visual areas is on the order of 20, and only some are retinotopic, and each may have several representations, acceptable values for the number of representations that comprise the output of early vision, that is M, are less than 20.

Additional Assumptions

Hexagonal images are assumed, packed with hexagonal pixels, of order N (i.e., N pixels per side). A hexagonal tiling of a hexagonal image was chosen for convenience, and for the resemblance to the retinal mosaic; however, much of the discussion is independent of the choice of image mosaic. Whenever the choice does have an impact on the results, it will be pointed out. The number of pixels in such a hexagonal image is $P_N = 3N^2 - 3N + 1$. In these hexagonal images pixels are uniformly distributed across the image. All receptive fields and prototypes are hardwired to the processors, and all data from a receptive field can be accessed simultaneously, as is also true for all data associated with a visual prototype. Connectivity will be considered in terms of other units, not synapses. All maps are assumed to be the same size.

A Sufficient Set of Optimizations

This section will develop a small, sufficient set of optimizations that will lead to a biologically plausible function relating P, PP, VP and M. The development can be likened to "a back of the envelope" computation, with estimates for variable values used only to guide the development. Once the function has been determined, conclusions will be drawn about the values of some of the variables.

Efficiency can be gained by attacking the prototype search through a process of successive refinement. This is quite a standard tactic in AI-"divide and conquer", has been used in wide varieties of AI systems. As used here, the idea is similar to the perceptual "20 questions game" described in (Richards, 1982). Assume that we can build a binary tree whose leaves are the prototypes of the knowledge base, and whose nodes are superclasses of prototypes. This is not unlike the specialization or decomposition hierarchies found in the knowledge representation literature. Note that although a binary tree search is serial in nature, the key here is the number of operations, and the search will be "parallelized" later in the paper. Therefore:

$$PP = 2^{P \times M} \times \log_2 VP \tag{3}$$

This is a very minor improvement. On its own, the standard technique employed by all knowledge-based vision systems, namely prototype organization, is at best a small, (but crucial) contributor in defeating the complexity problem of vision. Note that although a

Organized Visual Prototypes

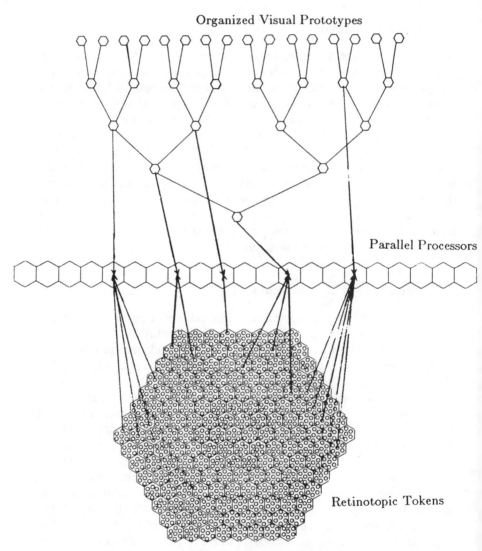

Parallel Processors

Retinotopic Tokens

Figure 2. Architecture #2. The visual prototypes are now organized into a tree, thus reducing prototype search time logarithmically.

hashing scheme would be even faster, it would not be sufficient on its own to make a difference, and further, it is not clear what biologically plausible mechanism could implement hashing. This architecture is shown in Figure 2.

A critical observation on the physical world is that it is not the case that all 2^P possible combinations of locations are meaningful and thus reasonable to consider. Objects are not spread arbitrarily in 3-space, and events are not spread arbitrarily in the time dimension. Their physical characteristics are also similarly localized. Assuming a hexagonal image of order N, and that only hexagonal contiguous regions of whole array elements are considered as processor receptive fields, then some simple geometry yields N^3 receptive fields over the whole image, or in pixels, approximately,

$$\frac{P^{1.5}}{3\sqrt{3}} + \frac{P}{2} + \frac{5\sqrt{P/3}}{8} \tag{4}$$

Only this number of receptive fields need be considered. Figure 3 illustrates the receptive field structure. The degree of speed-up func-

Figure 3. The hexagonal retinotopic stimulus representation. The hexagon is of order N, that is, N elements per side. The diameter of the hexagon is 2N − 1 elements. For three of the elements, the corresponding complete set of receptive fields that may centered on those elements are shown. Each element in the hexagon can be the center of a number of hexagonal receptive fields in this manner. In total, there are N^3 receptive fields.

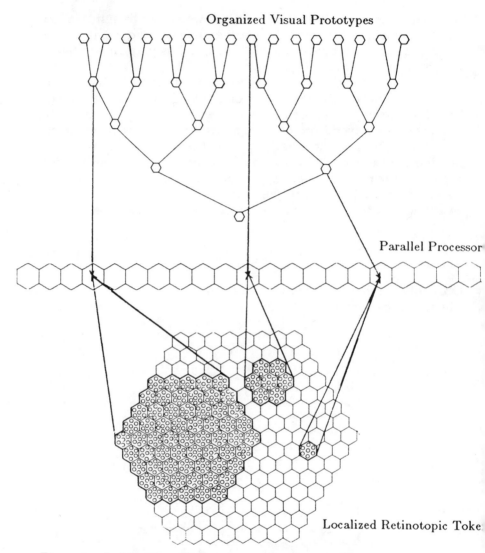

Figure 4. Architecture #3. Processors choose spatially contiguous receptive fields for input, as shown in the lower portion of this figure.

tion for this third architecture (shown in Figure 4), is dramatically different:

$$PP = N^3 \times (2^M - 1) \times \log_2 VP \qquad (5)$$

The powerset of maps still remains in the expression because a priori, it is not known which subset of maps is the correct one for a best image to prototype match, and therefore in the worst case, all subsets must be examined. The null set has been removed however, since it may have a numerical effect. Table 2 gives the value for PP resulting from this expression. Although there has been significant change in the estimated degree of speed-up, the values are still not close to biologically plausible values. One important consequence of the localization of receptive fields is that no comparative relations between receptive fields may be computed. This is not worrisome since it has been observed in humans that the determination of spatial relations requires serial processing, and is not a pre-attentive ability (Ullman, 1983), and thus this optimization leads to an implication consistent with the observations. Another side-effect of this particular receptive field structure is that it does not permit as fine a selection of tokens across the receptive field as the first expression (equation (1)). In equation (1), some of the subsets could indeed represent contiguous space, but the powerset of elements implied that over a contiguous space, each element could be a different type of parameter. The new definition of receptive field requires that tokens for each selected type of parameter are used for each location across the receptive field. This too is reasonable, since visual parameters display the same localization as the objects which exhibit them.

As a third potential optimization, we note that it is not the case that all visual stimuli involve all types of tokens. Let $\hat{M}$ represent the number of types of visual parameters that are relevant for a given input. Thus, the number of possible subsets of types is $2^{\hat{M}} - 1$. This could be implemented via a computation of *pooled response*, that is, an output associated with each map that signals whether or not the map has been activated. The idea is borrowed from (Treisman, 1985). A direct result is the logical segregation of types, an idea that arose in the "intrinsic image" theory of (Barrow & Tenenbaum, 1978) and also in the "feature integration" theory of Treisman. Physical separation of types into physically distinct maps follows if connectivity lengths are considered. Cowey presents this reason for the evolution of physically separate visual maps: units that compute similar quantities need to communicate with one other for consistency purposes and thus need to be connected to one another (Cowey, 1979). The

Table 2.   PP for Varying Values of P, M, and VP for Architecture 3.

PP	VP = 10^5		VP = 10^7	
P	M = 1	M = 12	M = 1	M = 12
13,000,000	$10^{12.68}$	$10^{16.29}$	$10^{12.82}$	$10^{16.43}$
1,000,000	$10^{9.5}$	$10^{13.12}$	$10^{9.65}$	$10^{13.26}$
65,536	$10^{7.73}$	$10^{11.35}$	$10^{7.88}$	$10^{11.49}$

connectivity lengths would be prohibitive if the units were separated. The architecture including logically separable maps is shown in Figure 5. The new expression for speed-up is:

$$PP = N^3 \times (2^{\hat{M}} - 1) \times \log_2 VP \qquad (6)$$

The values for $\hat{M} = 1$, the simplest input, are found in Table 2 in the $M = 1$ column. Even for the smallest image, the values of PP are barely biologically plausible. Therefore, since pooled response and

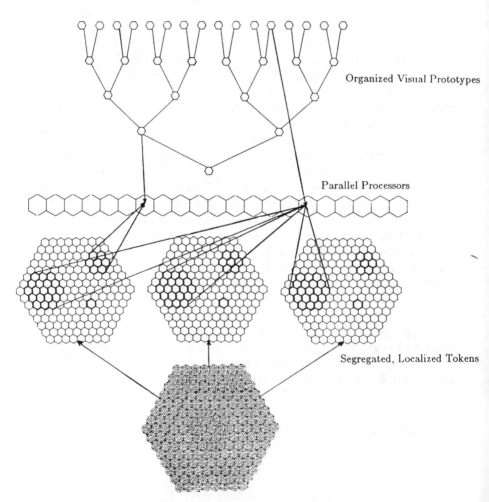

Figure 5.   Architecture #4. Maps are logically segregated and significant activity in a given map is signaled by a pooled response.

map segregation does not lead to savings for all possible images, one may speculate that the role is that of speeding up the computation for the simpler inputs (the simplest and therefore fastest situation being for $\hat{M} = 1$), thus minimizing response time for simple inputs, and may be considered a mechanism for the system's graceful degradation with increasing complexity of the input. The issue of the number of maps and their functionality will be further elaborated in a later section.

Further efficiency could be gained by trading off precision, and this constitutes the only optimization that causes a degradation in the fidelity of the incoming signal. This can be achieved by reducing the resolution of the visual image, and simultaneously, abstracting the input in order to maintain its semantic content. The "processing cone" representation of Uhr (Uhr, 1972) has the right flavor, but does not include the proper semantic abstraction. Abstraction implies that some data is lost, and thus, the "filter" of attention theories has a strong computational counterpart.

Let $\hat{N}$ be the size of the new abstracted array. What is the largest array that leads to complete inspection within the time constraint? The expression for degree of speed-up is changed to:

$$PP = \hat{N}^3 \times (2^{\hat{M}} - 1) \times \log_2 VP \qquad (7)$$

If VP is set to 10,000,000, $\hat{M}$ to 1, and PP to 1,000,000, then from this equation, $\hat{N}$ is 35, and $P_{\hat{N}}$ is 3571. For VP = 100,000, $\hat{N}$ is 39, and $P_{\hat{N}}$ is 4447. It is easy to see that variations in PP, $\hat{M}$ and VP lead to changes in $P_{\hat{N}}$, and that there are a great many possible configurations that lead to values of $P_{\hat{N}}$ that are less than 2100. This, then, is the satisfying architecture and is illustrated in Figure 6.

Exploring further, more insights can be obtained from equation (7). Figure 7 shows a family of curves of this relationship for $P_{\hat{N}}$ vs. $\log_{10}PP$ for values of $\hat{M}$ ranging from 1 through 10, and for VP = 100,000 through 10,000,000. Thus, the thick solid curves, one for each value of "$\hat{M}$, represent the family of curves for the same value of $\hat{M}$ for all values of VP between 100,000 and 10,000,000. Qualitatively, several conclusions can be drawn, that are also verified analytically. If these are the basic performance relationships, then the designer of the visual system is faced with a few choices and tradeoffs. First of all, there seems to be a 'hard complexity wall' on the number of processors. It is very cheap in terms of processors to incorporate a very large knowledge base of prototypes, as is clear from Figure 7. Changes in VP have a very small-effect on PP, as can be easily seen from the partial derivative, $\dfrac{\partial PP}{\partial VP} = PP \times \dfrac{\log_2 e}{VP \times \log_2 VP}$. It is more ex-

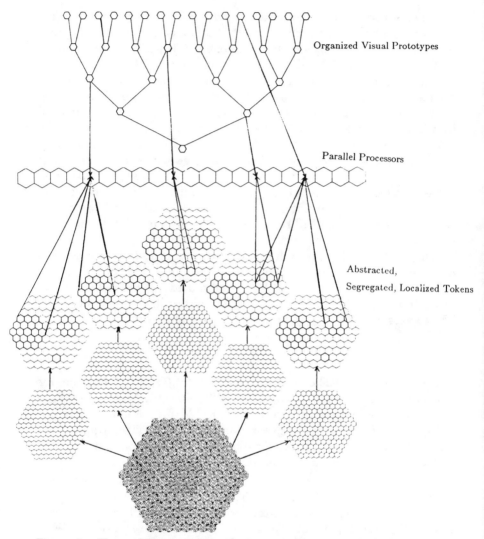

Figure 6. The architecture that satisfies the timing constraint for a single bottom-up processing pass. The binary tree knowledge base of prototypes is at the top. The processor layer is in the middle, each processor accessing one visual prototype and one receptive field from one or more maps. The lower half of the diagram depicts the early abstraction hierarchy, with logically separated maps, and decreasing resolution.

pensive to use larger maps, since $\dfrac{\partial PP}{\partial \hat{N}} = PP \times \dfrac{3}{\hat{N}}$. The largest expense is incurred for adding maps, because $\dfrac{\partial PP}{\partial \hat{M}} = PP \times \dfrac{2^{\hat{M}} \times \log_e 2}{2^{\hat{M}} - 1}$.

If, for example, VP = 10,000,000, PP = $10^{5.6}$, $\hat{M}$ = 1, an $\hat{N}$ = 26, then the derivative of PP for changes in $\hat{M}$ is 12 times steeper than for changes in $\hat{N}$ and 223,000,000 times steeper than for changes in VP. Thus, of the three variables, it is most critical that the value of $\hat{M}$ be set as low as possible.

Although equation (7) may lead to 'ballpark' figures, it still remains to determine reasonable estimates for the configuration of the visual system. The Maximum Power Principle can be used at this point to guide the search among all the reasonable values. A simple objective function may be formulated to embody some of the con-

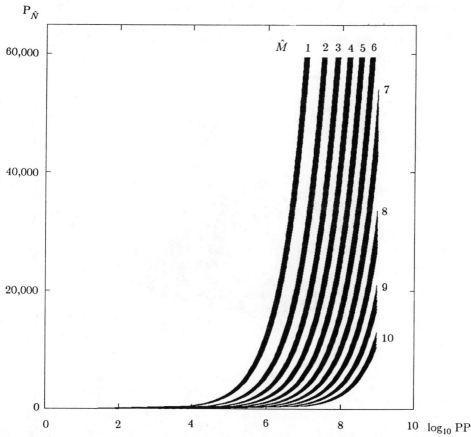

Figure 7. The family of curves generated using equation (7) for varying values of PP, VP, $\hat{M}$, and $P_{\hat{N}}$. $P_{\hat{N}}$, ranging from 0 to 60000, is plotted against $\log_{10}$ PP, ranging from 1 to 9. Each thick solid curve represents a value of $\hat{M}$, with $\hat{M}$ = 1 the leftmost, and $\hat{M}$ = 10 the rightmost. The thickness of each curve represents that fact that within it is the entire range of VP, from 100,000 to 10,000,000.

straints of the principle. A relationship is defined that expresses the amount of data that must be processed in a single bottom-up pass, in the worst case. A different measure is required here, rather than using the one that has been the basis of all the analysis so far, since by definition, each processor is allowed time to perform only one of the basic matching operations defined earlier. Also, this measure must be totally divorced from the algorithm or organization used by the vision system in order to be an appropriate metric for comparison purposes. The input to the system is an array of $P_{\hat{N}}$ elements, $\hat{M}$ types of tokens at each elements. There are $P_{\hat{N}} \times \hat{M} \times VP$ data 'chunks' in total for the worst case. Or, in other words, this is the amount of raw data that must pass through the processors in the worst case. The greater the amount of data that each processor can process, the greater the system's power. The system power, then, may be expressed as:

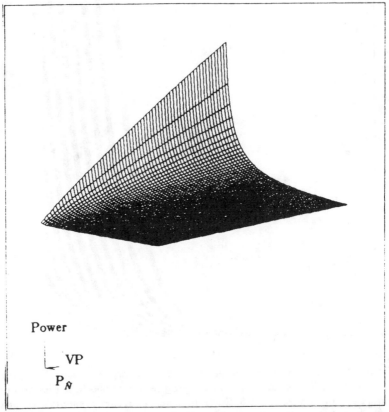

Figure 8.   The Power function: VP ranges from 10,000 to 640,000, and $P_{\hat{N}}$ ranges from 100 to 6400. $\hat{M} = 1$.

$$\text{Power} = \frac{P_{\hat{N}} \times \hat{M} \times \text{VP}}{\text{PP}} \tag{8}$$

where PP is defined in equation (7). A system can increase its power by increasing input array size, the number of maps, and/or the size of the knowledge base and/or decreasing the speed-up required to satisfy the time constraints. This is precisely what the maximum power principle requires. Figures 8, 9, and 10 show the power profiles in the three dimensions of interest. The best configuration satisfying the requirements of immediate perception is the one that maximizes the power principle. Unfortunately, there is no global maximum. (See Figures 8, 9, and 10.) System power increases with increasing values of VP, and thus it can be concluded that VP should be as large as possible. Power decreases with increasing values of $P_{\hat{N}}$ as it does with

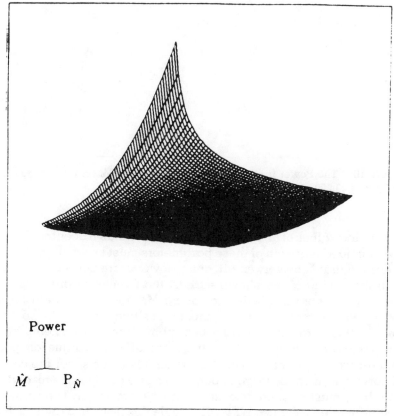

Power

$\hat{M}$ $\quad$ $P_{\hat{N}}$

Figure 9. The Power function: $\hat{M}$ ranges from 1 to 64, and $P_{\hat{N}}$ ranges from 100 to 6400. VP = 1,000,000.

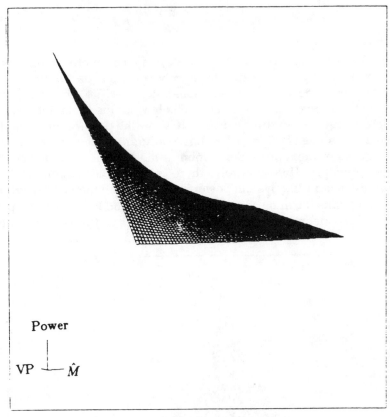

Power

VP ⊥ $\hat{M}$

Figure 10. The Power function: $\hat{M}$ ranges from 1 to 64, and VP ranges from 10,000 to 640,000. $P_{\hat{N}} = 1000$.

increasing values of $\hat{M}$, but much more slowly than it does for $\hat{M}$. It can be concluded that both of these parameters must be small. Further, it is critical that $\hat{M}$ be very small, since power decreases exponentially with increasing $\hat{M}$. As shown earlier, the number of processors increases exponentially with increasing $\hat{M}$, much faster than for the other two parameters. Since $\hat{M}$ must a positive integer, the best setting for this parameter in order to minimize the requirements for PP and to maximize power is $\hat{M} = 1$. It is more difficult to argue for specific settings of $P_{\hat{N}}$ since the constraints on this value so far are contradictory: a) $P_{\hat{N}}$ must be large enough to ensure good image resolution; and, b) $P_{\hat{N}}$ must be small to ensure high system power. The minimum cost principle for connectivity will decide this issue in a later section.

IMPLICATIONS AND PREDICTIONS

Using the basic conclusions of the previous section, a number of characteristics may be derived for the resulting architecture. Some are confirmed by the known neuroanatomy of primate vision, and others will stand as predictions. The architecture, guided by the philosophy of the complexity level and the maximum power / minimum cost principle, exhibits columnar processor organization, and inverse processor layer magnification if implemented on "flat" hardware. Connectivity constraints predict the average sizes of the retinotopic maps, the number of maps that comprise the output of early vision, and the degree of required speed-up due to parallelism.

Decreasing System Speed with Increasing Data Complexity

From equation (7), and the conclusions of the previous section, a statement may be made about the speed of processing for increasingly complex input data. The simplest input data situation, that is, $\hat{M} = 1$, also corresponds to the maximum power situation, and fixes the degree of parallelism for the system, to:

$$PP = \hat{N}^3 \times \log_2 VP \tag{9}$$

and thus, the time for computation is T_{min}, given by:

$$T_{min} = \frac{\hat{N}^3 \times \log_2 VP \times PS}{PP} = PS \tag{10}$$

As $\hat{M}$ increases, the time to compute increases exponentially with $\hat{M}$, up to a maximum given by T_{max}, when all maps (M) are active:

$$T_{max} = \frac{\hat{N}^3 \times (2^M - 1) \times \log_2 VP \times PS}{PP} = T_{min} \times (2^M - 1) \tag{11}$$

The progressive increase in base computation time for increasingly complex preattentive tasks has been experimentally documented in (Nakayama & Silverman 1986). However, the exponential relation is not immediately apparent. It is rather difficult to examine this from reaction time comparisons, and in fact, there may be other optimizations that play a role in reducing this exponential increase. Obviously, the expression in equation (11) gives the computation time for the worst case with respect to searching the entire set of map subsets. The average case, which is what the visual search experiments of Nakayama and Silverman expose, requires half the computation time. Thus, the average computation time, if $\hat{M}$ maps are active is given by:

$$T_{ave} = \frac{\hat{N}^3 \times (2^{\hat{M}} - 1) \times \log_2 VP \times PS}{2 \times PP} = T_{min} \times \frac{(2^{\hat{M}} - 1)}{2} \qquad (12)$$

A further optimization may be that map subsets are internally ordered, and this would also lead to a speedier response. Finally, there is the issue of discriminability along the stimulus quality dimension. The closer two tokens along a given type dimension are, the more difficult it will be to separate them perceptually. For example, pop-out is faster for a display where the target is red and every other element of the stimulus is black than for a display where the target is orange and the distractors are red or yellow. It seems that arbitrary amounts of computation are required to distinguish tokens along the same dimension depending on their perceptual similarity along that dimension. These are issues for further consideration and are not directly addressed by the theory in this paper.

Columnar Processor Organization

The question addressed by this section is "How are the processors connected to the retinotopic maps?" At each array element of the most abstract maps, we can define a *processor assembly*. A processor assembly contains, on average, $PP/P_{\hat{N}}$ processors. The number of processors, in the best configuration for immediate perception derived earlier, is given by setting $\hat{M}$ to 1 in equation (7). Using this, and the expression for $P_{\hat{N}}$ in terms of $\hat{N}$, the number of processors in an assembly is:

$$\frac{\hat{N}^3}{3\hat{N}^2 - 3\hat{N} + 1} \times \log_2 VP \qquad (13)$$

But, $\dfrac{\hat{N}^3}{3\hat{N}^2 - 3\hat{N} + 1}$ is the average number of processor receptive fields

at each location. Thus, there are $\log_2 VP$ processors for each receptive field at each location. Call this set of processors a *receptive field assembly*; this will be the basic processing unit for the remaining discussion. Each of the receptive field assemblies must be connected to their relevant retinotopic elements, and stacking the assemblies over the centers of their receptive fields minimizes connection length. The proof is straightforward. Assume a one-dimensional receptive field, whose center is at position Y, and whose rims are at positions Y + (K + 1)/2 and Y − (K + 1)/2. Thus, the diameter of the receptive field is K, an odd integer, and this is the number of units to which each processor must be connected. The total length of all connections for a single processor processor to this receptive field can be expressed by:

$$\sum_{x=Y-\frac{K+1}{2}}^{Y+\frac{K+1}{2}} \sqrt{1 + (loc - x)^2} \tag{14}$$

It is assumed that processors are unit distance above the stimulus array, but this does not affect the result. loc is the location of the processor and could take values between 1 and K. This function is minimized when loc = Y. Thus, in the one-dimensional case described above, placing the processor over the center of its receptive field minimizes total connection length for those connections. The same is true of the two-dimensional case, since the situation is circularly symmetric. Thus it follows that for one layer of processors, the configuration with minimal total connectivity is one where each processor is placed directly over the center of its receptive field. If there is more than one layer of processors, as is true in this situation, the same conclusion is reached. More than one processor cannot occupy the same physical space. If a layer is configured so that the processors are over the centers of their receptive fields, then the remaining processors must be placed above or below this layer. Then, the same argument applies—the minimum connection length for this next layer of processors is achieved if the processors are centered over their receptive fields. This procedure is applied until all processors have been allocated.

There is a column of processor assemblies for each retinotopic element (or pixel), and within the column there is a receptive field assembly for each of the receptive fields centered on that pixel. Figure 11 illustrates the organization of processor assemblies. This structure is not unlike that of Hubel and Wiesel's hypercolumns in an abstract sense. In principle, if the decision criteria for branching in the knowledge base search are known, and one branch decision does not depend on the previous decision, then the processors can categorize each receptive field, in parallel, in one time step, since

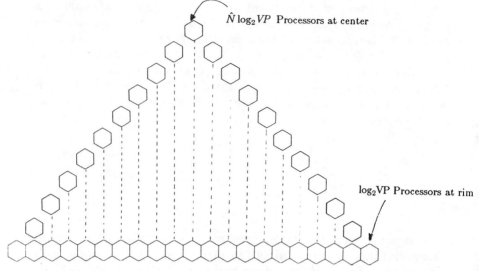

Figure 11. Processor columnar organization, into a cone of columns.

there is one processor for each of the $\log_2 VP$ branches, for each receptive field. The result of each receptive field match would be available at the outputs of the corresponding receptive field assembly, and the pattern of responses within a receptive field assembly points to the most appropriate prototype (matched pre-attentively). This is one way in which the serial nature of binary search is 'parallelized'. The center pixel requires $\hat{N} \log_2 VP$ processors (or $\hat{N}$ receptive field assemblies), while the pixels on the rim require $\log_2 VP$ processors, (or 1 receptive field assembly).

Processor Layer Inverse Magnification Factor

The cortex is flat, yet the columnar processor organization described above is conical, having a peak in the center. In order to implement such a structure on "flat" hardware, and maintain at least some of the connectivity benefits that arise through stacking the processors over their corresponding receptive field centers, the solution is to flatten the set of processors as if pressing down on the cone's peak, and redistributing the elements locally, as if the cone were made of putty. This leads to an inverse magnification, that is, more cortical area is devoted to central visual fields per unit than in the periphery. A magnification function may be obtained by dividing the number of processors at a given radius R in the conical model by the average number of processors at each location over the whole cone. The value of R at the center is 1, while its value at the rim of the stimulus array is $\hat{N}$. Thus, the area magnification function is:

$$\frac{2(\hat{N} - R + 1)\log_2 VP}{(\hat{N} + 1)\log_2 VP} = 2 - \frac{2R}{\hat{N} + 1} \tag{15}$$

Daniel and Whitteridge (Daniel & Whitteridge, 1961) measured cortical magnification factors for the monkey, and discovered an inverse relationship between the location of a receptive field in the cortex and the corresponding location on the retina. This relationship was measured in terms of the amount of distance across the cortex that must be traversed in order to achieve a one degree traversal in visual space on the retina. A fit to the data was done by Schwartz (1977), who found the following relationship:

$$M = \frac{6}{R^{0.9}} \tag{16}$$

where M is magnification in millimeters of cortex per degree of visual eccentricity and R is degrees of eccentricity. The predicted magnification function and the experimental one are shown in Figures 12 (a) and (b) respectively. These expressions are not directly comparable, since the dimensions are all different. Moreover, the model in this paper assumed uniform pixel distribution, whereas the retina does not have uniform photoreceptor distribution, nor is the distribution of receptive fields uniform. Nevertheless, it is interesting that receptive field localization, and conservation of connection lengths, alone, lead to the negative slope across a flat processor layer.

An even more interesting conclusion however is revealed if retinal receptor distribution is taken into account, and the predicted inverse magnification function (equation (15)) is used as a filter for the receptor distribution function, and the result compared to the experimental findings. Figure 13 portrays the distribution profiles of rods and cones across the retina, from zero to 70 degrees eccentricity. The predicted function was then used as a filter for the rod density, the cone density, and the sum of the two profiles. The filtered cone profile, shown in Figure 14, is virtually identical in shape to the experimental magnification curve in Figure 12 (b). The result for rods and for the sum of rods and cones is significantly different. One may speculate (probably pre-maturely) that the organization of receptive fields is dictated by the distribution of cones, and that rods play a subservient role to the cones.

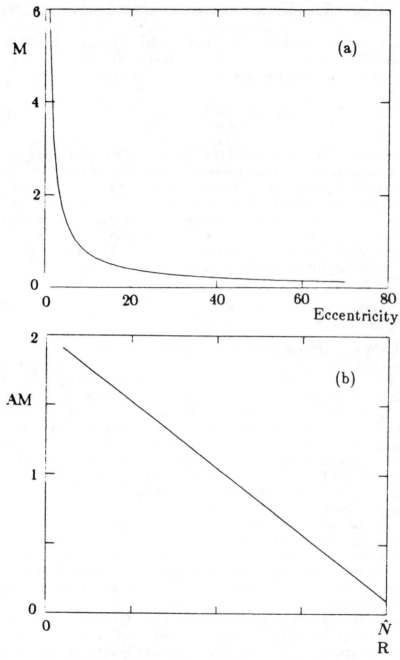

Figure 12. (a) The magnification curve determined by Schwartz, the ordinate representing magnification in millimeters of cortex from 0 to 6, and the abscissa representing degree of retinal eccentricity from 0 to 70 degrees. (b) The predicted magnification function, in equation (15), the ordinate representing area multiplicative factor from 0 to 2, and the abscissa representing radius of the processor array, from 0 to $\hat{N}$.

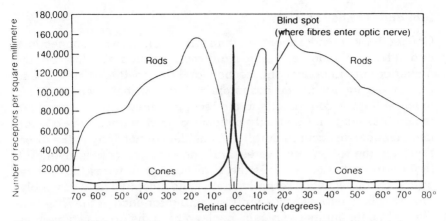

Figure 13. The spatial distribution of rods and cones in the retina.

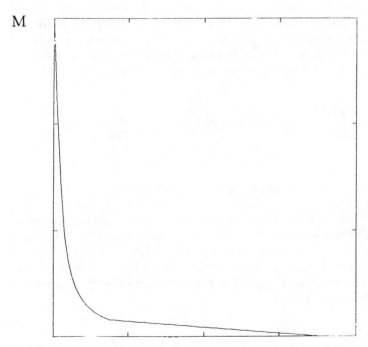

Eccentricity

Figure 14. The resulting cortical magnification obtained by filtering the cone distribution in the retina from Figure 13 through the predicted magnification due to processor organization in Figure 12 b). The units are not meaningful since the appropriate mappings from area to linear magnifications have not been done.

Size and Number of Maps

Connectivity considerations, both in terms of number of connections and in terms of lengths, lead to many more predictions. Note that the numerical predictions of this section (and this section alone) depend on the hexagonal image assumption. Each of the receptive fields must be hard-wired directly to the receptive field assemblies: there is no other way that a strictly bottom-up pass, without any a priori knowledge, can occur in parallel. Consider connectivity in the direction from the retinotopic representations to the processor layer. The first quantity to determine is the total number of wires that are required to connect each receptive field to its receptive field assembly. This will be computed by simply summing for each of the $\hat{N}^3$ receptive fields, its number of pixels. Each point in the array is a member of a ring of points, each point of which is the center for the same number and sizes of receptive fields. The number of elements of each ring, at radius i is given by $P_i - P_{i-1}$. The receptive fields that are centered by each member of the ring are of sizes 1 through $\hat{N} - i + 1$. The sum of the elements in each of these size receptive fields for each element of each possible ring then gives the total number of wires required to hard-wire all the receptive fields, independently of one another. The following expression accomplishes this:

$$\sum_{i=1}^{\hat{N}} \left\{ (P_i - P_{i-1}) \sum_{j=1}^{\hat{N}-i+1} P_j \right\} = \frac{\hat{N}}{10} (3\hat{N}^2 + 1)(\hat{N}^2 + 1) \tag{17}$$

Call this the area (A) of each map. The total number of wires is $A \times M$, and the average fan-out from the retinotopic representations is $(A \times M)/(P_{\hat{N}} \times M)$, or $A/P_{\hat{N}}$. At the receptive field assemblies, the average fan-in from the retinotopic representations is the total number of wires divided by the number of receptive field assemblies, or, $A \times M / \hat{N}^3$. Now if we use the biological constraint of fan-out and fan-in for neurons of approximately 1000, these two connectivity expressions yield predictions of:

$\hat{N} = 21$ or 22
$P_{\hat{N}} = 1261$ or 1387
$M = 7$ or 6
Average Fan-out = 975 or 1118
Average Fan-in = 929 or 874
PP: for VP = 10^7: PP = $10^{5.33}$ or $10^{5.4}$; for VP = 10^5 PP = $10^{5.19}$ or $10^{5.25}$

Each of these values is completely consistent with the biologically plausible ranges described earlier, for areas MT and V4. Since each map was assumed to be the same size, $P_{\hat{N}} \times M$ is probably a better prediction for total number of data items in the output of early vision. It is more correct, however, to note that the values for M and $P_{\hat{N}}$ are lower bounds only and that the values for PP are thus upper bounds only. There may in fact be other connectivity optimizations that are present that would permit a larger number of elements to be connected. A simple calculation would also reveal that connectivity considerations would predict that at least three layers are required in the input abstraction hierarchy to go from a retina of 130,000,000 receptors to a map of size 1300.

There are no connections from the processors to any of the larger maps in the input abstraction hierarchy. The number of such connections would be prohibitive. A back of the envelope calculation can help here as well. Suppose that the processors are to be connected to M maps of high resolution, say 1K by 1K. We can use the formulae developed earlier for receptive field fan-in to the processor layer, but this time, P = 1,000,000. The resulting additional number of connections per map would be approximately 10^{13}. The additional average fan-in at each receptive field assembly, if PP is on the order of 10^5, is on the order of 10^7. This calculation could be repeated for each of the layers of resolution as well. Given that the cortex contains 10^{10} neurons, with an estimated total connections of 10^{13}, this is clearly not how Nature implemented access to high resolution maps. If information is to be transmitted to the processors from the larger maps, then it must be done *through the input abstraction hierarchy*, "attentively", by tuning of the operators that compute the representation of the top-level maps. This conclusion supports the findings of Moran and Desimone (1985) by providing additional justification for a top-down tuning mechanism.

If the visual areas of the brain contain on the order of 10^9 neurons, then the prediction for PP implies that about 5000 neurons are required for each unit of incremental speed-up due to parallelism. What this analysis claims is that with connectivities of 1000 on average, it is not possible to hard-wire more than 7 maps, or 7 types of early vision output parameters, to a set of processors, and even then, the maps can contain only about 1300 elements (or a 36 × 36 image). Further, the degree of speed-up due to parallelism that is required is on the order of $10^{5.3}$ for knowledge bases of prototypes that may possibly be large enough to reflect human performance. Thus, these figures may be considered as the limits on the capacity of early vision schemes.

The prediction of 6 or 7 maps in total now predicts the response time for the slowest pre-attentive worst case performance of the system, using equation (11). However, the prediction of 6 or 7 maps, intuitively, seems small, given that it was earlier stated that perhaps many more parameters may be needed to completely characterize each point in visual space. Given that the uncertainty principle must play a role in the measurement of signal properties, some amount of inseparability seems necessary on those grounds too. If each type is really along more than one dimension, then it is possible to have many more actual values, implying a coarse-coded representation. A coarse-coded representation at this level would certainly allow many more actual values to be extracted, thus leading to a visual system capable of much richer interpretations of the visual world (see Hinton, 1981, and Ballard, Hinton & Sejnowski, 1983 for discussions on coarse-coded representations for vision).

Re-expanding the Collapsed Layer of Processors

The architecture presented assumes a single processor layer and a complete separation of data from processors. It is known that at one level of description, biological vision systems are composed of many layers of neurons, with complex interconnections and functionalities. We propose that the fuzziness associated with the neural and columnar functionality, and with the boundaries between retinotopic and nonretinotopic areas is not an artifact of Nature. Suppose that the architecture presented thus far, that is, in Figure 6, is "blurred" vertically. That is, the input abstraction hierarchy is extended upwards, the prototype hierarchy is extended downwards and the processor layer is extended in both directions, such that there is no distinction any longer between processor and data. The result is that a number of layers of processing units are created. Each unit of each layer has an integrated function of computing a portion of the algorithm for token extraction, representing a portion of a prototype, and matching. Since all sizes of receptive fields were included in the original definition of receptive field structure shown in Figure 3, the blurring process mentioned could result in receptive fields at the early levels of the hierarchy being smaller than those at the later stages of the hierarchy. At the early stages of the hierarchy, processors are dominated by a signal measurement function, while towards the top of the hierarchy, processors have a mostly interpretive function. A lower bound on the number of layers would be $\log_2 VP$, from the prototype hierarchy, and Feldman and Ballard's 100-step rule provides an upper bound. Data passes through the system, and is partially transformed at each layer serially in time, yet spatially parallel. The layers at the bottom of the hierarchy represent visual space exclusively retinotopically, while at the top, nonretinotopically. Search constraints dictate that any non-retinotopic representation must be very small, since spatio-temporal localization of receptive fields cannot be used to assist in reducing the space of possibilities that need be considered. Finally, the resulting network is a most plausible way of parallelizing the tree search included in the optimizations described earlier.

The visual world is inseparably spatio-temporal, not purely spatial or temporal. All units involved in the network must necessarily involve both spatial and temporal abstraction. Neurons are selective for certain specific inputs (response decreases dramatically as input differs slightly from the specific tuning) and sensitive to others (response increases or decreases roughly monotonically as the value of a particular parameter increases or decreases). This allows for a very wide range of possible degrees of spatio-temporal selectivities and sensitivities, not all of them semantically meaningful. In order to maintain semantic integrity, single unit spatio-temporal abstraction requires that: (a) there must be a semantically meaningful relationship in space and in time among all input units; (b) the bottom-up inputs must be synchronized, i.e., must be derived from the same spatio-temporal input *wave*. Lateral and top-down inputs must also conform to similar constraints.

An obvious implication is that there are no units that compute static information, divorced from time. Even stationary edges exist during a time interval. In fact, it is not necessarily the case that the output of any unit can be described using a simple semantic label, such as edge or flow. Each level of computation *specializes* the abstraction computed at the previous level, and further, at each level there may be several different specializations occuring in parallel. Even combinations which occur must do so in a synchronized manner. Yet, they can still be regarded as specialization since in effect two separate responses intersect. The flow of data is constrained by temporal integration times at all units, and thus it is not necessarily true that each unit completes processing, passes on information to the next level, and then takes on new input. Rather, it seems that the entire network is busy at work on several time slices of events, so that neural elements are not merely computation steps, but more correctly, introduce delay into processing, and compute continuously.

DISCUSSION

An interesting comparison can be drawn between the use and implications of connectivity constraints in this paper with that of Feldman and Ballard (Feldman & Ballard, 1982). In that paper, Feldman and Ballard correctly point out that conservation of connections is important, and they propose a number of "tricks" (their term) that would assist in this task. The first is functional decomposition, and indeed, that is used here as well in a slightly different manner, namely that of knowledge base search optimization. The second optimization is limited resolution computation. They claim that only the resolution required for the computation should be used, and no larger. How exactly this is to be determined in a bottom-up a priori situation is unspecified, and it seems that the utility of this idea may come in only after the first bottom-up processing pass. They also claim the utility of coarse coding; in this paper, the details of the utility of coarse coding in connectivity are demonstrated. The fourth optimization is that of tuning, and here their prediction is right on the money. They claim that larger units could be tuned to respond to a single saturated smaller unit within their input range, thus reducing connections required. Moran and Desimone, as described earlier, discovered this in 1985. However, the winner-take-all implementation of Feldman and Ballard and the details of the functionality are unfortunately incorrect. In this paper, a different view on the functionality is described, consistent with the experimental findings, and it is used to provide a new explanation for the pop-out effect and the visual search paradigm. Finally, they claim the importance of spatial coherence. The most serious problem requiring conservation of connections is claimed to be that of the representation of complex concepts. Only properties that are spatially coincident can activate concepts, thus one can factor out location from representations. Of course, they are right; but this is very different from my use of spatio-temporal localization. It is important to note that Feldman and Ballard's work did not address the specific complexity issues of visual perception, did not provide a quantitative demonstration of why their particular set of conserving techniques was sufficient, and did not propose an integrated approach to time and space complexity issues, that is, the consideration of connectivity complexity was separate from the consideration of other space and time constraints.

A NEW EXPLANATION FOR THE "POP-OUT" PHENOMENON

An important conclusion of this research is that pre-attentive vision, as defined by Neisser (1967) and by Treisman (1985), can be shown to be just a special case of the entire process of visual perception, and not a process distinct from attentive vision. The reason is complex and involves careful consideration of the functionality of maps and the matching process required for finding targets in a display.

Treisman used two basic types of displays, conjunctive and disjunctive. In a conjunctive display, the target is defined by conjoining two stimulus qualities, such as shape with color. There is more than one element in the stimulus pattern that displays the same conjunction of types of stimulus qualities and each of these is called a distractor. There is only one target with the pre-specified conjunction of tokens (for example, red with round). An increase in the number of distractors leads to the observed positive linear slope in response time, and for such displays where there is no target (called conjunction negatives), the slope is approximately twice as large leading to the conclusion that a self-terminating search is taking place. On the other hand, a disjunction display requires a target defined by a single stimulus quality, such as shape. However, the definition of distractors is different and here means all other elements in the display. Only one element of the stimulus has this quality, and the response time curve is flat with respect to the number of elements in the display. The disjunction negative case does exhibit a positive linear slope with number of elements in display, perhaps pointing to the need for a exhaustive search that verifies that no target is present.

The explanation Treisman and her colleagues propose is that for conjunction displays, separate feature maps must be brought into register using a spotlight of attention. This spotlight is necessarily serial, and thus a positive linear slope in response time versus distractors is found due to a self-terminating search. Unfortunately, spatial registration cannot be the reason for the phenomenon as claimed by Treisman—if all elements of the maps are hard-wired, registration is quite easily solved by the neural hardware. In the disjunction or pop-out case, Treisman claims that map activity is sufficient to signal that the target is present. This view assumes that there is a map for each feature type—unfortunately there are a great many more types used by Treisman than the number of maps in the brain, both those that are known as well as the number predicted by the theory of this paper. And moreover, why then is a linear search required to verify the disjunction negative case? Could not the same cue just as easily be used to signify the lack of a target? Response time slopes for both display types seem to be related to ease of stimulus quality discriminability. Why should this be the case for disjunction negatives? Finally, the response time for a conjunction display with only one element, namely the target, is larger than for a disjunction display of only the target. Treisman's explanation does not account for this. A different explanation for both conjunction as well as disjunction performance is required.

To this point, global decisions on image contents have not been discussed. Global decisions are dependent on the goals of the recognition task. As presented, the architecture in Figure 6 includes $\hat{N}^3$ receptive fields, and there is sufficient processing time to compute the best association for each with a visual prototype from the knowledge base. Each of these associations is a candidate for satisfying the goals of the visual task, and may wholly or only partially identify the visual entity in the corresponding receptive field. Thus, there are $\hat{N}^3$ separate outputs, or candidates for matching with the goals of the recognition task. Assuming some kind of cooperative process within columns, at best there will be $P_{\hat{N}}$ candidates, rather than $\hat{N}^3$ candidates. Let the number of goals or a priori visual targets be G. In the worst case, $P_{\hat{N}} \times G$ matches of candidates to targets are needed. Goals are dynamic, and are usually too few to try and organize, nor do they necessarily share properties on which organization can be done. Thus, the optimizations used earlier cannot be applied here. Candidates, however, could be ordered by "goodness of fit" criteria, using strength of response. Thus, it seems that the best this architecture can accomplish is a serial, self-terminating search. This is what is observed, and in this view, the positive slope is due only to the number of candidates and goals.

What is the number of candidates that are extracted to be matched against the goals or targets, by the subjects in the visual search experiments? Moran and Desimone (1985) have discovered, in monkeys, that single neurons as early as V4 (as well as in IT, but not in V1) can be tuned so that separate stimuli within the same receptive field can be individually attended, via top-down control, depending on spatial location and/or stimulus quality (Moran & Desimone, 1985). Effective and ineffective stimuli were determined for each neuron, and then each presented in different portions of the receptive field simultaneously. Attention to the effective stimulus lead to a strong response, while attention to the ineffective stimulus lead to a significant attenuation of response, even though the effective stimulus was still in the field. The tuning was observed to be changeable from trial to trial. They claim that unwanted information is filtered from the receptive fields of neurons in extrastriate cortex as a result of selective attention. If the findings of Moran and Desimone are also true for human vision, then over a course of several trials, subjects can tune out nongoal responses at the map level. The tuning in the case of Treisman's experiments would not be on location, but only on stimulus quality. So for example, if the targets are either a brown "T" or a letter, for a given display, subjects could tune out nonbrown, non-T, and nonletter stimuli at very early processing levels. In a disjunction display, there would always be only one candidate remaining for matching against goals because of the definition of the display. In the conjunction case, the number of candidates would be exactly the number of distractors plus the target.

Earlier in the paper, it was concluded that the best solution for immediate perception was for $\hat{M} = 1$. Yet, it is was also shown that up to seven maps could be accomodated using connectivity constraints. There are two possibilities. Firstly, it could be the case that only one physical map is active and relevant. The evidence seems to go against this view however, due to the inseparable nature of neural processing. Thus, it would indeed be very special circumstances in which a single neuron would yield information about only one physical parameter (for example, color without shape, depth or motion information). These are not the kinds of stimuli that are used to demonstrate pop-out. For the second explanation, recall that $2^{\hat{M}} - 1$ is the number of subsets of maps, not the actual number of maps. Since the receptive field assemblies can be directly connected to all maps, and pooled response could point to the active subset, then all tokens could conceivably be integrated in parallel, one after another. The resulting time to compute all these parameters would vary as in equation (12). Of course, stimulus quality discriminability also plays a role, and this affects both the determination of the candidates and the sequential selection of candidates. Earlier, it was demonstrated that there was a need for a coarse-coded representation at the map level. Thus, it would be possible only under special stimulus conditions, that consideration of a single subset of types leads to single visual parameters that lead to unambiguous target-stimulus matches, and thus pop-out. If those special conditions do not exist, then more subsets must be considered. There is no spotlight required for this, just more work. The spotlight is required only for selecting from among competing candidates. This explanation predicts that the response time for pop-out displays will increase as the number of stimulus qualities increases. This increase in base computation load is most apparent in the results of (Nakayama & Silverman, 1986), who have tried a more complex set of disjunction experiments than Treisman, thus confirming the prediction (color with motion, for example). If on the other hand, it had been assumed that the maps were independent, then the extraction of all the data needed for the candidate-to-goal matching would occur in constant time regardless of the number of stimulus qualities. Using Treisman's experimental parameters, if we had assumed the independence of maps, the theory presented in this paper would predict that the slope of the curve for response time vs. number of distractors in the display for conjunction negatives would be the same as that for feature negatives, and this slope would be twice that for conjunction positives. The disjunction positive curve would still be flat, yet the conjunction positive curve would intersect with the disjunction positive curve for size of display of one element. This is not what is observed. On the other hand, if the maps are not independent, the observed results are predicted: serial self-terminating search for conjunctions, flat response time for disjunction positives, a higher response time for conjunction displays with one element than for disjunction displays with one element, and a disjunction negative response slope that is smaller than for conjunction negatives.

The second issue is that of visual features or primitives. With the view presented above, Treisman's claimed features represent only what is "tunable"—and is much larger than the set of "default" features or primitives that the system computes. In the terminology of this paper, Treisman discovers tokens but not types. However, if Treisman's visual search paradigm rejects a feature, then it is indeed not even tunable. The conclusion is that visual search can be used only to reject candidate stimuli as features, and not to prove their existence. Using visual search in this mode only would be a very time-intensive task as well offering only indirect evidence for particular features.

CONCLUSIONS

The development of theories of visual perception lacks guiding principles, that is, a set of fundamental considerations that can both direct the creation of a theory, and that can test its validity. Two such principles are proposed in this paper, the "complexity level" of analysis and the Maximum Power / Minimum Cost Principle. This research has demonstrated that significant conclusions about the architecture of biologically plausible visual systems can result by the faithful application of these principles.

The implications for computer vision are clear and quite important. The reason that many of the high level vision proposals have not been entirely satisfactory (see Tsotsos, 1987b for a comprehensive overview), is that a strong argument for the computational need for high level processing has never been presented. That need must be in terms of the basic computational inadequacies of spatially parallel, bottom-up visual architectures. The capabilities of such architectures have been derived in this paper for biologically motivated designs (and still greatly apply for nonbiologically motivated designs, but not entirely). The argument for high level vision, and indeed, for computational modeling of human vision, is now on a solid foundation, and the results of this paper point to a very different style of high level vision research than currently practised.

It has been shown that in addition to spatial parallelism, the other characteristics of a visual processing architecture, that satisfies the timing constraints, are:

- hierarchical organization through abstraction of prototypical visual knowledge, in order to cut search time at least logarithmically
- localization of receptive fields, noting that the physical world is spatiotemporally localized and that objects and events, and their physical characteristics, are not arbitrarily spread over time and space
- maps are summarized via a pooled response, using the observation that not all visual stimuli require all possible parameter types for interpretation, and thus leading to separable, logical maps
- hierarchical abstraction of the input token arrays, in such a way as to maintain semantic content yet reducing the number of retinotopic elements.

These optimizations may be considered as sufficient, but not necessary, conditions to satisfy the time complexity constraint for the architecture of a visual system with performance comparable to human pre-attentive visual performance.

Applying connectivity constraints to this architecture, that is, determining how all the elements are to be connected, and the resulting cost of connection, many further characteristics of primate visual systems are implied and several others predicted:

- processor columnar organization;
- inverse magnification within the processor layer with respect to the retinotopic array
- tokens of visual parameters at high resolution cannot be directly accessed, rather must be obtained by tuning of computing units and through the input abstraction hierarchy
- token coarse coding
- physical separation of some maps
- predictions for the best architecture for immediate perception
- predictions for the overall configuration of the visual system in terms of the size and number of maps
- a top-down control mechanism, in the spirit of the experimental

findings of Moran and Desimone. This top-down control can be driven both by location and/or stimulus quality. This would intuitively require the matching of visual prototypes with abstracted retinotopic descriptions, the determination of differences and similarities between successful matches and perceptual goals, and if goals are unsatisfied, determination of which computations to tune and how to tune them. In addition, the visual routines of (Ullman, 1983) necessarily play a role. The overall control may proceed in much the same fashion as in (Tsotsos, 1980), (Tsotsos, 1985).

Another conclusion of this work is that, contrary to current psychological theories, pre-attentive vision is shown to be simply a special case of the visual process, and not a component separable from attentive vision. Put simply, if a single bottom-up pass yields an unambiguous immediate match to the goals of the perceptual task, then one has the pop-out phenomenon observed in pre-attentive vision. In non-pop-out situations, it is not the case that a different sort of mechanism takes over. More time is required for integration of more types of parameters, and subsequent serial search is required to select candidates for matching with the targets. This leads to the serial nature observed for attentive vision tasks. Since in typical perceptual experiments, subjects are told what to expect in the stimuli, and in most experiments have a substantial amount of training on the stimulus set, the first bottom-up pass may be tuned to attenuate the responses to nontarget stimuli, as pointed to by the results of Moran and Desimone.

REFERENCES

Allman, J., Miezin, F., & McGuinnis, E. (1985). Stimulus specific responses from beyond the classical receptive field: Neurophysiological mechanisms for local-global comparisons in visual neurons. *Annual Review of Neuroscience, 8,* 407–430.

Ballard, D. (1986). Cortical connections and parallel processing: Structure and function. *The Behavioral and Brain Sciences, 9–1,* 67–90.

Ballard, D., Hinton, G., & Sejnowski, T. (1983). Parallel visual computation. *Nature, 306–5938,* 21–26.

Barlow, H. (1981). Critical limiting factors in the design of the eye and the visual cortex. *Proceedings of the Royal Society of London, B212,* 1–34.

Barrow, H., & Tenenbaum, J. M. (1978). Recovering intrinsic scene characteristics from images. In A. Hanson & E. Riseman (Ed.), *Computer vision systems* (pp. 3–26). New York: Academic Press.

Beck, J. (1982). Textural segmentation. In J. Beck (Ed.), *Organization and representation in perception* (pp. 285–318). Hillsdale, NJ: Erlbaum.

Brooks, R. (1981). Symbolic reasoning among 3-dimensional models and 2-dimensional images. *Artificial Intelligence, 17,* 285–348.

Bushnell, M., Goldberg, M. & Robinson, D. (1981). Behavioral enhancement of visual responses in monkey cerebral cortex I: Modulation in posterior parietal cortex. *Journal of Neuroscience, 46–4,* 755–772.

Churchland, P. (1986). *Neurophilosophy.* Cambridge, MA: MIT Press / Bradford Books.

Cook, S. (1971). The complexity of theorem-proving procedures. Proceedings of the 3rd Annual ACM Symposium on the Theory of Computing, 151–158, New York.

Corbeil, J. C. (1986). *The Stoddart visual dictionary.* Toronto: Stoddart Publishing Co.

Cowey, A. (1979). Cortical maps and visual perception. *Quarterly Journal of Experimental Psychology, 31,* 1–17.

Crick, F. & Asunama, C. (1986). Certain aspects of the anatomy and physiology of the cerebral cortex. In D. Rumelhart & J. McClelland (Ed.), *Parallel Distributed Processing* (pp. 333–371). Cambridge, MA: MIT Press.

Crick, F., Marr, D. & Poggio, T. (1980). An information processing approach to understanding the visual cortex, MIT AI Memo 557, Cambridge, MA.

Daniel, P. & Whitteridge, D. (1961). The representation of the visual field on the cerebral cortex in monkeys. *Journal of Physiology, 159,* 203–221.

Desimone, R., Chein, S., Moran, J. & Ungerleider, L. (1985). Contour, color, and shape analysis beyond the striate cortex. *Vision Research, 25–3,* 441–452.

Dobson, V. & Rose, D. (1985a). Application of an explicit procedure for model building in the visual cortex, In D. Rose & V. Dobson (Ed.), *Models of the visual cortex* (pp. 546–560). Chichester, Great Britain: John Wiley & Sons.

Dobson, V. & Rose, D. (1985b). Models and metaphysics: the nature of explanation revisited. In D. Rose and V. Dobson (Ed.), *Models of the visual cortex* (pp. 22–36). Chichester Great Britain: John Wiley & Sons.

Feldman, J. & Ballard, D. (1982). Connectionist models and their properties. *Cognitive Science, 6,* 205–254.

Feldman, J. (1985a). Four frames suffice: A provisional model of vision and space. *The Behavioral and Brain Sciences, 8–2,* 265–289.

Feldman, J. (1985b) (Special Issue Editor). Connectionist models and their applications. *Cognitive Science, 9–1,* 1–169.

Fleet, D., Hallett, P. & Jepson, A. (1985). Spatio-temporal inseparability in early visual processing. *Biological Cybernetics, 52,* 153–164.

Garey, M. & Johnson, D. (1979). *Computers and intractability: A guide to the theory of NP-completeness.* New York: W. H. Freeman and Co.

Hildreth, E. & Hollerbach, J. (1985). The computational approach to vision and motor control, MIT AI Memo 846, Cambridge, Massachusetts.

Hinton, G. (1981). Shape representation in parallel systems, Proceedings International Joint Conference on Artificial Intelligence, 1088–1096, Vancouver.

Hubel, D. & Wiesel, T. (1977). Functional architecture of macaque visual cortex. *Proceedings of the Royal Society of London, B 198*, 1–59.

Hummel, R. & Zucker, S. (1980). On the foundations of relaxation labeling processes, TR-80-7, Department of Electrical Engineering, McGill University.

Julesz, B. (1978). Perceptual limits of textural discrimination and their implications to figure-ground separation. In E. Leewenberg and H. Buffart (Ed.), *Formal theories of visual perception*. New York: John Wiley & Sons.

Kandel, E. & Schwartz, J. (Ed.) (1981). *Principles of neural science*. New York: Elsevier/North Holland.

Kirousis, L. & Papadimitriou, C. (1985). The complexity of recognizing polyhedral scenes, 26th Annual Symposium on Foundations of Computer Science, Portland, Ore.

Lynch, J., Mountcastle, V., Talbot, W. & Yin, T. (1977). Parietal lobe mechanism for directed visual attention. *Journal of Neurophysiology, 40–2*, 362–389.

Mackworth, A. (1978). Vision research strategy: Black magic, metaphors, mechanisms, miniworlds, and maps. In A. Hanson & E. Riseman (Ed.), *Computer vision systems*, (pp. 53–60). New York: Academic Press.

Mackworth, A. & Freuder, E. (1985). The complexity of some polynomial network consistency algorithms for constraint satisfaction problems. *Artificial Intelligence, 25*, 65–74.

Mansfield, R. (1982). Role of the striate cortex in pattern perception in primates. In D. Ingle, M. Goodale and R. Mansfield (Eds.), *Analysis of visual behavior* (pp. 443–482). Cambridge, MA: MIT Press.

Marr, D. (1982). *Vision*. San Francisco: W. H. Freeman.

Maxwell, N. (1985). Methodological problems of neuroscience. In D. Rose and V. Dobson (Ed.), *Models of the visual cortex* (pp. 11–21). Chichester, Great Britain: John Wiley & Sons.

Moran, J. & Desimone, R. (1985). Selective attention gates visual processing in the extrastriate cortex. *Science, 229*, 782–784.

Nakayama, K. & Silverman, G. (1986). Serial and parallel processing of visual feature conjunctions. *Nature, 320–6059*, 264–265.

Neisser, U. (1967). *Cognitive psychology*. New York: Appleton-Century-Crofts.

Poggio, T. (May 1982). Visual algorithms, MIT AI Memo 683, Cambridge, MA.

Poggio, T. (March 1984). Vision by man and machine, MIT AI Memo 776, Cambridge, MA.

Richards, W. (1982). How to play twenty questions with nature and win, MIT AI Memo 660, Cambridge, MA.

Rose, D. & Dobson, V. (Eds.) (1985). *Models of the visual cortex*. Chichester Great Britain: John Wiley & Sons.

Rumelhart, D. & McClelland, J. (1986a). PDP models and general issues in cognitive science. In D. Rumelhart & J. McClelland (Ed.), *Parallel distributed processing* (pp. 110–146). Cambridge, MA: MIT Press.

Rumelhart, D., & McClelland, J. (Eds.) (1986b). *Parallel distributed processing.* Cambridge, MA: MIT Press.

Stensaas, S., Eddington, D. & Dobelle, W. (1974). The topography and variability of the primary visual cortex in man. *Journal of Neurosurgery, 40,* 747–755.

Stone, J., Dreher, B. & Leventhal, A. (1979). Hierarchical and parallel mechanisms in the organization of the visual cortex. *Brain Research Reviews, 1,* 345–394.

Schwartz, E. (1977). Spatial mapping in the primate sensory projection: Analytic structure and relevance to perception. *Biological Cybernetics, 25,* 181–194.

Treisman, A. (1986). Features and Objects in Visual Processing. *Scientific American, 255–5,* 114B–125.

Treisman, A. (1985). Preattentive processing in vision. *Computer Vision, Graphics and Image Processing, 31,* 156–177.

Tsotsos, J. (1980). *A framework for visual motion understanding,* PhD Thesis, also, CSRI-TR-114, Department of Computer Science, University of Toronto.

Tsotsos, J. (1985). Knowledge organization and its role in the interpretation of time-varying data: The ALVEN system. *Computational Intelligence, 1–1,* 16–32.

Tsotsos, J. (1986). Connectionist computing and neural machinery: examining the test of 'timing'. *The Behavioral and Brain Sciences, 9–1.* 106–107, (commentary on (Ballard 1986)).

Tsotsos, J. (1987a). Representational axes and temporal cooperative processes. In M. Arbib and A. Hansen (Eds.), *Vision, brain and cooperative computation* (pp. 361–418). Cambridge, MA: MIT Press / Bradford Books.

Tsotsos, J. (1987b). Image understanding. In S. Shapiro (Ed.), *The encyclopedia of artificial intelligence* (pp. 389–409). New York: John Wiley & Sons.

Ullman, S. (1983). Visual routines, MIT AI Memo 723, Cambridge, MA.

Ungerleider, L. & Mishkin, M. (1982). Two cortical visual systems. In D. Ingle, M. Goodale & R. Mansfield (Eds.), *Analysis of visual behavior* (pp. 549–586). Cambridge, MA: MIT Press.

Uhr, L. (1972). Layered 'recognition cone' networks that preprocess, classify and describe. *IEEE Transactions on Computers,* 758–768.

van Essen, D., Newsome, W. & Maunsell, J. (1984). The visual field representation in striate cortex of the macaque monkey: Asymmetries, anisotropies and individual variability. *Vision Research, 24–5,* 429–448.

van Essen, D. & Maunsell, J. (September 1983). Hierarchical organization and functional streams in the visual cortex. *Trends in Neuroscience,* 370–375.

van Essen, D. & Zeki, S. (1978). The topographic organization of rhesus monkey prestriate cortex. *Journal of Physiology, 277,* 193–226.

Zeki, S. (1978). Uniformity and diversity of structure and function in rhesus monkey prestriate visual cortex. *Journal of Physiology, 277*, 273–290.

Zucker, S. (1985). Does connectionism suffice? *The Behavioral and Brain Sciences, 8-2*, 301–302, (commentary on (Feldman 1985a)).

IV
Recognition and Representation of Form and Object

13
Preattentive Processing in Vision*

Anne Treisman

University of British Columbia, Vancouver, British Columbia,
Canada

Psychologists, perhaps other scientists too, tend to like dichotomies. We should, I suppose, be suspicious of this preference for simplification, but there does seem to be a striking one in the visual system, at least as a first approximation. Some discriminations appear to be made automatically, without attention and spatially in parallel across the visual field. Other visual operations require focused attention and can be performed only serially. These two kinds of visual processing have been attributed to different levels, originally by Neisser in 1967. He distinguished a stage at which simple features are preattentively registered, which determines texture segregation and figure ground grouping, from a second stage at which objects with their complex combinations of features are identified. In computational vision, a similar distinction has been drawn by a number of people, for example, between Marr's (1982) primal sketch and his model-based recognition, or between Barrow and Tenenbaum's intrinsic images (1978), again followed by model-based recognition. This dichotomy is one starting point for the research that I will describe.

Another is the accumulating evidence, both physiological and psychological, that early stages of visual processing are analytic,

* This research was supported by a grant from the National Scientific and Engineering Research Council of Canada. The author held a fellowship from the Canadian Institute of Advanced Research. Thanks are due to Daniel Kahneman for helpful criticism and suggestions, and to Hilary Schmidt, Janet Souther, and Stephen Gormican who collaborated in running many experiments. This chapter was originally published in *Computer Vision, Graphics and Image Processing, 31*, 156–177, 1985. It is reprinted here with the permission of the publisher, Academic Press.

that is, that they decompose the physical array of light along a number of separate dimensions. These seem to be mapped into different areas in the brain, each of which is specialized to analyze a different property; for example, neural units selectively responsive to particular orientations are found in one area, those responsive to stereo depth in another, those to color, movement, and so on in yet other functionally separate areas (Cowey, 1979; Hubel & Wiesel, 1968; Zeki, 1981). Differences in processing time may supplement differences in localization as evidence for specialized analysis. Thus, in humans, one can measure on the scalp evoked responses which pool the activity from populations of cells in different areas of the brain. By varying the discriminations required by the task and showing different evoked response latencies, one can infer that some properties are extracted before others (Towle, Harter, & Previc, 1980). One can also demonstrate adaptation which is selective to particular properties. If one stares at a waterfall and then looks at the bank, it appears to flow in the opposite direction. This may reflect selective adaptation of detectors for a particular direction of movement, which respond independently of what is moving. The bank looks very different from the water in color and texture, but it still shows the aftereffects of movement. Finally, one can attend selectively to different attributes. For example, a speeded classification task in which cards must be sorted on the basis of color shows little interference from irrelevant variations in shape (Garner, 1974). (See also Treisman (1982) for a more detailed review of psychological evidence for feature analysis).

If we accept that there is some kind of early decomposition along different dimensions, we are confronted with two different problems for psychological research. One is to define which features or properties are the basic elements or visual primitives in this language of early vision. The second concerns how they are put together again into the correct combinations to form the coherent world that we perceive. I have been interested in both these questions. I will focus on two kinds of psychological evidence about early visual processing. The first is texture segregation. This is likely to reflect early stages of analysis because it is a prerequisite for figure ground separation which sets up the candidates for subsequent object recognition. The other source of evidence is visual search. Subjects are asked to look for a particular target in displays containing varying numbers of distractor items. If the target is defined by a simple visual feature, detection appears to result from parallel processing; the target "pops out" of the display regardless of how many distractors surround it. The spatially parallel processing suggests that the task depends on early vision, before the stage at which attention becomes involved. I will start with a brief discussion of evidence from texture segregation, and then devote most of the paper to some recent research using visual search to explore the nature and the limits of preattentive processing.

Texture segregation and grouping appear to reflect what Julesz (1975) described as immediate, effortless perception, without scrutiny, of distinct areas in the visual field, the initial "parsing" of the world into the objects and backgrounds that may later be identified. Some time ago, Beck (1967) showed that texture segregation is easy when it is based on simple properties such as color (a red area will segregate clearly from a green area); brightness (a dark area will segregate from a light one); and line orientation (an area containing tilted Ts segregates well from an area containing vertical Ts). However, if two areas differ only in line arrangements (one contains Ts and the other Ls), the boundary is much more difficult to detect. Beck suggested that texture segregation depends on features that are easily discriminable with peripheral vision and with distributed attention. More recently, Beck (1982) also suggested that texture segregation results from a computation of local *differences*, which is performed automatically across the visual field. Any discontinuity where adjacent elements differ in one or more features sets up a boundary between adjacent regions.

Our experiments have confirmed that segregation is easy when areas differ in simple properties like shape and color (Figure 1a and b; Treisman & Gelade, 1980). However, a boundary which is defined solely by a conjunction of properties (e.g., green triangles and red circles on the left and red triangles and green circles on the right), is much harder to find (Fig. 1c). Latencies to decide whether the boundary is vertical or horizontal are much slower than when either shape alone or color alone are sufficient to divide the display into discrete areas. Similar results hold with parts of shapes as well as with values on different dimensions. Segregation is easy, for example, when the boundary divides Ps and Os from Rs and Qs (Figure 1d). The letters on one side have a diagonal slash whereas those on the other side do not. If the areas mix Ps with Qs and Rs with Os (Fig. 1e), there is no single distinguishing element, and texture segregation is again much harder. Some of our results, however, seem to conflict with Beck's suggestion that local differences are computed automatically, and that texture segregation depends on the detection of local boundaries only. For instance, if you look quickly at Figure 2a you may miss the two odd items. On the other hand, if you look at their local contexts alone (Figure 2b and 2c) without the sur-

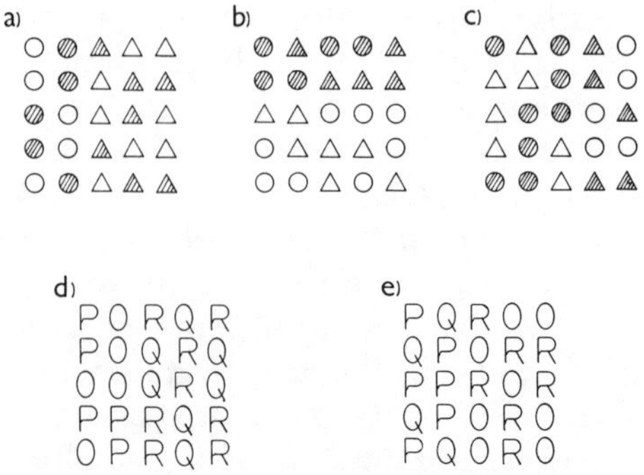

a)
```
O ⊘ ▲ △ △
O ⊘ △ ▲ △
⊘ O △ △ △
⊘ O ▲ △ △
O ⊘ △ ▲ ▲
```
b)
```
⊘ ▲ ⊘ ⊘ ▲
⊘ ⊘ ▲ ▲ ▲
△ △ O O O
O △ △ △ O
O O △ O △
```
c)
```
⊘ △ ⊘ ▲ O
△ △ ⊘ ▲ O
△ ⊘ ⊘ O △
△ ⊘ △ O O
⊘ ⊘ △ ▲ ▲
```

d)
```
P O R Q R
P O Q R Q
O O Q R Q
P P R Q R
O P R Q R
```
e)
```
P Q R O O
Q P O R R
P P R O R
Q P O R O
P Q O R O
```

Figure 1. Examples of easy and of difficult texture segregation: (a) salient vertical boundary between circles (curved shapes) on left and triangles (straight and angular shapes) on right; (b) salient horizontal boundary between red shapes above and green shapes below; (c) no salient boundary between conjunctions of green triangles and red circles on left and green circles and red triangles on right; (d) salient vertical boundary between letters without diagonal line on left and letters with diagonal line on right; (e) no salient boundary between Ps and Qs on left and Rs and Os on right.

rounding groups, the odd ones are immediately salient. The "masking" effect depends on the presence elsewhere of other elements sharing the locally distinctive property; unique elements are not masked in the same way. I ran an experiment showing that subjects scan each group serially when the locally different property is pre-

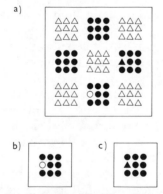

Figure 2. (a) Preattentive "masking" of locally salient items by the presence of their locally distinctive properties elsewhere in the display. Unless attention is directed to the relevant group, the odd items are likely to be missed. (b) and (c) the same items are immediately obvious when attention is restricted to their local context.

sent elsewhere, but not when it is unique in the whole display (Treisman, 1982). The result seems inconsistent with the idea that local discontinuities are automatically detected.

Another demonstration may also cast doubt on the local processing suggestion: In Figs. 3a and b the immediate impression is of three different types of element, but if one looks quickly at Fig. 3c without focusing attention on the central column, one tends to see two groups rather than three. The central group shares one property with the group on the left and one property with the group on the right. When texture segregation depends only on preattentive processing, it seems to occur within a color representation or within a shape representation, but not within a global representation which combines both color and shape. If attention is distributed over the whole display, the central group is absorbed either into the white group or into the circles. At the preattentive level the white circles appear to have no independent and separate existence (Treisman, 1982). For Neisser (1967), Beck (1982), Marr (1982), and Julesz (1984), I believe that the spatially parallel, preattentive representation or primal sketch is a single representation, a global map showing the boundaries formed by any number of simple properties. For Julesz the visual primitives that mediate texture segregation (which he calls "textons") include conjunctions of different dimensional values, such as "a vertical elongated red blob in a blue surround." The experiments I have described perhaps cast some doubt on these views, and suggest instead that several different representations may be set up preattentively within each dimension separately, and

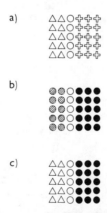

Figure 3. Three groups are perceptually salient when defined by different features of shape in (a) and of color in (b), but the central group may be camouflaged preattentively, as in (c), when it shares one feature (color) with the group to the left and another feature (shape) with the group to the right.

that they are accurately combined only at a later stage through focused attention. This view is more consistent with Barrow and Tenenbaum's idea (1978) of separate intrinsic images, and with the modular organization that physiologists describe. Clearly the information from the different feature maps must be pooled before it reaches conscious awareness. Perhaps Beck's observations relate more to this level of integrated response of which we are subjectively aware; in this sense, his results are not necessarily incompatible with mine. However, when we overload or divert attention, the earlier stages of spatially parallel analysis seem to occur within separate representations for different dimensions.

The remainder of this paper will be concerned with visual search. If a single blue letter is embedded in a display of brown and green letters of other colors, it "pops out." Detection appears to be spatially parallel. There is no need to check each of the brown and green letters before the blue one can be found. Similarly, if a single curved letter (e.g., "S") is presented in a display of straight or angular letters (e.g., "X" and "T"), it is also immediately salient. But if the task is to find a target which conjoins two properties (e.g., green and "T"), each of which is present in other distractors (e.g., green "X"s and brown "T"s), the search is much more difficult. Latencies increase linearly with the number of distractors, as if attention had to be focused on each item in turn (1980). The slope relating search time to display size (number of distractors) is twice as steep on trials where the target is absent as on trials where it is present, which suggests a serial self-terminating search. The area or visual angle of the display has little effect, provided that acuity limits are not exceeded. Kraus and I have shown more recently that linear search functions are also obtained with very brief displays in which eye movements would be ineffective. The serial scan appears to depend on movements of the "mind's eye" rather than (or as well as) movements of the physical eyes.

As a tentative conclusion, I suggest that parallel pop-out may be diagnostic of the presence of a unique feature which is analyzed in early vision. To link this idea with the hypothesis of multiple feature maps, we might hypothesize that pop-out occurs when the target produces activity in a separate feature map, which is unaffected by the distractors. The strategy would be to consult the particular feature map that uniquely responds to the target and, if there is any activity present, to respond "yes." Thus, when looking for a red target, you check the map for red and see if there is any activity there. On the other hand, if you are looking for a conjunction of properties, such as green and *T* among green distractors and *T* distractors, there is no uniquely activated feature map. Each item must be located and checked individually.

The role of attention can be tested more directly by giving subjects a cue in advance, which tells them where the target will occur if it is present at all. Now if attention is needed to detect conjunction targets, the precue should eliminate the serial checking phase. On the other hand, with targets defined by a single feature, the cue should have very little effect: separate features are registered in parallel anyway. We used displays like those in Figure 4a containing shapes that varied in shape, size, color, and whether they were outline or filled. The target was defined either by a conjunction or properties, for example, a large brown outline triangle, or by a single property like red or large or outline. We precued the location at which the target would occur if it was present at all, by flashing a pointer to that location 100 ms before presenting the display. The precue was valid on 75% of the trials on which the target occurred. It was invalid on 25% of trials: in these cases the target occurred somewhere other than cued location. On invalid trials attention would be directed to the wrong location rather than distributed across the whole display.

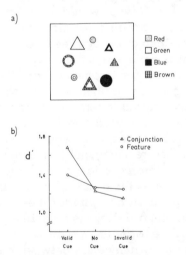

Figure 4. (a) Example of a display of multi-dimensional stimuli used either for feature search or for conjunction search. The feature target for this display might be a red item and the conjunction target might be a large brown outline triangle. An advance cue, consisting of a bar marker outside the display pointing at the location to be occupied by one of the eight items, was given on most trials. (b) Mean accuracy (expressed as the signal detection measure d') in the different cue conditions for feature and for conjunction targets.

An invalid cue should therefore give rise to costs rather than benefits relative to the condition with no cue. On 25% of trials no advance information was given about the target location, although a temporal warning signal was given to equate the general level of preparation. We matched the accuracy of performance for feature targets and for conjunction targets by presenting the display for a longer duration for the conjunction targets (though never more than 150 ms, to minimize the effect of eye movements). The question we asked was whether the effect of the cue would be greater for conjunction than for feature targets. Figure 4b shows the results: for conjunction targets, there was a substantial benefit from a valid cue, while for feature targets the cue had very little effect. In fact the whole improvement for feature targets with the valid cue was due to better detection of the small size target when it was precued. Small targets were missed more often when they were in invalid locations. Thus there appears to be little effect of attention when a target is defined by a single property, but there is a large effect of knowing where to attend when the target is defined by a conjunction of properties.

If attention is necessary for conjoining features, we should predict errors of conjunction when attention is divided or overloaded. The preattentive levels of early vison would represent only the presence of various features and not their interrelations. We briefly presented displays which contained a row of three colored letters in the middle and two black digits at each end of the array (see Figure 5a; Treisman and Schmidt (1982). In order to ensure that attention would not be focused on any single letter, subjects were asked to report first the digits, and then any colored letters that they were reasonably sure they had seen. Subjects reported almost as many "illusory conjunctions" in which the color of one letter was recombined with the shape of another as they did correct conjunctions. Some of these conjunction errors were made with high confidence and appeared to be genuine perceptual illusions. Does the occurrence of illusory conjunctions of shape and color depend on how similar the stimuli are on other dimensions? For example, are subjects more likely to switch colors between two small outline triangles, because they are the same shape, size and style, than to switch colors between a small, blue, outline circle and a large, red, filled triangle (see Fig. 5b)? If the color is represented in a wholistic, analog form which preserves the quantity and the spatial distribution of the color as well as the hue, saturation, and lightness, one might expect fewer exchanges to occur when they would violate these constraints. In fact, it seems that subjects are quite happy to fill in, for example, whatever amount of blue is necessary to represent the shape, size, and style they have coded as be-

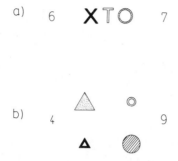

Figure 5. (a) Example of displays used to demonstrate the occurrence of illusory conjunctions when attention cannot be focused on each letter in turn. If solid letters were blue, outline letters green and speckled red, then subjects, given this display, might report, for example, a blue *T* or a red *X*. (b) Example of displays used to investigate possible similarity constraints on illusory conjunctions. Subjects were as likely to attribute the color of the large filled triangle to the small outline circle as to attribute to it the color of the small outline triangle (or even of another small outline circle, when two were included in the display).

longing with it. It is as if, when attention is overloaded or divided, the presence of one or more values on each dimension is coded separately and categorically; the codes are then combined, whether correctly or incorrectly, and consciously experienced in whatever form fits the set of specified codes. If the color blue is assigned to a large, filled triangle, then the amount and the spatial distribution of the blue will be appropriate to that set of descriptions.

The final evidence suggesting that features are preattentively detected without being conjoined whereas conjunctions require attention comes from an analysis of interdependencies between the accuracy of target identification and the accuracy of target localization. If we do conjoin only by attending to an object, we should be forced to locate a conjunction target in order to identify it correctly, whereas this would not be necessary for a feature target. Subjects were shown displays like those in Figure 6 and were asked both to identify the target and to say where it was (Treisman & Gelade, 1980). In the feature task, they had to decide on each trial whether there was an orange letter or an *H*. In the conjunction task they had to decide whether there was a blue *O* or a red *X*. We analyzed the conditional probability that the identity was correct given that the location was wrong. We expected this probability to be above chance for feature targets, and it was. Thus subjects could sometimes tell correctly that there was an orange letter in the display

$$P(I_{correct} / L_{error})$$

Matched Performance	Matched Accuracy

OO X H XO
X O X X O X   0.68   0.75

X X O O X O
O O X O X X   0.50   0.45

Figure 6.   Examples of displays used to investigate the dependence of accurate identification on accurate localization. (a) The target was either an *H* or an orange letter (among red and blue distractors). Forced choice identification was significantly better than chance even when the target was mislocalized by more than one position in any direction. (b) The target was either a red O or a blue *X* (among blue O and red *X* distractors). These targets differ from the distractors only in the way their properties are conjoined. Forced choice identification was at chance when localization was incorrect.

while mislocating it by two or more squares. For conjunction targets, on the other hand, correct identification was completely dependent on correct localization, as it should be if attention must be focused on a location in order correctly to combine the features it contains.

It seems, then, that information about features can be "free-floating" or indeterminate in location, but information about conjunctions is available only through accurate localization. If attention is overloaded, free-floating features may recombine at random when their associations were originally arbitrary. In a natural scene, of course, many conjunctions are ruled out by our prior knowledge. We seldom come across blue bananas or furry eggs. There are top-down constraints which may prevent us from seeing too many conjunction illusions in everyday life.

The remainder of this paper will describe our current attempts, to discover more about the features that form the functional elements in the language of early vision. The visual system is likely to have developed ways of detecting structure in the real world, the regularities that are nonaccidental because they are caused by physical objects. Surfaces in the real world tend to be homogeneous rather than randomly colored or textured; lines, curves, or boundaries are continuous; edges are often parallel; texture gradients change smoothly within objects and discontinuously between objects or between object and ground. Witkin and Tenenbaum (1983) suggest that vision scientists should look for "an alphabet soup of descriptive chunks that are almost certain to have some fairly direct semantic interpretation." The experiments I have described so far used uncontroversial properties like orientation and color which are almost certain to be separately coded. I used these to identify a set of behavioral "symptoms" that these features are likely to show in perceptual tasks. They mediate effortless texture segregation; they allow "pop-out" or parallel search when a single feature is present to define the target; they may recombine to form illusory conjunctions; they can be identified without being accurately localized. We can now try to use symptoms from this feature syndrome both to define more precisely the features we have already classified, such as orientation, and also to discover or diagnose new features which are more theoretically controversial. For example, we can use behavioral tests to help determine whether closure is a perceptual feature or not.

In the next experiment, I used the different pattern of results expected in feature and in conjunction search to explore how the visual system codes orientation. Is the same representation formed for the orientation of solid lines and for the orientation of dot pairs, or are there different codes for orientation depending on the nature of the carrier? Marr (1982) suggested that vision "goes symbolic" at an early stage; that orientation, for example, is coded abstractly as the orientation of a "virtual" line whose ends can be marked by any kind of discontinuity or change in intensity. If that is the case, the same symbolic code may be used by the visual system to represent the orientation of dot pairs and of lines. We have seen that orientation is separable from color. The question now is whether orientation is integral with the substance that carries it or whether it is also separable from its carrier.

The test we devised was to look for evidence of conjunction problems in a search task, which would suggest that the visual system uses interchangeable codes for line orientation and dot orientation. Subjects searched in one condition for a target line tilted right among distractor lines that were tilted left (see Figure 7a). In another condition they searched for a tilted line amongst dot pairs that were tilted in the same direction (see Figure 7b). In both conditions the target "popped out" regardless of the number of distractors (see Figure 8a and 8b). Thus we have evidence that both the orientation of lines and the difference between lines and dot pairs are preattentively discriminable, separate features. In the critical condition, the target was defined by the conjunction of an orientation and a carrier; subjects searched for a tilted line among dot pairs tilted in the same direction and lines tilted in the opposite direction (see Figure 7c). If

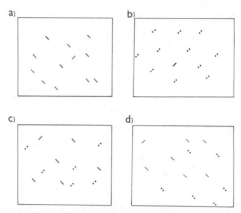

Figure 7. Examples of displays used to test the nature of the code used for orientation. In (a) and (b) the target differs from the distractors in a single feature (line orientation and line versus dot, respectively); in (c) the target differs only in the conjunction of orientation with carrier (line versus dot). (d) Control condition to see whether mixed line and dot displays are inherently more difficult. In each case the target is a line tilted to the right.

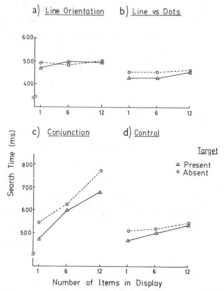

Figure 8. Search times to decide whether the target was present or absent in displays like those shown in Figure 7a to d: (a) Line orientation; (b) line versus dots; (c) conjunction; (d) control. The target pops out in every case except the conjunction search, where the latency appears to reflect serial item by item checking at the rate of about 20 ms per item.

the same representation is used for the orientation of dot pairs and of lines, we would expect search to be serial, as in fact it appears to be (see Figure 8c). A control condition, using both dots and lines but with the target line orientation unique did not show the same effect of display size (see Figure 7d and 8d), so the difficulty of the conjunction condition does not simply reflect the fact that the mixture of dots and lines is confusing.

The result is consistent with Marr's suggestion that orientation is coded abstractly, in the sense that a common representation is used for the orientation of dot pairs and of target lines. However, this abstraction could still be implemented by cells with oriented receptive fields, like those described by Hubel and Wiesel (1968). The activity from the two dots would be pooled if both fell within a receptive field and could activate the cell in the same way as an oriented line. In order to test this possibility, we will repeat the experiment using black and white dot pairs on a grey background to see whether orientation is abstracted even if the two dots contrast in opposite directions from the background. Simple cells would not respond to dot pairs of opposite contrast; their effects would cancel out. But Marr's place-markers should be established in the same way as for two black dots.

We looked next at other possible visual primitives. I have suggested that pop-out may be mediated by activity in separate feature maps. The idea is that we access directly a pooled response for each feature map—for example, a map for red, a map for vertical, a map for movement to the right. The pooled response would tell us whether there is activity in a given map, and perhaps also how much activity there is, but it would not tell us where it is, nor how it is arranged spatially, nor how it relates to activity in other maps. We can think of the visual system coding a certain number of simple and useful properties (not too many, one hopes), which are separately detected and represented in orderly arrays. The properties may be ordered systematically across a stack of maps, coding continuously changing orientations in one three-dimensional cube, color in another, the direction of movement in another. These maps, the physiologists tell us, are usually spatiotopic (Cowey, 1979); they do preserve spatial relations. However, the spatial information may not be directly accessible in parallel. In particular, the spatial information which links one map to another may be accessed only through a serial scan. In Figure 9, I show the serial scan as controlled through links to a single master map of locations, in which the presence of any discontinuities in intensity or color is registered without specifying what the discontinuities are. Focused attention acts through this

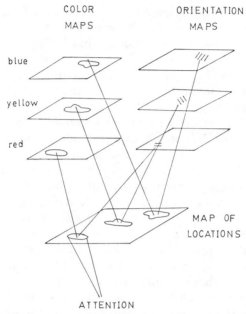

COLOR
MAPS

ORIENTATION
MAPS

blue

yellow

red

MAP OF
LOCATIONS

ATTENTION

Figure 9. General framework relating hypothesized feature maps within and across dimensions to a master map of locations, through which focused attention serially conjoins the properties of different objects in the scene.

master map, simultaneously selecting, through its links to all the separate feature maps all the features that are currently present in the selected location. This would be one mode of perceptual processing—the focused attention mode, which automatically retrieves and conjoins all features currently in the spatially limited attention spotlight, and does this serially for each object in turn.

In addition, I suggest that, independently of attention, each feature map also directly signals the presence of its particular feature, but without specifying where it is. There could be some kind of pooled signal conveying the information that, for example, red is present in the scene, in small or large amounts, at any given moment. The pooled response loses position information; it is equivalent to a simple form of Hough transform. It may tell us how much red is present, provide a means of linking separate areas sharing the same color, and also help to direct attention to the locations in the master map that are linked to a currently salient or relevant feature.

This is certainly a crude framework, but it does predict an interesting phenomenon. There should be a marked asymmetry between search for the *presence* of a feature and search for the *absence* of the same feature. Beck (1982) reports such an asymmetry for long lines in a background of short lines and for complete triangles in a background of incomplete triangles compared to the inverse displays. He attributes the asymmetries to differences in discriminability in peripheral vision with distributed attention—a kind of masking by neighboring stimuli which affects shorter more than longer lines. The prediction can be derived more directly from the framework proposed above. Suppose that the target has feature "X" and the distractors do not. For example, the target is a circle with a vertical slash across it and the distractors are regular circles. Now in order to decide whether the target is present, one could simply check categorically whether there is activity in a feature map for a straight line (or for vertical orientation, or for any other feature that characterizes the slash but not the circles). Now suppose that all the distractors have the feature "X" and the target does not; in this case, there is no feature map which will tell one whether the target is present or not. The only difference between the effects of displays containing the target and displays without it would be a slight reduction in activity in the feature maps coding the slash when the target was present compared to when it was absent. So when the target was the circle without the slash, subjects might be forced to scan serially with focused attention in order to find the one circle that did not have a slash. On the other hand, if the target was the single circle with the slash, its presence would be signalled by the presence of activity in the relevant feature map or maps. We tested this prediction (Treisman & Souther, 1985), using circles with and without slashes as a clear example of feature presence or absence (see Figure 10). As predicted, we found a very large asymmetry, shown in Figure 11. Search appears to be serial for the circle without a slash among circles with slashes, while search is parallel for the circle with a slash among circles without. Notice that this is a somewhat strange and counter-intuitive prediction, because the same discrimination between the same two stimuli is involved in each case. One might think that the only relevant factor would be how discriminable the circle with a slash is from a circle without a slash; but in fact, performance looks completely different in the two cases. The result is consistent with the idea of a pooled signal conveying the presence but not the absence of a distinctive feature.

We can now apply this test to other hypothetical features to see which pop out when they are present but not when they are absent. Note that the primitives extracted by the visual system need not (and in fact are unlikely to) correspond to the simple dimensions defined by physicists. The relevant features are those that best characterize

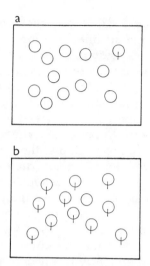

Figure 10. Examples of displays used to test the efficiency of search for a target defined by the presence or by the absence of a single feature. The target in (a) is a circle with an intersecting vertical line. The target in (b) is a circle without an intersecting line.

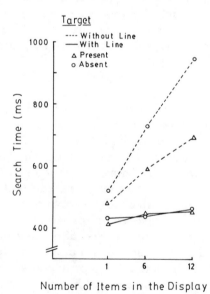

Figure 11. Search times for displays like those shown in Fig. 10. Search appears to be parallel for the presence of the intersecting line and serial for its absence.

objects and surfaces in the real world. These may be relational or "emergent" features, created by the combinations or arrangements of components. They would nevertheless be directly coded by specialized functional detectors at the preattentive level (see Pomerantz, Sager, & Stoever, 1977). We tried a number of different candidates. We did find pop-out for filled versus outline circles (see Figure 12a); however, we also found it for outline versus filled circles. No asymmetry was present, suggesting that both are coded preattentively as visual primitives. We found pop-out for blue among green distractors and also for green among blue (see Figure 12b). Again, both colors are presumably coded as separate features. We tested the supposedly more abstract property of "inside" versus "outside." Ullman (1984) has suggested that this distinction is difficult to extract computationally; however, it appears not to be hard for people, at least in the form in which we tested it. We used a circle containing a central dot contrasting with a circle with one dot at the same distance from the circumference but outside the circle. Both targets popped out independently of display size (see Figure 12c). However, the uniformity of the container shapes may make this a special case

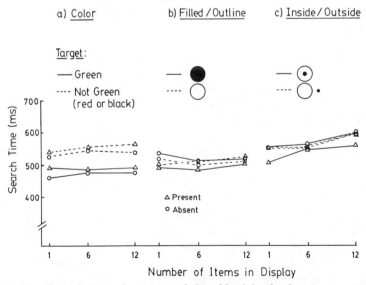

Figure 12.   Search times for targets defined by (a) color (green vs not green), (b) filled vs outline shape, and (c) inside vs outside dot. In each case the roles of targets and distractors were reversed in the paired condition. Search appeared to be unaffected (or hardly affected) by the number of distractors in all these tasks, suggesting that all six targets are characterized by a preattentively detectable feature which is absent from the distractors.

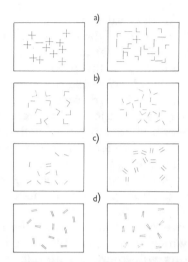

Figure 13. Examples of displays used to test search for targets defined by the arrangement or number of straight lines: (a) The target was in one case a plus and in the other case a separate line; (b) the target was in one case a right angle and in the other two separated orthogonal lines; (c) the target was in one case a pair of parallel lines and in the other a single line; (d) the target was in one case a pair of parallel lines and in the other a pair of converging lines.

of the inside-outside distinction—one which possesses an additional emergent feature. The bull's eye pattern might even be innately coded as part of a facial recognition system. A better test of the inside-outside discrimination would use varied shapes as containers and various dot positions within the shapes. We are now exploring this version of the inside-outside distinction.

Next we tried some features which may characterize line arrangements. The first was intersection, using displays such as those in Figure 13a. Notice that the stimuli without intersections shared as many other properties with the intersecting lines (the plus) as possible; in particular we removed the cue of global size which could have been used if the only stimuli were a plus and an "L" or "T" with lines of the same length as the plus. We did not find pop-out for intersection (see Figure 14a). There was no asymmetry in search latencies and search appeared to be serial in both cases. Julesz and Bergen (1983) suggested that intersection is a preattentive feature (or "texton"). However, texture discrimination in their displays could have been mediated by the global outline of the shapes; the pluses were more compact than the "L"s and this may have been preattentively detected.

Another potential feature formed by the arrangement of lines is juncture. If two lines are joined to form a right angle, as opposed to separated (see Figure 13b), would either the joined or the separate pairs pop out as a target among distractors? The answer is no: we did find an asymmetry, but neither target popped out. The joined lines gave slower search rates than the unjoined lines, but in both cases there was a substantial effect of the number of distractors (see Figure 14b). Next, we looked at Julesz's suggestion (1984) that the preattentive system can "count" or discriminate numbers at least up to four or five. The two stimuli we tested were a single line versus a pair of parallel lines (see Figure 13c). We found a clear asymmetry (see Figure 14c). The pair of lines pops out in a background of single lines, but the single line does not when embedded among pairs. It seems, unlikely, however, that the discrimination depends on counting, since the number one should be as discriminable from two as two is from one. It is possible that the pair of lines creates another "emergent" feature; two possible candidates might be parallelism and a form of "closure" in the sense of a partly enclosed space. The single lines possess no emergent feature relative to the paired lines and search is therefore serial. Finally, to distinguish whether the pairs pop out because they are parallel or because they form a partly enclosed area, we tested search with pairs of lines that were either

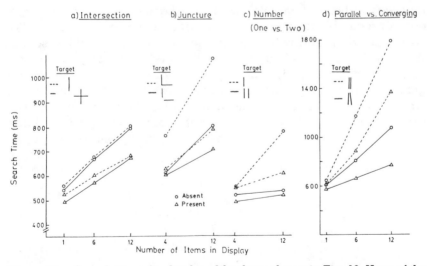

Figure 14. Search times for displays like those shown in Fig. 13. None of the targets appears to "pop out" of the displays, except the pair of parallel lines among separate single lines: (a) intersection; (b) juncture; (c) number (one vs two); (d) parallel vs converging.

parallel or converging (see Fig. 13d). Neither type of target popped out, although there was an asymmetry (see Fig. 14d). Serial search for the converging lines was faster than serial search for the parallel lines.

Closure is an interesting and controversial feature. Most subjects show good pop-out for a triangle in a background of angles and lines, but not for an arrow, made up of the same component lines (Treisman & Paterson, 1984; see Figure 15). The triangle contains an enclosed space surrounded by a connected contour, and we attributed the parallel detection to the presence of this emergent feature, which is absent both from the angles and lines and from the arrows. Julesz (personal communication) suggested as an alternative possibility that triangle pop-out is mediated by the absence of line ends or terminators rather than by closure. The arrow, on the other hand, has three terminators and therefore differs less than the triangle from the right angle and line distractors, which each have two. Julesz and Bergen (1983) tested texture segregation with stimuli like those in Fig. 16 and found no salient preattentive boundary, despite the presence of closure on one side of the display. Clearly, further experiments were needed.

The test we devised was to use circles with and without gaps (Treisman & Souther, 1985). The gaps produce line ends which are absent from closed circles. Subjects were asked to search for a target that was either a complete circle in a background of circles with gaps or a circle with a gap in a background of complete circles. We varied discriminability by making the gap larger or smaller, so that it was ⅛, ¼, or ½ the circumference. We predicted that if closure is a feature, it should mediate good pop-out for the complete circle targets. Similarly if line ends are features, we predict pop-out for the circles with gaps. Figure 17 shows examples of all the displays. The results were very clear and quite unexpected. There was a large asymmetry in search latencies (see Figure 18): circles with gaps pop

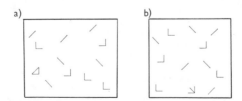

Figure 15. Examples of displays used to test search for a triangle in (a) and an arrow in (b). The triangle pops out of the display while the arrow, for many subjects, appears to require serial search.

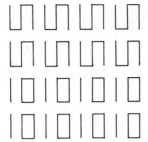

Figure 16. Example of stimuli used by Julesz and Bergen (1983) to test texture segregation based on closure. The boundary is not immediately apparent, suggesting that closure is not preattentively detected.

out and closed circles do not. Another interesting point to note is that there was no effect of discriminability when the target was the circle with a gap. Either line ends are present or they are not; their separation does not matter (once they are above the acuity threshold). On the other hand, if the target is a connected circle, the size of the gap makes a large difference. The asymmetry suggests that performance is controlled by a different feature in the two different target conditions. If subjects are looking for a closed circle, the degree of closure matters; if they are looking for terminators, their presence is deter-

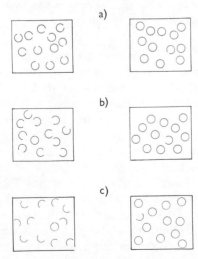

Figure 17. Examples of displays used to test search for targets defined either by a connected contour (closure) in the displays on the left, or by a gap (line ends) in the displays on the right; (a), (b), and (c) differ in the size of the gap.

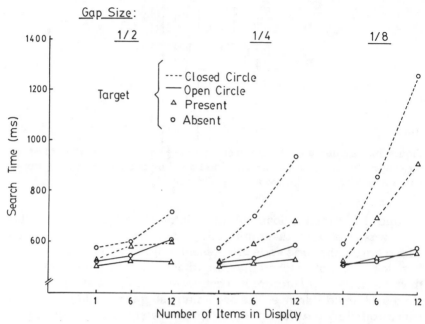

Gap Size:

Figure 18. Search times for complete circles and for circles with gaps in displays like those shown in Fig. 17. The size of the gap was half the circumference, or one quarter, or one eighth. The closed circle appears in each case to require serial search, although the rate varies with gap size, while the open circle with a gap usually pops out.

mined categorically. Julesz was right that closure, in the sense of a connected contour with no gap, is not (by this test) a categorically detected feature at the preattentive level. On the other hand, the absence of terminators does not allow pop-out, so Julesz's explanation for our triangle data was probably wrong. Triangles must have some other distinctive feature differentiating them from angles and lines; or perhaps they have closure by a different definition, where it means not connectedness but some sort of "blobness" or convex surround, where the closure can vary in degree (Rosenfeld, personal communication). The circles with gaps are closed (by this definition) to different degrees. Like the completely closed target, they activate the feature map for closure when they are the distractors, but they do so quantitatively less than the complete closed circle.

Another controversial feature is curvature. Neurophysiologists have not yet reported cells which respond selectively to curvature, but they may not have looked systematically for them. Behaviorally, we can ask whether curves pop out of straight line distractors, or whether straight lines pop out of curves. It is sometimes suggested that curves are coded as conjunctions of straight lines with changing orientations. If this were correct, they should not pop out of lines in mixed orientations like those in Figure 19. But in fact, these displays gave another surprising result. There was a large asymmetry which favored curves as the visual primitives, rather than straight lines (see Figure 20). Again there was a large effect of discriminability (in this case the degree of curvature) when the target was a straight line, but much less when the target was a curve. If we apply the presence/absence diagnostic for visual primitives, then we would have to infer that straightness is coded not as a positive feature, preattentively detected by the visual system, but simply as the absence of curvature. In other words, it might be the neutral point on the curvature dimension, while curves (or departures from straightness) are what the nervous system detects.

Line orientation is an obvious candidate for a visual primitive, and we showed earlier that it can mediate pop-out. In this case, there is considerable physiological evidence for early coding (Hubel & Wiesel, 1968). Beck (1967) showed long ago that orientation mediates easy texture segregation. We would therefore expect pop-out for any target and distractor pair that activate differently tuned orienta-

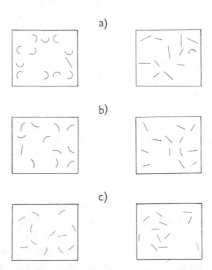

Figure 19. Examples of displays used to test search for a straight line and for a curved line. The curves in (a)–(c) differ in degree of curvature and therefore in discriminability from the straight line targets or distractors. The target in the displays on the left is a straight line and the target in the displays on the right is a curved line.

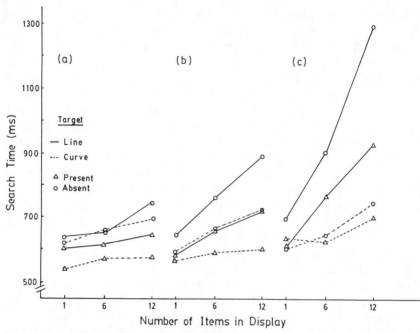

Figure 20. Search times for displays like those shown in Fig. 19. Curved targets appear to pop out of displays of straight lines, but straight line targets require serial checking to find them in displays of curved lines. Conditions (a)–(c) differ in the degree of curvature of the curves.

tion channels, and we would not expect any search asymmetry. We tested subjects with vertical targets among tilted lines and tilted targets among vertical lines (see Fig. 21a) and again obtained a surprising result (see Fig. 22a). The tilted line pops out, and the vertical line does not. Again, the presence/absence logic would suggest that the tilted lines have a feature that is preattentively coded and the vertical lines do not. Perhaps vertical, like straight, is coded as a neutral point on the orientation dimension, simply as the absence of tilt.

Perhaps, however, another explanation is possible. Rather than the target being the one item that is identified, perhaps the visual system encodes the whole display, so that the ease or difficulty of search is determined more by the nature of the distractors than by the nature of the target. It may be the fact that all the distractors are vertical and not the fact that the target is tilted that makes the tilted target easier to detect. Similarly, the displays with the curved target could be easy because all the distractors are straight lines rather than because the target is a curved line. We tested this account of

a)

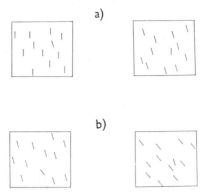

b)

Figure 21. Examples of displays used to test search for a target line differing from the distractor lines in orientation. (a) The target is a tilted line among vertical distractors on the left and a vertical line among tilted distractors on the right. (b) The target is a line that is more tilted than the distractors on the left and less tilted than the distractors on the right.

the orientation results by trying a search task which did not involve vertical lines. All the lines were tilted and the target was either more or less tilted than the distractors (see Figure 21b). Neither of these conditions should give pop-out if parallel processing depends on the distractors being vertical; but in fact they both do (see Figure 22b). The difficulty, therefore, does seem to arise when the target has the standard value rather than when the distractors do not.

Can we further define the nature of this standard value in the case of orientation? Is it vertical on the retina, or vertical with respect to the surrounding framework, which was a vertical and horizontal aperture in the tachistoscope? We tested this by putting an additional outline frame, which was either vertical or tilted, around the displays (see Figure 23b). If the difficulty arises when the target matches the frame, we would expect the tilted target to be difficult to find in the tilted frame. On the other hand, if the difficulty arises when the target is vertical on the retina, the frame should make no difference. The results showed a large effect of the frame (see Figure 24); there may in addition have been an effect of retinal (or gravitational) vertical, which reduced the effect of the frame, but the main difficulty in both cases arose when the target orientation matched the frame. The relevant codes determining performance seem, then, to be relative ones rather than those activating a particular subset of physiological feature detectors. Search is controlled by further operations relating different populations of feature detectors, and not by a single cell shouting! We certainly suspected this; pop-out in search

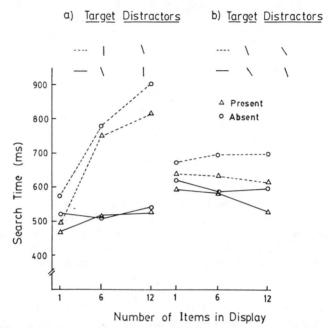

Figure 22. Search times for displays like those shown in Fig. 21. (α) The tilted target appears to pop out of displays of vertical lines, but the vertical target among tilted distractors does not. (b) Both the more tilted and the less tilted target pop out of tilted distractors.

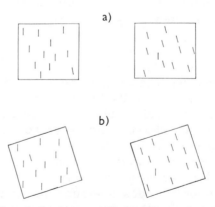

Figure 23. Examples of displays used to test search for vertical or tilted targets in displays (α) with a vertical frame, or (b) with a tilted frame. The targets is a tilted line in the displays on the left and a vertical line in the displays on the right.

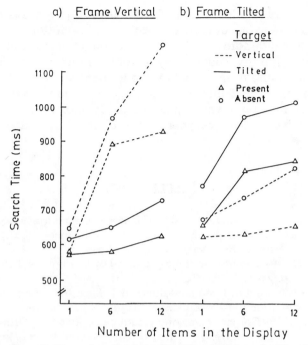

Figure 24. Search times for displays like those shown in Fig. 23. Search times are slower when the target orientation matches the frame rather than when the target orientation is vertical. (a) Frame vertical. (b) Frame tilted.

cannot be used as a psychological electrode inserted in single cells of the various visual areas. We are exploring the perceptual properties that are extracted from patterns of activity across different populations of cells, the functional codes that are presumably produced by combining the responses of many units.

Very briefly, what conclusions seem to emerge from these search experiments? I think they are consistent with the idea that early visual analysis results in separate maps for separate properties, and that these maps pool their activity across locations, allowing rapid access to information about the presence of a target, when it is the only item characterized by a particular preattentively detectable feature. Second, the pop-out test has so far uncovered rather few such features, which, I think, is encouraging. We would not want to assume that there are hundreds of visual primitives, each of which is separately represented in a map in the brain. Color, brightness, line ends or terminators, blobness or closure (in the sense of convexity), tilt and curvature appear to be among the visual primitives, whereas our data cast some doubt on other candidates such as intersection, juncture, number and connectedness. Several properties seem to be coded relative to one or more neutral points; deviations are signalled whereas the standard is not; we code curved not straight, and tilted not vertical. Finally, coding may be relative rather than absolute, as suggested by the frame effect on the vertical and tilted targets. The main contribution of this research may be to add some converging operations to other psychological methods of exploring early preattentive vision and of deciphering the elementary codes of perception.

REFERENCES

Barrow, H. G., & Tenenbaum, J. M. (1978). Recovering intrinsic scene characteristics from images. In A. Hanson & E. Riseman, (Eds.), *Computer Vision Systems*. New York: Academic Press.

Beck, J. (1967). Perceptual grouping produced by line figures. *Perception and Psychophysics, 2,* 491–495.

Beck, J. (1982). Textural segmentation. In J. Beck (Ed.), *Organization and Representation in Perception*. Hillsdale, N.J.: Erlbaum.

Cowey, A. (1979). Cortical maps and visual perception, The Grindley Memorial Lecture. *Quarterly Journal of Experimental Psychology, 31,* 1–17.

Garner, W. R. (1974). *The Processing of Information and Structure*. Potomac, MD: Erlbaum.

Hubel, D. H. & Wiesel, T. N. (1968). Receptive fields and functional characteristics of monkey striate cortex. *Journal of Physiology, 195,* 215–243.

Julesz, B. (1975). Experiments in the visual perception of texture. *Scientific American, 232,* 34–43.

Julesz, B. (1984). Toward an axiomatic theory of preattentive vision. In G. Edelman, M. Cowan, & E. Gall (Eds.), *Dynamic Aspects of Neocortical Function*. New York: Wiley.

Julesz, B., & Bergen, J. R. (1983). Textons, the fundamental elements in preattentive vision and perception of textures. *Bell System Technical Journal, 62,* 1619–1645.

Marr, D. (1982). *Vision: A Computational Investigation into the Human Representation and Processing of Visual Information*. San Francisco: Freeman.

Neisser, U. (1967). *Cognitive Psychology*. New York: Appleton-Century-Crofts.

Pomerantz, J. R., Sager, L. C., & Stoever, R. G. (1977). Perception of wholes and their component parts: Some configurational superiority effects. *Journal of Experimental Psychology: Human Perception and Performance, 3,* 422–435.

Towle, V. L., Harter, M. R., & Previc, F. H. (1980). Binocular interaction of orientation and spatial frequency channels: Evoked potentials and observer sensitivity. *Perception and Psychophysics, 27,* 351–360.

Treisman, A. (1982). Perceptual grouping and attention in visual search for features and for objects. *Journal of Experimental Psychology: Human Perception and Performance, 8,* 195–214.

Treisman, A. (1986). Properties, parts, and objects. In K. Boff, L. Kaufman, & J. Thomas (Eds.), *Handbook of Perception and Human Performance.* New York: Wiley.

Treisman, A., & Gelade, G. (1980). A feature integration theory of attention. *Cognitive Psychology, 12,* 97–136.

Treisman, A., & Paterson, R. (1984). Emergent features, attention and object perception. *Journal of Experimental Psychology: Human Perception and Performance, 10,* 12–31.

Treisman, A., & Schmidt, H. (1982). Illusory conjunctions in the perception of objects. *Cognitive Psychology, 14,* 107–141.

Treisman, A., & Souther, J. (1985). Search asymmetry: A diagnostic for preattentive processing of separable features. *Journal of Experimental Psychology: General, 12,* 3–17.

Ullman, S. (1984). Visual routines. *Cognition, 18,* 97–159.

Witkin, A. P., & Tenenbaum, J. M. (1983). On the role of structure in vision. In J. Beck, B. Hope, & A. Rosenfeld (Eds.), *Human and Machine Vision.* New York: Academic Press.

Zeki, S. M. (1981). The mapping of visual functions in the cerebral cortex. In Y. Katsuki, R. Norgren, & M. Sato, (Eds.), *Brain Mechanisms of Sensation.* New York: Wiley.

14
Aspects and Extensions of a Theory of Human Image Understanding*

Irving Biederman
State University of New York at Buffalo

Humans have a remarkable ability in object recognition. The ability is termed "remarkable" in that an object can typically be accurately and quickly recognized even when its image has never been previously experienced, or the image is extensively degraded, or the object is viewed from a novel orientation. Moreover, most often only a single, brief fixation is all that is required to achieve understanding of the image.

A recent proposal for how object recognition is achieved, Recognition-by-Components (RBC), assumes that an image of an object is segmented at regions of deep concavity into an arrangement of simple convex generalized cone primitives, such as cylinders, bricks, wedges, and cones (Biederman, 1987a). The central assumption of the theory is that the members of this particular set ($N = 36$) of primitives, called *geons* (for geometrical ions), are distinguishable on the basis of dichotomous or trichotomous contrastive properties of image edges, such as symmetry, curved vs. straight, and cotermina-

* This research was supported by the Air Force Office of Scientific Research (Grant 86-0106). I would like to express my deep appreciation to Thomas W. Blickle, Ginny Ju, H. John Hilton, Elizabeth A. Beiring, and Deborah Gagnon for their invaluable contributions to the research described in this article. I also thank Irvin Rock for several helpful suggestions on an earlier paper. The Overview section is modified from Biederman (1987a).

Correspondence about this paper should be addressed to Irving Biederman, who is now at the Department of Psychology, University of Minnesota, Elliott Hall, 75 East River Road, Minneapolis, Minnesota 55455.

tion of segments. These image properties can be determined from a general viewpoint and are highly resistant to degradation. Consequently, the geons, which are derived from these edge contrasts, themselves will be determinable under degradation and variations in viewpoint. A derivation of the theory is that the basic level classification of most single visual entities can be achieved from a specified arrangement of only two or three geons.

An update and two extensions of the theory—one to scene perception and the other to expert visual classification—are presented in this paper. In addition to various recent developments, results, and refinements of the theory, a general analysis of the problem of image understanding is presented. This analysis is useful, if not required, for articulating the *goal* of the computation in pattern recognition and it has important methodological implications for the kinds of data that will likely prove relevant for human image understanding.

AN ANALYSIS OF THE GOAL OF THE COMPUTATION FOR IMAGE UNDERSTANDING

There are at least three aspects of specifying the goal of any model of human image understanding that need specification: (a) the performance characteristics for which the proposed processes are relevant, (b) the class of images for which the presumed processes are operative, and (c) the processing environment.

Performance Characteristics

The fundamental problem addressed by RBC is how the edges extracted from an image of an object can be matched—in real time—to an appropriate representation of that object in memory.[1]

The core of the problem in specifying the goal of the computation for image understanding stems from the fact that an image projected from a real-world object generally provides a *redundant* description with respect to that object's classification. Consequently, it is possible to filter that image in various ways, demonstrate successful classification, and conclude that what was eliminated was not necessary for recognition. Presumably, the information that was passed was critical, or at least sufficient, for classification.

For example, in 1954, Attneave published a picture of a cat that

---

[1] Although the term *matching* an image to a representation will often be used in this chapter, it is not meant to exclude a parallel distributed model (e.g., Rumelhart, McClelland, & The PDP Research Group 1986) by which an image might *activate* a network that best represents that object.

had been modified by connecting, with straight lines, the points of maximum curvature (Figure 1, upper left). This image could be identified. But so could an image of Attneave's cat drawn by connecting the *midpoints* between points of maximum curvature as shown in Figure 1, lower right (Lowe, 1984). Marr (1982) invited his readers to note that the stylized silhouettes in Picasso's *Rites of Spring* could also be recognized. Kolers (1970) presented a page of pictures of somewhat unusual models of chairs and noted that each of the individual pictures could be classified as a chair.

From Attneave's (1954) and Lowe's (1984) cats, are we to conclude that curves have no effect on the time to identify an object? Are we to make a similar conclusion about the absence of internal contours and surface gradients and the distortion of silhouette contours in the *Rites of Spring*? Are we to conclude that all chairs are equally identifiable?

If the goal of the computation is to account for a possible basis for the *ultimate* classification of an image, then these demonstration suffice. Indeed, the information isolated from such demonstrations might serve as the basis for the design of an engineering effort in robot vision. But if the goal of the computation is to model the *real-time performance of human observers*, then such demonstrations can be misleading. The issue in real-time modeling is not whether some classification can ultimately be achieved by some subset of the information in the image, but whether the speed, accuracy, and attentional demands are affected by the filtering of a particular aspect of the image.[2] Let us refer to these two computational goals as ultimate and real-time classification. To a large extent, they are the goals for machine and human vision, respectively.

This somewhat lengthy exposition about the domain of recognition is motivated by a phenomenon that became apparent to us the very first time we viewed slides of objects at brief exposure durations: *Objects that can be identified with little subjective effort when viewed casually can differ markedly in their identifiability when presented at brief exposures.* The variation was not a consequence of a difference in size or contrast (which have their expected effects) but what might be called *typicality of depiction.* Typicality refers to the similarity of an image to a prototype for a given object class. Some objects, near the prototype, could be identified quickly at near perfect accuracy at a 100-msec. masked exposure duration, but others, dissimilar to the prototype, would have error rates of 30-40 percent

---

[2] Pisoni, Nusbaum, and Greene (1985) have been employing such *performance based* criteria in evaluating the intelligibility of particular systems for generating artificial speech.

1078   (17)

689   (0)

845   (42)

697   (0)

939   (39)

714   (0)

Figure 1.   Six cats with their naming reaction times and error rates (in parenthesis) from a 100 msec. masked exposure. *Upper left.* The original version of Attneave's (1954) cat constructed by connecting points at extrema of curvature. *Middle left:* Removal of the eye slit resulted in markedly higher error rates. *Lower left:* Lowe's (1984) cat constructed from Attneave's by connecting points midway between points at extrema of curvature. *Right side:* Three cats were curved regions retained. These could be identified perfectly with markedly lower reaction times. The middle cat on right is from Snodgrass and Vanderwart (1980).

373

and long naming reaction times even with a doubling of the exposure duration.

Biederman and Hilton (1987a) recently completed an experiment where we explored some of these differences in a controlled design. Subjects saw a series of slides of line drawings of common objects, each of which could be readily named with a basic-level term. The slides were shown for 100 msec. followed by a mask. The mean correct naming reaction times (RTs) and error rates for simple "standard" line drawings of cats in various postures averaged 700 msec. and 0% errors. Attneave's cat (Figure 1) required 1078 msec. and had a 17% error rate. Merely removing the segment that represented that cat's eye (Figure 1, right middle) increased the error rate to 42%! Although with an RT of 939 msec. Lowe's cat could be named somewhat faster than Attneave's, its error rate was considerably higher in that it was missed on 39% of its exposures. An office chair, made up only of simple volumetric parts, could be named in 671 msec., with no errors. Kolers' Bentwood rocking chair required 1129 msec. and could not be identified on 52.8% of the trials!

The approximately 300 msec. increase in naming RTs for the modified or stylized images, though representing an increase of 43% over the RTs for the original images, actually represents a considerably greater increase when only the time for recognition is considered. Approximately 600 msec. of the 700 msec. mean naming RTs for the standard or original versions of the objects is used for the selection and execution of an overt naming response, as assessed by techniques demonstrating that targets can be detected when embedded in a train of images, each presented for approximately 100 msec. (Intraub, 1981). The 300 msec. increase in RTs for the stylized images then represents a tripling of the time devoted to perceptual recognition. Even this value would have to be increased considerably to bring the extremely high error rates to levels comparable to the prototypes.

The problem here is that the increased time for the stylized or modified images may allow them to be recognized as symbols or through inference or through several iterations of a more basic process, rather than through an initial direct mapping of a representation of image contours to a stored representation of an object. The goal of the computations for ultimate identifiability as applied to human vision thus becomes a most elusive one, in that there are likely a number of paths, potentially obeying different constraints, by which an image might receive classification. The important methodological point from our experimental results is that the casual viewing of images is unlikely to offer much insight to the operation of the various real-time mechanisms for human vision.

Class of Images

The issue here is that different kinds of images might be ultimately identified by different processes. Is the model designed to account for the recognition of objects as they might appear in their real-world configuration? Textured surfaces? Arbitrary symbols? Any visual input? RBC only purports to account for the recognition of objects with specifiable boundaries. These entities are typically designated with *count* nouns in English. For such objects we can use the indefinite article or number, as in "a giraffe" or "four chairs." In English, we use *mass* nouns to refer to concrete entities that do not have specifiable boundaries, such as "water," "snow," or "sand." We conjecture that the latter class of objects are recognized through their surface properties and position in a scene.

Processing Environment

There are two aspects of the processing environment that are critical in evaluating image understanding efforts: (a) the number of possible objects, and (b) the presence of contextual constraints.

Many robot vision models (e.g., Knoll & Jain, 1986) are designed to identify a single object from a small class of a dozen or so well-specified objects. One can then capitalize on unique contours of the objects to distinguish each from the others. However, these contour descriptions might not be those employed when attempting to identify an object from a larger and unspecified class of other possible objects. Another problem with such schemes is that the diagnostic contours are rarely those that would be employed by a human. Consequently, interactions with such systems are difficult in that neither errors nor the effects of altering the set of objects are transparent to human intuitions.

The most obvious contextual constraints are those provided by real-world scenes which often allow highly occluded or degraded objects to be recognized (Biederman, 1987a). Other kinds of contextual constraints arise from mere repetition: ambiguous stimuli are often identified as instances of whatever appears to be repeated. Artists have capitalized on this effect for centuries. In painting an army or forest of thousands of men or trees, it is unnecessary to paint each and every man or tree. Only the individual entities in the foreground need be accurately depicted and the rest are inferred through a texture gradient. It is likely that the apparent ease of identifying any one of Kolers' chairs was facilitated by the presence of other chairs on the same page.

RBC is directed toward modeling the identification of unanticipated, isolated objects. This allows us to defer study of (a) those case where subjects employ a limited set of cues in *discriminating* one object from another object (or a small set of other objects) known beforehand, and (b) the constraints furnished by a scene or display context.

Primal Access

Based on these considerations, we defined the goal of our modeling to account for what we term *primal access:* The initial classification of an unanticipated, isolated concrete count-noun object.

The *initial classification* of an object allows us to concentrate on perceptual recognition rather than the *consequences* of perceptual recognition. For example, in looking at a chair one might remember where it was purchased and its cost and that it is a member of the class FURNITURE. But our primary interest is in how the image was originally classified as CHAIR. Often, but not always, the initial categorization of an isolated entity will be at a *basic level*, as when we know that a given object is a typewriter, banana, or giraffe (Rosch, Mervis, Gray, Johnson, & Boyes-Braem 1976). Much of our knowledge about objects is organized at this level of categorization—the level at which there is typically some readily available name to describe that category (Rosch *et al.*, 1976). RBC predicts, however, that when the componential description of a particular, familiar subordinate differs substantially from a basic-level prototype, categorizations will initially be made at the subordinate level. Thus, we might know that a given object is a floor lamp, penguin, sports car, or dachshund, more rapidly than we know that it is a lamp, bird, car, or dog (e.g., Jolicoeur, Gluck, & Kosslyn, 1984). [For both theoretical and expository purposes, these familiar non-prototypical members (subordinates) of basic level categories will also be considered basic level in this article.][3]

Implications of primal access for experimental research strategy. Although our goal is to study the perception of objects in their original appearance, our experimental strategy often requires altering an object's appearance in some specified manner in order to study

---

[3] It is possible that what J. J. Gibson (1966) termed *affordances* can also be included as part of the initial classification. Thus the configuration of a chair can be said to *afford sitting* or that a ball can *afford rolling*, in that the potentiality of certain stimulus or response properties might be directly elicited from the appearance of the object. This aspect of RBC has not yet been explored, but the stimulus descriptions provided by RBC would appear to allow a broad range of relatively direct functional implications.

primal access. If the image modification does not affect identification speed and accuracy compared to the original image, we assume that the modification was not a controlling factor in primal access but might serve as a backup route for identification. If we can identify a giraffe from a line drawing as quickly as we can from a full-color image, then we might conclude that color and texture were not controlling primal access. If we are ultimately able to identify a giraffe from a color patch of its coat, this identification will not (most likely) be as rapid as a line drawing of the intact giraffe. The color and texture is then furnishing only a secondary route for recognition. Classification through such secondary routes need not be under voluntary or strategic control. More likely is the possibility that activation starts from several routes simultaneously. When looking at a color photograph of a giraffe we might be processing both its contours and its surface properties. When the most direct routes for classification are unsuccessful, as when we only have a patch of fur, then identification will be achieved by a secondary route. In such cases, the time for identification should be slower than when recognition can be achieved by the primary routes. For this reason, the time for identification assumes a central role in our modeling of recognition.

BASIC PHENOMENA OF OBJECT RECOGNITION

Independent of laboratory research, the phenomena of everyday object identification provide strong constraints on possible models of this process. In addition to the fundamental phenomenon that objects can be recognized at all (not an altogether obvious conclusion from just a statement of the problem and the study of the brain), at least five facts are evident: Typically, an object can be recognized

1. Rapidly
2. When viewed most from novel orientations
3. Under moderate levels of visual noise
4. When partially occluded
5. When it is a new exemplar of a category

Implications

The preceding five phenomena constrain theorizing about object interpretation in three ways.

Implication 1: Access to the mental representation of an object should not be dependent on absolute judgments of quantitative detail, because such judgments are slow and highly error-prone (Miller, 1956; Garner, 1962). The degree of curvature of a segment can be highly diagnostic as to an object's identity and serves as the basis of a number of robot vision schemes (e.g., Knoll & Jain, 1986). Whatever value curvature might possess for a robot's eye, however, humans will typically require more time to distinguish among just several levels of curvature than that required for the identification of the object itself. Consider, for example, Figure 2, which shows seven edges differing in curvature. Imagine an experiment in which a subject would study these curves and then be presented with one curve in isolation, such as the one shown in Figure 3. The subject's task is to provide the number of the curve that matched the test curve. The time required to perform this type of absolute judgment task will typically exceed, by an order of magnitude, the time required to actually name an object. Consequently, such quantitative processing cannot be the controlling factor by which object recognition is achieved.

Implication 2: The information that is the basis of recognition should be relatively invariant with respect to orientations in depth and noise. Varying the orientation or length of the test curve in Figure 3 or interrupting its contour would increase the difficulty of that task far more than the resultant disruption on the identification of an object.

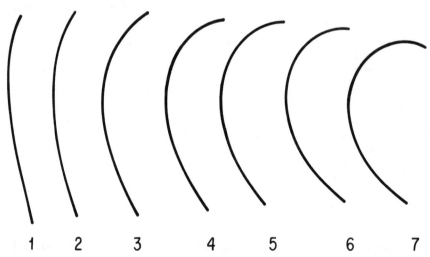

Figure 2.   A set of stimuli in an absolute judgment task of segment curvature.

Test Stimulus

?

Figure 3.   A test stimulus for the stimulus set in Figure 2. The correct an-swer is "4." The time required to make such absolute judgments is substan-tially longer than the time required to identify an object.

Implication 3: Partial matches should be computable. A theory of object interpretation should have some principled means for comput-ing a match for occluded, partial, or new exemplars of a given catego-ry. We should be able to account for the human's ability to identify, for example, a chair when it is partially occluded by other objects, or when it is missing a leg, or when it is a new model. As will be argued below, it is important to distinguish the deletion of parts of an object's contours when produced by noise, as specified in Implication 2, from conditions where an object is either missing parts or when parts are occluded, as specified in Implication 3. These two conditions of con-tour deletion have different consequences for perception.

RECOGNITION-BY-COMPONENTS: AN OVERVIEW

RBC bears some relation to several prior schemes for representing objects by parts or modules (e.g., Binford, 1971; Brooks, 1981; Guzman, 1971; Marr, 1977; Marr & Nishihara, 1978; Tversky & Hemen-way, 1984). RBC's contribution lies in its proposal for a particular vocabulary of components derived from perceptual mechanisms and its account of how an arrangement of these components can access a representation of an object in memory.

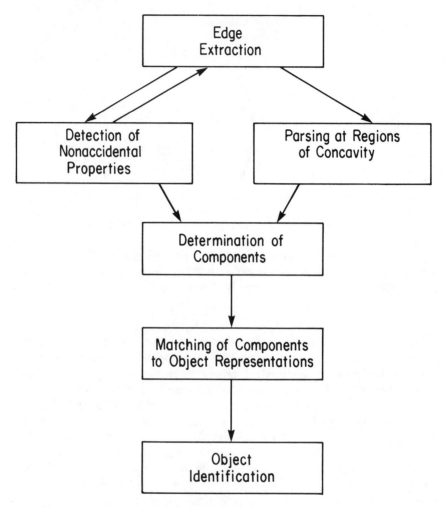

Figure 4.   Presumed processing stages in object recognition. From Biederman (1987a), Fig. 2, p. 118.

Stages of Processing

Figure 4 presents a flowchart of the presumed subprocesses by which an object is recognized. The stages need not proceed in strict sequential fashion in that it would be possible, if not likely, for the stages to be arranged in cascade whereby partial activation (processing) at one level is sufficient to initiate activation at the next (McClelland, 1979).

An early edge extraction stage, responsive to differences in surface characteristics, viz., sharp changes in luminance or texture, provides an edge-based description of the object. Edge extraction is assumed to be accomplished by a module that can proceed independently of the later stages, save for top-down influences that are likely to be in evidence under conditions where the image is difficult or ambiguous. No particular edge extraction scheme is assumed by RBC. Nonaccidental properties of the image's edges (e.g., collinearity, symmetry) are determined simultaneously with a parsing of the image. The parsing is performed primarily at matched concave regions. The nonaccidental properties of the parsed regions provide critical constraints on the identity of the components. Within the temporal and contextual constraints of primal access, the stages up to and including the identification of components are assumed to be bottom-up.[4] A delay in the determination of an object's components should have a direct effect on the identification latency of the object.

The arrangement of the components is then matched against a representation in memory. It is assumed that the matching of the components occurs in parallel, with no loss in capacity when matching objects with a large number of components. Partial matches are possible with the degree of match assumed to be proportional to the similarity (described below) in the componential descriptions of a representation of the image and the memorial representation. This stage model is presented to provide an overall theoretical context. The focus of this article is on the nature of the representation.

Parsing. RBC assumes that a representation of the image is segmented—or parsed—into separate regions at points of deep concavity. Support for parsing at a given region could be variable (see Connell, 1985) but is undoubtedly strongest at matched deep concavities (Marr & Nishihara, 1978), particularly where there are discontinuities in minima of negative curvature (cusps). In general, matched concavities will arise whenever convex volumes are arbitrarily joined, a principle that Hoffman & Richards (1985) term *transversality*. Such segmentation conforms well with human intuitions about the boundaries of object parts and does not depend on

---

[4] The only top-down route shown in Figure 4 is an effect of the nonaccidental properties on edge extraction. However, as with the nonaccidental properties, a top-down route from the component determination stage to edge extraction could precede independent of familiarity with the object itself. It is also likely that other top-down routes, such as those from expectancy, object familiarity or scene constraints (e.g., Biederman, Mezzanotte, & Rabinowitz, 1982) will be observed at a number of the stages, for example, at segmentation, component definition, or matching, especially if edges are degraded.

familiarity with the object, as can be demonstrated with the non-sense object in Figure 5 (see also Connell, 1985; Bower & Glass, 1976).

Geons. Each segmented region is then approximated by one of a possible set of convex or singly concave components, called *geons*, that can be modeled by generalized cones (Binford, 1971; Marr, 1977, 1982). A generalized cone is the volume swept out by a cross-section moving along an axis (as illustrated in Figure 9 below). Marr (1977, 1982), showed that the contours generated by any smooth surface could be modeled by a generalized cone with a convex cross section. The cross-section is typically hypothesized to be at right angles to the axis. Secondary segmentation criteria (and criteria for determining the axis of a component) are those that afford descriptions of volumes that maximize symmetry, axis length, and constancy of the

Figure 5.   There is strong consensus in the segmentation of this configuration and in the description of its parts despite its overall unfamiliarity. From Biederman (1987a) Fig. 1, p. 116.

size and curvature of the cross-section of the component. Of these, axis length often provides the most compelling subjective basis for selecting subparts, perhaps because computations of symmetry could then be determined over a narrower extent (see Brady, 1983; Brady & Asada, 1984; Connell, 1985).

The fundamental perceptual assumption of RBC is that the components can be differentiated on the basis of dichotomous or trichotomous contrasts of properties of edges in the 2-D image that are readily detectable and relatively independent of viewing position and degradation, such as curved vs. straight, parallel vs. non-parallel. These 2-D image properties allow strong inference as to the existence of these same properties in the 3-D edge projecting them and have been termed "nonaccidental" by Lowe (1984) but many were noted earlier by Rock (1975) as a tendency of the visual system to "avoid coincidence." These terms reflect the possibility that the 2-D image property in question could have been produced by an "accidental" or "coincidental" alignment of image and eye. For example, an image (2-D) of a straight edge could have been produced by an alignment of a curved edge (3-D) where the axis of curvature was exactly aligned with the viewpoint. The visual system appears to disregard consideration of such rare possibilities when interpreting an image of an object. Moreover, even when victimized by an accident, the damage is likely to be brief and minor in that slight alterations of viewpoint will readily correct any error.

Relations among the components. Although the components themselves are the focus of this paper, as noted previously the arrangement of primitives is necessary for representing a particular object. Thus an arc *side-connected* to a cylinder can yield a cup as shown in Figure 6. Different arrangements of the same components can readily lead to different objects, as when an arc is connected to the top of the cylinder to produce a pail in Figure 6. Whether a component is attached to a long or short surface can also affect classification as with the arc producing either an attache case or a strongbox in Figure 6.

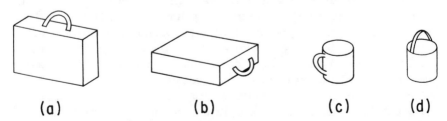

(a)   (b)   (c)   (d)

Figure 6.   Different arrangements of the same components can produce different objects. From Biederman (1987a) Fig. 3, p. 119.

The identical situation between primitives and their arrangement exists in the phonemic representation of words, where a given subset of phonemes can be rearranged to produce different words.

The representation of an object would thus be a structural description that expressed the relations among the components (e.g., Winston, 1975). A suggested (minimal) set of relations is described in a later section (Table 1), and would include specification of the relative sizes and positions of the components and their points of attachment.

Noncomponential Indexing

The model shown in Figure 4 provides a basis for modeling primal access based on a matching of a structural description of an arrangement of components of the image to a like representation in memory. Not shown in the figure is a route by which the representation of an object might be partially indexed directly by other aspects of the image, such as color or aspect ratio. It is my guess that such indexing has its effect through bottom-up inhibition of object representations that are inconsistent with image features rather than through activation of object representations that are consistent. In this sense, they would exert effects similar to that proposed by McClelland and Rumelhart (1981) for inconsistent features on letter identification. For example, the detection of a highly elongated image or the color red might inhibit the representations of TABLE and MUSHROOM, respectively, but not necessarily activate PEN and KETCHUP. These inhibitory indexing routes differ from the kind proposed by McClelland and Rumelhart (1981) in that they would bypass the componential analysis described in Figure 4 to work directly on the thresholds for the various object representations. In word perception (or reading), it may be the case that word *length* ultimately receives similar theoretical treatment (cf. Rayner, 1975).

Color and texture. By the preceding account, after the extraction of image edges, surface characteristics such as color and texture will typically have only secondary roles in primal access. This should not be interpreted as suggesting that the perception of surface characteristics *per se* is delayed relative to the perception of the components but merely that in most cases the surface characteristics are generally less efficient routes for accessing the classification of a count object. That is, we may know that a chair has a particular color and texture simultaneously with its componential description, but it is only the volumetric description that provides efficient access to the mental representation of CHAIR.

Silhoutte vs. internal contours. No special status is accorded the silhouette (or bounding contour) of an image. Often, however, the silhouette will provide sufficient contour to activate the representation of an object. This could be accomplished with information relevant to the geons' identity or through indexing the object's representation through its aspect ratio and axis structure. As Figure 7 illustrates, images can readily be found in which the silhouette is a less effective image for recognition than the internal contours.

RBC AND PERCEPTUAL ORGANIZATION

The nonaccidental relations include several that have traditionally been thought of as principles of perceptual organization, such as good continuation and symmetry (Kanizsa, 1979; Rock, 1983). When combined with the parsing principle of segmenting visual patterns at cusps, the nonaccidental properties will produce the Gestalt phenomenon of Good Figure or Pragnanz. RBC thus provides a principled account of the relation between the classic phenomena of perceptual organization and pattern recognition: although objects can be highly complex and irregular, the units by which objects are identified are simple and regular. The constraints toward reg-

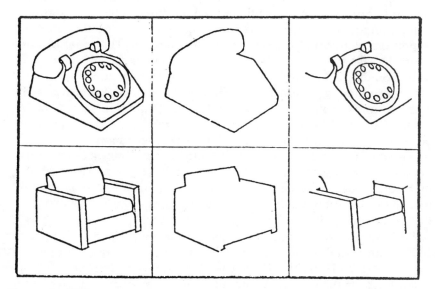

Figure 7. Silhouettes do not always result in better identification than internal contours.

ularization (Pragnanz) are thus assumed to characterize not the complete object but the object's components. In many respects entities like chairs, airplanes, and elephants are "Bad Figures." As will be discussed later in the section on the perception of partial objects, RBC predicts that such complex patterns will be more readily identifiable than simple ones, such as a flashlight or spoon.

NONACCIDENTAL PROPERTIES: A PERCEPTUAL BASIS FOR A COMPONENTIAL REPRESENTATION

Recent theoretical analyses of perceptual organization (Binford, 1981; Lowe, 1984; Witkin & Tenenbaum, 1983; but see Rock, 1983 for an earlier discussion) suggest a perceptual basis for RBC. As noted earlier, the central organizational principle is that certain *nonaccidental* properties of edges in a 2-D image are taken by the visual system as strong evidence that the edges in the 3-D world contain those same properties. For example, if there is a straight line in the image (*collinearity*), the visual system infers that the edge producing that line in the 3-D world is also straight. The visual system ignores the possibility that the property in the image is merely a result of a (highly unlikely) accidental alignment of eye and a curved edge. Smoothly curved elements in the image (*curvilinearity*) are similarly inferred to arise from smoothly curved features in the 3-D world.

If an image region is symmetrical (*symmetry*), we assume that the region projecting that image is also symmetrical. The order of symmetry is also preserved: Images that are symmetrical under both reflection and 90° increments of rotation, such as a square or circle, are interpreted as arising from objects (or surfaces) that are symmetrical under both rotation and reflection. [Although skew symmetry is often readily perceived as arising from a tilted symmetrical object or surface, there are cases where skew symmetry is not readily detected (Attneave, 1982).] When edges in the image are *parallel* or *coterminate* we assume that the real world edges also are parallel or coterminate, respectively. These five nonaccidental properties and the associated 3-D inferences are illustrated in Figure 8 (modified from Lowe, 1984).

Witkin and Tenenbaum (1983, see also Lowe, 1984, and Pentland, 1986) argue that the leverage provided by the nonaccidental relations for inferring a 3-D structure from 2-D image edges is so great as to pose a challenge to the effort in computational vision and perceptual psychology that assigned central importance to variation in local *surface* characteristics, such as luminance gradients, from which surface cur-

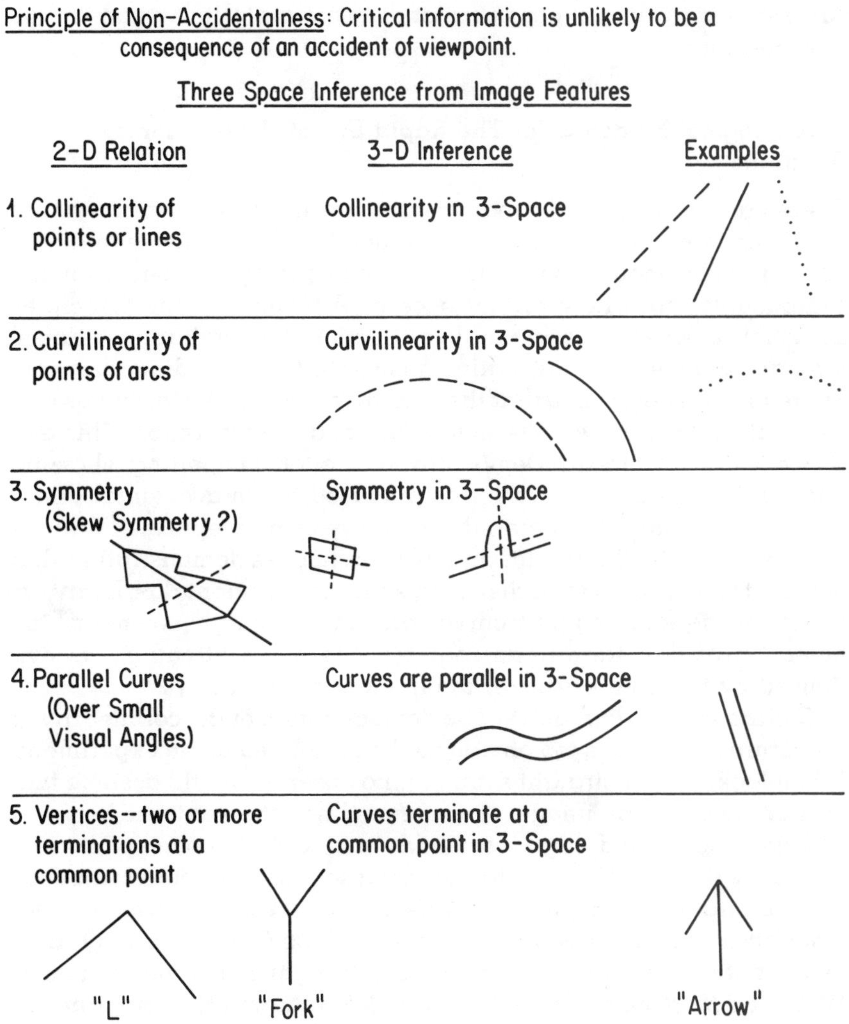

Principle of Non-Accidentalness: Critical information is unlikely to be a
consequence of an accident of viewpoint.

Three Space Inference from Image Features

2-D Relation	3-D Inference	Examples
1. Collinearity of points or lines	Collinearity in 3-Space	
2. Curvilinearity of points of arcs	Curvilinearity in 3-Space	
3. Symmetry (Skew Symmetry ?)	Symmetry in 3-Space	
4. Parallel Curves (Over Small Visual Angles)	Curves are parallel in 3-Space	
5. Vertices--two or more terminations at a common point	Curves terminate at a common point in 3-Space	

"L"   "Fork"   "Arrow"

Figure 8. Five nonaccidental relations [Adapted from Lowe, 1984.]

vature could be determined (as in Marr, 1982; Knoll & Jain, 1986). Although a surface property derived from such gradients will be invariant over some transformations, Witkin and Tenenbaum (1983) demonstrate that the suggestion of a volumetric component through the
shape of the surface's silhouette can readily override the perceptual
interpretation of the luminance gradient. The psychological literature, summarized in the next section, provides considerable evidence supporting the assumption that these nonaccidental properties

can serve as primary organizational constraints in human image interpretation.

Psychological Evidence for The Rapid Use of Nonaccidental Relations

There can be little doubt that images are interpreted in a manner consistent with the nonaccidental principles. But are these relations used quickly enough so as to provide a perceptual basis for the components that allow primal access? Although all the principles have not received experimental verification, the available evidence strongly suggests an affirmative answer to the preceding question. There is strong evidence that the visual system quickly assumes and uses collinearity, curvature, symmetry, and cotermination. This evidence is of two sorts: (a) Demonstrations, often compelling, showing that when a given 2-D relation is produced by an accidental alignment of object and viewpoint, the visual system accepts the relations as existing in the 3-D world, and (b) search tasks demonstrating that when a target differs from distractors in a nonaccidental property, as when one is searching for a curved arc among straight segments, the detection of that target can require little or no attention, as evidenced by minimal effects of the number of distractors.

Collinearity vs. curvature. The demonstration of the collinearity or curvature relations is too obvious to be performed as an experiment. When looking at a straight segment, no observer would assume that it is an accidental image of a curve. That the contrast between straight and curved edges is readily available for perception was shown by Neisser (1963). He found that a search for a letter composed only of straight segments, such as a Z, could be performed faster when in a field of curved distractors, such as C, G, O, and Q, then when among other letters composed of straight segments such as N, W, V, and M. More recently Treisman (1988) showed that preattentive discrimination was possible for curved segments among straight segments. Pentland (1986) reported a similar effect for shapes with a curved axis when embedded in fields of shapes with straight axes.[5]

Symmetry and parallelism. Many of the Ames demonstrations (Ittleson, 1952), such as the trapezoidal window and Ames room, derive from an assumption of symmetry (and parallelism). Treisman (1988)

---

[5] Often marked asymmetries are found in the detectability of one pole of the contrast, when the other serves as distractor. Almost always it is the standard that is the *more* difficult target. Thus, ellipses pop out from circles, but not circles from ellipses; nonparallel pairs of segments pop out from parallel pairs but not parallel from non-

showed that an ellipse can be detected preattentively among circles. King, Tangney, Meyer, & Biederman (1976) demonstrated that a perceptual bias towards symmetry contributed to apparent shape constancy effects. Garner (1974), Checkosky & Whitlock (1973), and Pomerantz (1978) provided ample evidence that not only can symmetrical shapes be quickly discriminated from asymmetrical stimuli, but the degree of symmetry was also a readily available perceptual distinction. Thus stimuli that were invariant under both reflection and 90° increments in rotation could be rapidly discriminated from those that were only invariant under reflection (Checkosky & Whitlock, 1973).

Once we have evidence that an asymmetrical surface is extended in depth, we might be *more* likely to perceive it as symmetrical (King, Tangney, Meyer, & Biederman, 1977). The reason for this, I believe, is that viewpoint and object slant are subject to error. Consequently, if there is a viewpoint within the possible slant values that would render a tilted surface symmetrical, we readily see it that way. By allowing some variation in depth that is subject to error, there may be more tolerance with which to allow a regularization bias.

Cotermination. The "peephole perception" demonstrations, such as the Ames chair (Ittleson, 1952) or the physical realization of the "impossible" triangle (Penrose & Penrose, 1958), are produced by accidental alignment of the ends of noncoterminous segments to produce—from one viewpoint only—L, Y, and Arrow vertices.

With polyhedra (volumes produced by planar surfaces), the Y, Arrow, and L vertices allow inference as to the identity of the volume in the image. The Y vertex is produced by the cotermination of three segments, with none of the angles greater than 180°. The tangent Y

---

parallel; curves popout from straight edges but straight edges do not popout from curves; magenta pops out from red but red does not pop out from magenta (Treisman, In press). A possible explanation for this is that the "standard," e.g., parallelism, is detected by a more broadly tuned detector so that a nonparallel pair of edges with just moderate convergence, would activate not only a detector for convergence (say) but also one for parallelism. The standard would activate only the parallel detector. Consequently, when a parallel pair of edges is the target, the nonparallel distractors also excite the parallel detectors and the detection must be made on the basis of a difference in degree of activation. When a nonparallel pair of edges is the target, the subject merely has to check a location where there is activation of a nonparallel detector. This speculation then accounts for why it is the standard (vs. the nonstandard) that appears to be the one that is most difficult to detect. It also provides an explanation of the biases favoring the standard, for example in the trapezoidal window and Ames room in the case of parallelism, in the image interpretation.

vertex present in a cylinder can be distinguished from the Y or Arrow vertices in that the termination of the segments in the curved Y are tangent (Chakravarty, 1979). An arrow vertex, also formed from the cotermination of three segments, contains an angle that exceeds 180°. An L vertex is formed by the cotermination of two segments. The presence of vertices composed of three segments, such as the Y and Arrow (and their curved counterparts) are important in that a given region is part of a volumetric body, rather than planar. Planar components lack three pronged-vertices. As shown in Figure 11 (below), the structural description of a geon specifies the type and location of the image's vertices.

The T vertex represents a special case in that it is not a locus of cotermination (of two or more segments), but only the termination of one segment on another. Such vertices are important for determining occlusion and thus segmentation (along with concavities), in that the edge forming the (normally) vertical segment of the T cannot be closer to the viewer than the segment forming the top of the T (Binford, 1981). By this account, the T vertex might have a somewhat different status than the Y, Arrow, and L vertices, in that the T's primary role would be in segmentation, rather than in establishing the identity of the volume.[6]

COMPONENTS GENERATED FROM CONTRASTS IN NONACCIDENTAL PROPERTIES AMONG GENERALIZED CONES

The particular set of nonaccidental properties shown in Figure 8 may constitute a principled basis for the generation of a set of perceptually plausible components. Any primitive that is hypothesized to be the basis of object recognition should be rapidly identifiable and invariant over viewpoint and noise. These characteristics would be attainable if differences among components were based on contrasts in nonaccidental properties. Although additional nonaccidental properties exist, there is empirical support for rapid perceptual access to the five described in Figure 8. In addition, these five relations reflect intuitions about significant perceptual and cognitive differences among objects.

From variation over only two or three levels in the nonaccidental

---

[6] The arrangement of vertices, even for curved objects, offers constraints on "possible" interpretations of lines as convex, concave, or occluding, Malik, 1987.

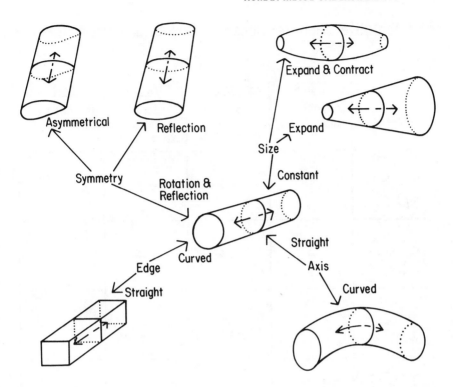

Figure 9.   An illustration of how variations in three attributes of a cross-section (curved vs. straight edges; constant vs. expanded vs. expanded and contracted size; mirror and rotational symmetry vs. mirror symmetry vs. asymmetrical) and one of the shape of the axis (straight vs. curved) can generate a set of generalized cones differing in nonaccidental relations. Constant-sized cross-sections have parallel sides; expanded or expanded and contracted cross-sections have sides that are not parallel. Curved vs. Straight cross-sections and axes are detectable through Collinearity or Curvature. The three values of cross-section Symmetry (Symmetrical under Reflection & 90° Rotation; Reflection only; or Asymmetrical) are detectable through the symmetry relation. Shown here are the neighbors of a cylinder. The full family of generalized cones has 36 members. From Biederman (1987a) Fig. 6, p. 122.

relations of four attributes of generalized cylinders, a set of 36 components can be generated. A subset is illustrated in Figure 9.

Six of the generated volumes (and their attribute values) are shown in Figure 10. [Additional volumes are shown in Biederman (1985, 1987a).] Three of the attributes describe characteristics of the cross section: Shape of edge (Straight vs. Curved); Symmetry (Invariant under Reflection and 90° Rotation; vs. Invariant under Reflection

Partial Tentative Geon Set Based on Nonaccidentalness Relations

CROSS SECTION

Geon	Edge Straight + Curved –	Symmetry Rot & Ref ++ Ref + Asymm –	Size Constant ++ Expanded – Exp & Cont – –	Axis Straight + Curved –
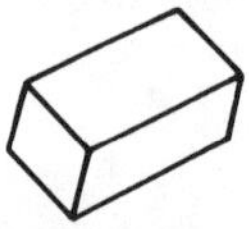	+	+ +	+ +	+
	–	+ +	+ +	+
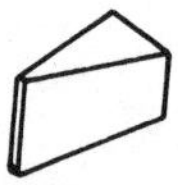	+	+	–	+
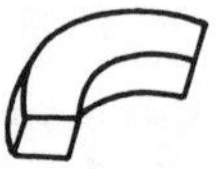	+	+ +	+	–
	–	+ +	–	+
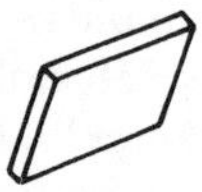	+	+	+	+

Figure 10. Proposed partial set of volumetric primitives (Geons) derived from differences in nonaccidental properties. Modified from Biederman (1987a) Fig. 7, p. 122.

vs. Asymmetrical); Size of cross section (Constant [parallel edges] vs. Expanded [nonparallel edges] vs. Expanded and Contracted [Nonparallel edges with a point of maximum convexity]). The fourth attribute describes the shape of the axis (Straight vs. Curved).

Nonaccidental 2-D Contrasts Among the Components

As indicated in Figure 9, the values of the four generalized cone attributes can be directly detected as contrastive differences in nonaccidental properties: straight vs. curved, symmetrical vs. asymmetrical, parallel vs. nonparallel (and if nonparallel, whether there is a point of maximal convexity). Cross-section edges and curvature of the axis are distinguishable by collinearity or curvilinearity. The constant vs. expanded size of the cross-section would be detectable through parallelism; a constant-sized cross-section would produce a generalized cone with parallel sides (as with a cylinder or brick); an expanded cross-section would produce edges that were not parallel (as with a cone or wedge). A cross-section that expanded and then contracted would produce an ellipsoid with nonparallel sides and extrema of positive curvature (as with a lemon). Such extrema are invariant with viewpoint (e.g., Hoffman & Richards, 1985) and actually constitute a sixth nonaccidental relation. The three levels of cross-section symmetry are equivalent to Garner's (1974) distinction as to the number of different stimuli produced by increments of 90° rotations and reflections of a stimulus. Thus a square or circle would be invariant under 90° rotation and reflection; but a rectangle or ellipse would be invariant only under reflection, as 90° rotations would produce another shape in each case. Asymmetrical figures would produce eight different figures under 90° rotation and reflection.

The Structural Description of a Geon

Specification of the nonaccidental properties of the three attributes of the cross-section and one of the axis, as described previously, is sufficient to uniquely classify a given arrangement of edges as one of the 36 volumes. These would be matched against a structural description for each *component* which implicitly specified the values of these four nonaccidental image properties. Although there is a psychological plausibility to the generating function, it is necessary to translate the generating function for each geon into a description of its diagnostic nonaccidental image features. This was done, for example, in characterizing a change in size of the cross-section as it is swept along its axis as producing nonparallel edges to the geon.

When the contrasts in the generating functions are translated into image features we find that there typically exists a much larger set of distinctive nonaccidental image features than the four that might be expected from a direct mapping of the contrasts in the generating function. Figure 11 shows some of the nonaccidental contrasts distinguishing a brick from a cylinder. The silhouette of a brick contains a series of six vertices, which alternate between *Ls* and *Arrows*, and an internal *Y* vertex, as illustrated in Figure 11.

The vertices of the silhouette of the cylinder, by contrast, alternate between a pair of Ls and a pair of tangent Ys. The internal Y vertex is not present in the cylinder (or any geon with curved cross sections). These differences in image features would be available from a general viewpoint and thus could provide, along with other contrasting image features, a basis for discriminating a brick from a cylinder.

These image features can be cast as elements in a structural description that is unique for each geon. There are two levels in this description. One is the arrangement and classification of vertices and edges in the image of the geon. The second is the characterization of the edges of the image as symmetrical or parallel.

Consideration of the featural basis for the structural descriptions for each component suggests that a similarity measure can be defined on the basis of the common vs. distinctive image features for any pair of components. The similarity measure would permit the promotion of alternative components under conditions of ambiguity, as when one or several of the image features were undecidable.

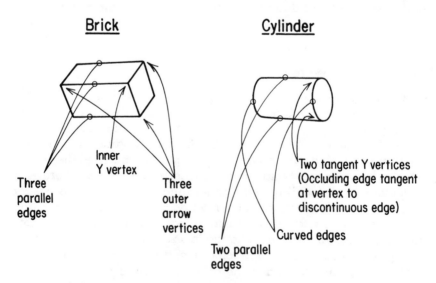

Figure 11.   Some differences in nonaccidental properties between a cylinder and a brick. From Biederman (1987a) Fig. 5, p. 121.

Unresolved Issues

Several general issues are still unresolved about the form of a model of human object recognition.

Geons or image features? It is possible that despite the subjective componential interpretation given to the arrangement of image features as simple volumes, it may be that it is the image features themselves, in specified relationships, that mediate perception, such as those shown in Figure 11. This issue may be empirically testable by determining whether image features of a geon that are missing in one viewpoint (typically through occlusion) can mediate perception of an object when it is projected from another viewpoint.

Is geon identification 2-D or 3-D? This issue is related to the prior one in that the image features are specifiable in 2-D, though a three-segment vertex, such as a fork or an arrow, typically induces a 3-D interpretation. Although the 36 components have a clear subjective volumetric interpretation, it must be emphasized they can be uniquely specified from their 2-D image properties. Consequently, recognition need not follow the construction of an "object centered" (Marr, 1982) 3-D interpretation of each volume.

Is there a need for asymmetrical geons? Given that a convex volume is parsed from matching adjacent concavities, it may not be necessary to assume geons with asymmetrical cross-sections. There are two possible reasons for this. First, it has been recently demonstrated (Leyton, 1987) that the convex 2-D region between two concavities will *necessarily* contain an axis of *local* symmetry (see also Brady & Asada, 1984). If this axis mediates perception *and* it can be demonstrated that these results hold in the 3-D case then geons with asymmetrical cross sections need not be posited. The problem with this is that it is not clear that people are particularly responsive to local (vs. global) symmetry.

Thus it is the second reason that may provide the strongest arguments against positing asymmetrical geons. To my knowledge, there are no cases where basic level classification requires the presence of an asymmetrical cross section. This does not mean that a component of an exemplar could not have an asymmetrical cross-section, but that primal access need not depend on the preservation of this asymmetry in the image. Empirically if it could be demonstrated that object classification performance did not depend on preservation of geon asymmetry in the image, a good case for dispensing with asymmetrical geons could be made. We are currently exploring this possibility.

The possible regularization of asymmetrical cross-sections raises the more general case for regularization of volumes that are generated by complex cross-sections, such as those with both straight and curved edges. When recognizing objects, these volumes could be matched to the closest regular geon in a manner analogous to the way phonemes in natural speech are related to their target. Alternatively, each volume could receive a unique representation in memory, perhaps in a schema-plus-correction fashion (Bartlett, 1932).

Additional Sources of Contour and Recognition Variation

RBC seeks to account for the recognition of an infinitely varied perceptual input with a modest set of idealized primitives. A full account would have to consider a number of subordinate and related problems such as metric variation, alternatives for component termination, asymmetrical cross-sections, axis selection, curved axis, and planar components. These are discussed more fully in Biederman (1987a). The interested reader is encouraged to consult that reference.

THE RELATIONS AMONG GEONS

RBC assumes that objects can be represented as a structural description which specifies both the geons and their relations. A tentative list of possible relations is presented in Table 1. Like the components themselves, the relations in Table 1 are *nonaccidental* in that they can be determined from a general viewpoint, are preserved in the 2-D image, and are categorical, requiring the discrimination of only two or three levels. The specification of only these five relations is likely conservative in that (a) it is certainly a nonexhaustive set in that other relations can be defined, and (b) the relations are only specified for a pair, rather than triples, of components. Let us consider these in order of their appearance in the table:

1. *Relative Size.* For any pair of geons, G_1 and G_2, G_1 could be much greater than, smaller than, or approximately equal to G_2.
2. *Verticality.* G_1 can be above, below, or side connected to G_2. A top-bottom relation is defined for at least 80% of the objects, by the author's estimate. Thus giraffes, chairs, and typewriters have a top-down specification of their components but forks and knives do not. The third level of this relation is *side connected*, as with the handle of a cup.

Table 1.   Geon Relations and the Number of Possible Two-Geon Objects

36	First Geon, G_1
X	
36	Second Geon, G_2
X	
3	Size [$G_1 \gg G_2$, $G_2 \gg G_1$, $G_1 = G_2$]
X	
2.4	G_1 top, bottom or side-connected (represented for 80% of the objects)
X	
2	Nature of Join [End-to-End (off center) or End-to-Side (centered)]
X	
2	Join at long or short surface of G_1
X	
2	Join at long or short surface of G_2
=	74,649 possible two-Component objects

3. *Centering.* Marr (1977) noted that the connection between any pair of joined components can be end-to-end (and of equal sized cross-section at the join), as the upper and forearms of a person, or end-to-side, producing one or two concavities, respectively, as illustrated in Figure 12. When components are joined arbitrarily, two-concavity joins will be far more common in that it will be rare that two arbitrarily joined end-to-end components will have equal-sized cross sections. This latter condition might be more profitably analyzed as a *bend* in a single component, as in a geon with a curved axis, rather than as a different type of relation. A more general distinction might be whether the end of one component in an end-to-side join is *centered* or *off centered* at the side of the other component. The end-to-end join might represent only the limiting, albeit special, case of off-centered joins. In general, the join of any two arbitrary volumes (or shpaes) will produce two concavities, unless an edge from one volume is made to be joined and collinear with an edge from the other volume.

4 and 5.   *Relative Size of Surfaces at Join.* Other than the special cases of a sphere and a cube, all primitives will have at least a long and a short surface. The join can be on either surface. The cup and the pail in Figure 6 differ, in part, as to whether the handle is connected to the long surface of the cylinder (to produce a cup) or the short surface (to produce a pail). In considering only two values for the relative size of the surface at the join, we are conservatively estimating the relational possibilities. Some volumes such as the wedge have as many as five surfaces, with

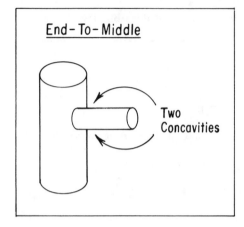
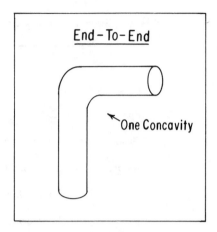

Figure 12. Two kinds of joins: End-to-middle and end-to-end.

no two necessarily of the same size. Table 1 (below) presents 57.6 combinations by which two geons can be joined by the five relations.

REPRESENTATIONAL CAPACITY OF RBC

I have argued that there is sufficient representational power in a set of 36 components and the five types of relations to express the human's capacity for primal access (Biederman, 1985, 1987a). In this section I will summarize these arguments and emphasize an important implication as to the number of components that will typically be sufficient to uniquely specify any object.

Two estimates are needed to determine the sufficiency of the 36 geons posited by RBC: (a) the number of readily available perceptual categories acquired by humans, and (b) the number of possible objects that could be represented by 36 components and the five kinds of relations. Obviously, the value for (b) would have to be greater than the value for (a) if 36 components are to prove sufficient.

How Many Readily Distinguishable Objects Do People Know?

A *liberal* estimate for this value would be in the order of 30,000 objects. Several sources suggest that there are words for approximately 1,500 readily identifiable and readily distinguishable basic level categories. A six-year-old child can name, at a level of competence that rivals that of an adult, just about every object encountered in his or her everyday world, including those that appear on television, books, and the movies. An average vocabulary for a six-year-old is approximately 10,000 words, about 10 percent of which are basic level, concrete count nouns.[7] A count of the number of such words in a dictionary sample taken by the author also supported this estimate (as did the estimates of several linguists). If we assume that this estimate is too small by a factor of two, allowing for idiosyncratic categories, such as our own laboratory apparatus or the weapons and vehicles shown on Saturday morning television, and errors in the estimate, then we can assume potential classification into approximately 3,000 basic level categories.

RBC assumes that perception is based on a particular componential configuration rather than the basic level category so we need to estimate the mean number of readily distinguishable componential configurations per basic level category. Almost all natural categories, such as elephants or giraffes, have one or only a few instances with differing componential descriptions. Dogs represent a rare exception for natural categories in that they have been bred to have considerable variation in their descriptions. Person-made categories vary in the number of allowable types, but this number often tends to be greater than the natural categories. Cups, typewriters, and lamps have just a few (in the case of cups) to perhaps 15 or more (in the case of lamps) readily discernible exemplars.[8] Let's assume (liberally) that the mean number of *types* is 10. This would yield an estimate of

---

[7] The kinds of descriptions postulated by RBC may play a central role in the child's capacity to acquire names for objects. It is my belief that the child is predisposed to employ different labels for objects that have different geon descriptions. When the perceptual system presents a new description for an arrangement of large geons, the absence of activation can readily result in a "what's that?" Among the last of the animal identifications mastered by children are those of lion and tiger. These, of course, have similar geon descriptions.

[8] It might be thought that faces constitute an exception to the estimate of a ratio of 10 examplars per category presented here, in that we can recognize thousands of faces. But individual faces may not be recognized as rapidly as we recognize differences among basic level categories. Another possible exception to the exemplar/category ratio presented here occurs with categories such as lamps which could have an arbitrarily large number of possible bases, shade types, etc. But these variations may actually serve to hinder recognition. In a number of experiments in our laboratory (e.g., Biederman & Hilton, 1987a), we have noted that highly stylized or unusual exemplars of a category are extremely difficult to identify under brief exposures (and out of context). The elements producing the variation in these cases may thus be acting as noise (or irrelevant components) in the sense that they are present in the image but not present in the mental representation for that category. These potential difficulties in the identification of faces or objects may not be subjectively apparent from the casual perusal of objects on a page, particularly when they are in a context that facilitates their classification, as noted on page 372.

30,000 readily discriminable objects (3,000 categories × 10 types/category).

Thirty thousand objects would require that the six-year-old learn an average of 13.5 objects per day from birth or about one per waking hour. By way of comparison, the maximum rates of word acquisition, reached during the ages of 2–6 years, is 4.5 words per day (Carey, 1978). By this criterion of plausible learning rate, 30,000 would appear to be a reasonable, if not liberal estimate of the number of readily distinguishable, known object exemplars.

How Many Objects Could Be Represented by 36 Components?

An estimate for two-component objects is summarized in Table 1. The 1,296 different pairs of the 36 volumes (i.e., 36^2) when multiplied by the number of relational combinations, 57.6 (the product of the various values of the five relations) gives us 74,649 possible two-component objects. If a third component is added to the two, then this value has to be multiplied by 2,073 pairs of possible components, (36 components × 57.6 ways in which the third component can be related to one of the two components) to yield 154 million possible three-component objects. This value, of course, readily accommodates the liberal estimate of 30,000 objects actually known.

The extraordinary disparity between the representational power of two or three components and the number of objects in an individual's object vocabulary means that there is an extremely high degree of redundancy in the filling of the 154 million cell component-relation space. Even if our estimate of the number of objects estimated to be known by an individual was too low by a factor of five so that 150,000 objects were actually known, we would still be using less than one tenth of one percent of the possible combinations of three components (i.e., 99.9% redundancy).

There is a remarkable consequence of this redundancy if we assume that objects are distributed homogeneously throughout the object space. A sparse, homogeneous occupation of the space means that, on average, it will be rare for an object to have a neighbor with the same structural description (viz., geons and relations)[9] Because the space was generated by considering only the number of possible two- or three-component objects, we have actually derived a con-

---

[9] Not only is it rare for different basic level objects to have the same structural description, informal demonstrations suggest that it is unusual to find different basic level objects differing by only a single component or relation. When a single component or relation of an object is altered only with rarity is a recognizable object from another category produced. The case of the cup and the pail in Figure 7, in which only a single relation was changed, represents one of these rare occurrences.

straint on the estimate of the likely number of components *per object* that will be sufficient for unambiguous identification. If objects were distributed relatively homogeneously among combinations of relations and components, *then only two or three components, in their specified relations, would be sufficient to unambiguously represent most objects!*

AN ANALOGY BETWEEN SPEECH AND OBJECT PERCEPTION

The geon-based account of object perception offered by RBC bears a close resemblance to phoneme-based accounts of speech perception (eg., Marslen-Wilson, 1980). Whatever the theoretical account, the computational problem is the same: How to map an infinitely variable input to eye or ear onto approximately 10^6 semantically relevant entities (objects or words). As shown in Figure 13, the computational strategy for solving this problem may be identical for speech and objects. In both cases, the input is mapped onto a small set (<10) of contrastive features. Because the contrasts are only dichotomous or trichotomous the human's limited capacities for absolute judgment are not taxed. A modest number (<60) of primitives—phonemes or geons—is generated from combinations of these features. The representational power of the system derives from its permissiveness in allowing relatively free combinations of its primitives so that approximately 10^6 semantic entities—words or objects—can be readily coded.

The particular properties of edges that are postulated to be relevant to the generation of the volumetric primitives have the desirable properties that they are invariant over changes in orientation and can be determined from just a few points on each edge. Consequently, they allow a geon to be extracted with great tolerance for variations of viewpoint, occlusion, and noise. Moreover, the nonaccidental properties supporting one geon of an object would not be expected to interact with the nonaccidental properties from another part of the object for another geon. In the speech domain although some of the particular features of the speech signal are invariant with the immediate linguistic environment[10], the presence of coar-

---

[10] Examples of categorized speech features would be whether the transitions are "feathered" (a cue for voicing) or the formants "murmured" (a cue for nasality). That these features have only two or three levels allows the recognition system to avoid the limitations of absolute judgment in the auditory domain (Miller, 1956). It is possible that the limited number of phonemes derives more from this limitation for accessing memory for fine quantitative variation than it does from limits on the fineness of the commands to the speech musculature.

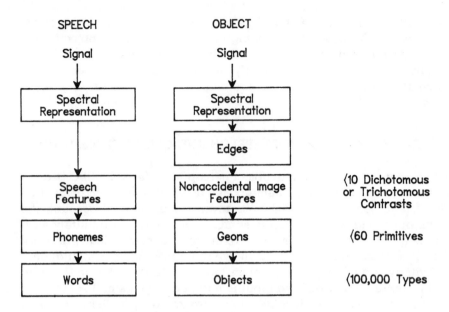

Figure 13.   An analogy between speech and object recognition.

ticulation effects presents a complication to the analogy. But the complication may be minor. Only a subset of the phonemes, the stopped consonants, require transitional mappings. All the others can be activated from invariant features. Perhaps it is not surprising that it is the transitional cues that best support the "speech is special" argument. Whatever the coarticulatory processing, even the transitional phonemes can be readily and subjectively decomposed into a string of phonemes which is then used to access memory. Perhaps a more significant difference between object and speech perception is that in the visual case we have a clear and powerful principle for parsing an object into geons—matched concavities. In the speech case the basis for parsing is still indeterminate but at least partly includes aspects of the signal, such as stress.

The feature pattern is then mapped onto a modest number (less than 60) of primitive phonemes in speech or geons in vision. We only need about 44 phonemes to code all the words in English, 15 in Hawaiian, 55 to represent virtually all the words in all the languages spoken on Earth. Lexical access to approximately 10^6 words during speech perception can be successfully modeled as a process mediated by the identification of a sequence of phonemes.

Just as the relations among the phonemes are critical in lexical access—"bat" and "tab" have the same phonemes but are not the same words—the relations among the volumes are critical for object recognition: Two different arrangements of the same components could produce different objects. In both cases, the representational power derives from the enormous number of combinations that can arise from a modest number of primitives. The relations in speech are limited to left-to-right (sequential) orderings; in the visual domain a richer set of possible relations allows a far greater representational capacity from a comparable number of primitives.

EMPIRICAL SUPPORT FOR A COMPONENTIAL REPRESENTATION

As noted earlier, one reason *not* to posit a representation system based on fine quantitative detail, for example, many variations in degree of curvature, is that such *absolute judgments* are notoriously slow and error prone unless limited to the 7 ± 2 values for a single attribute (Miller, 1956). This limitation on our capacities for making absolute judgments of physical variation, when combined with the dependence of such variation on orientation and noise, makes quantitative shape judgments a most implausible basis for object recognition. RBC's alternative is that the perceptual discriminations required to determine the primitive components can be made categorically, requiring the discrimination of only two or three viewpoint-independent levels of variation.

Our memory for irregular shapes shows clear biases toward "regularization" (e.g., Woodworth, 1938). Amply documented in the classical shape memory literature was the tendency for errors in the reproduction (and recognition) of irregular shapes to arise from the omission of slight deviations from symmetrical or regular figures. Alternatively, some irregularities were emphasized ("accentuation"), typically by the addition of a *regular* subpart. What is the significance of these memory biases? By the RBC hypothesis, these errors may have their origin in the mapping of the perceptual input onto a representational system based on regular primitives. The memory of a slight irregular form would be coded as the closest regularized neighbor of that form. If the irregularity was to be represented as well, an act that would presumably require additional time and capacity, then an additional code (sometimes a component) would be added.

RECENT EXPERIMENTAL ANALYSES

According to the RBC hypothesis, the preferred input for accessing object recognition is that of the volumetric components. In most cases, only a few appropriately arranged volumes would be all that is required to uniquely specify an object. Rapid object recognition should then be possible. Neither the full complement of an object's components, nor its texture, nor its color, nor the full bounding contour (or envelope or outline) of the object need be present for rapid identification. The problem of recognizing tens of thousands of possible objects becomes, in each case, just a simple task of identifying the arrangement of a few geons out of the set of 36.

Overview of Experiments

Several object naming reaction time experiments have provided support for the general assumptions of the RBC hypothesis, although none have provided tests of the specific set of components proposed by RBC or even that there might be a limit to the number of components. A general summary of the experiments can be found in Biederman (1987a). In all experiments, subjects named or quickly verified briefly presented pictures of common objects.

Surface vs. edged-based description. That an edge-based description may provide a sufficient account of object recognition was supported by experiments showing that simple line drawings, such as

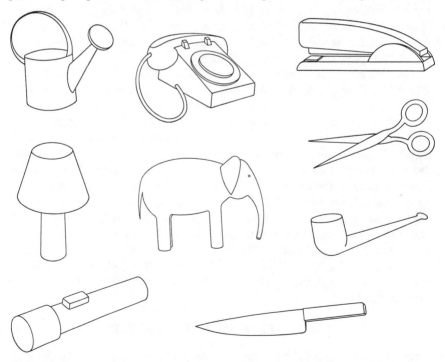

Figure 14. Nine of the experimental objects. From Biederman (1987a), Fig. 11, p. 130.

those shown in Figure 14, constructed to convey only the object's major geons, were identified about as rapidly as full colored, detailed, textured slides of the same objects (Biederman & Ju, 1988). There was no evidence, moreover, that an object with a characteristic color, such as a fork, mushroom, or camera, would derive any additional advantage when shown as a color slide compared to an object without a diagnostic color, such as a chair, mitten, or pen.

This failure to find a color-diagnosticity effect, when combined with the finding that simple line drawings could be identified so rapidly as to approach the naming speed of fully detailed, textured, colored photographic slides, supports the premise that the earliest access to a mental representation of an object can be modeled as a matching of an edged based representation of a few simple components. Such edged-based descriptions are thus *sufficient* for primal access.

Some objects, such as a broom or a zebra, require both a volumetric and a surface texture description to appear complete. A recent experiment in our laboratory (Biedeman & Hilton, 1987b) revealed that such objects were identified less accurately at brief exposures and with longer naming times than controls that were closely matched in volumetric structure and bounding contour. For example, a shovel and a horse served as controls for the broom and zebra, respectively. If the texture region was functioning as another component, then performance should have been facilitated through the additional geon, in that complex objects can be identified more rapidly than simple ones, as described in the next section. Alternatively, the detailed processing required to specify the texture field might not be completed in a brief exposure duration so such objects might prove to be less recognizable. The results supported the latter alternative and are consistent with the previously reported secondary status of surface features.

Partial objects and the effects of complexity. Biederman, Ju, & Clapper (1985) studied the perception of objects that were missing some of their components, such as those shown in Figure 15. Complex objects, defined as those requiring six or more components to appear complete, could be identified perfectly from only two or three of their geons, as long as subjects were not stressed to respond quickly. Under speed stress and with brief (100 msec.) exposures, both reaction times (RTs) and error increased with the removal of additional components. But even under these conditions, complex objects with less than half their components were accurately named on 75 percent of the trials. Importantly, for the complete versions of the objects, there was no evidence that simple objects (those requiring only two or three components to appear complete) were identi-

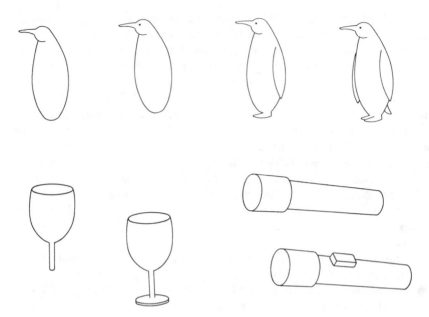

Figure 15. Illustration of the partial and complete versions of two three component objects (the wine glass and flashlight) and a nine component object (the penguin). From Biederman (1987a), Fig. 12, p. 131.

fied any more rapidly than the complex objects. Such a result would appear to be inconsistent with a tracing of contour, as in Ullman's (1984) visual routines.

Contour deletion. That the contours specified by RBC may provide a necessary account of object recognition was supported by a demonstration that degradation (contour deletion), if applied at the regions that prevented recovery of the components, rendered an object unidentifiable (Biederman & Blickle, 1987). Figure 16 shows some sample stimuli from that experiment. Those in the rightmost column, from which contour was removed to prevent recovery of the components were, essentially, unrecognizable, even with as long as five seconds of viewing. An equal amount of contour was removed from the control stimuli, shown in the middle column, but in midsegment so as to allow filling in from collinearity or smooth curvature. The disruption from the contour deletion on these objects was considerably smaller and they could be identified at near perfect accuracy at long exposures.

Objects with midsegment deletion (those in the middle column of Figure 16) did show a considerable loss in speed and accuracy of identification when compared to the intact objects shown in the left column. Additional research has established that the processes by

Figure 16. Example of five stimulus objects in the experiment on the perception of degraded objects. The left column shows the original intact versions. The middle column shows the recoverable versions. The contours have been deleted in regions where they can be replaced through collinearity or smooth curvature. The right column shows the nonrecoverable versions. The contours have been deleted at regions of concavity so that collinearity or smooth curvature of the segments bridges the concavity. In addition, vertices have been altered, from Ys to Ls, and misleading symmetry and parallelism introduced. From Biederman (1987a), Fig. 16, p. 135.

which contour is recovered through collinearity or smooth curvature is surprisingly slow and dependent on an image actually being present (as opposed to being done in memory) (Biederman, 1985). The effect of the contour deletion is object-centered, being directly related to the proportion of an object's contour that is deleted and independent of the actual size of the gaps in terms of degrees of visual angle (Biederman & Blickle, 1987). These characteristics of contour deletion allowed us to explore the arrangement of stages in the next investigation.

Perceiving degraded vs. partial objects. The model of RBC illustrated in Figure 4 can be partitioned into two critical stages: (a) those processes leading to and including the determination of the geons, and (b) those processes involved in the matching of an arrangement of geons to memory.

Consider Figure 17 which shows, for some sample objects, one version in which whole components are deleted so that only three (of six or nine) of the components remain and another version in which the same amount of contour is removed, but in midsegment distributed over all of the object's components (similar to those shown in the middle column in Figure 16). For objects with whole components deleted, missing components cannot be added prior to recognition. Logically, one would have to know what object was being recognized to know what parts to add. Instead, the activation of a representation most likely proceeds in the absence of the parts, with weaker activation the consequence of the missing parts. With the midsegment deletion, components can be determined from processes employing collinearity or smooth curvature.

The two methods for removing contour may thus be affecting different stages. Deleting contour in midsegment affects processes prior to and including those involved in the determination of the components (Figure 4). The removal of whole components (the partial object procedure) is assumed to affect the matching stage, reducing the number of common components between the image and the representation and increasing the number of distinctive components in the representation.

The two stages can be regarded as being arranged in cascade, with an earlier geon determination stage relaying activation to the object matching stage. Figure 18 shows the expected activation functions from the two procedures for deleting contour. Deleting contour in midsegment results in an initial slow growth in activation as the relatively slow processes for smooth contour continuation are required to restore the deleted contours. Once the restoration is complete there is a rapid growth in activation. By contrast, there is an

Complete

Component
Deletion

Midsegment
Deletion

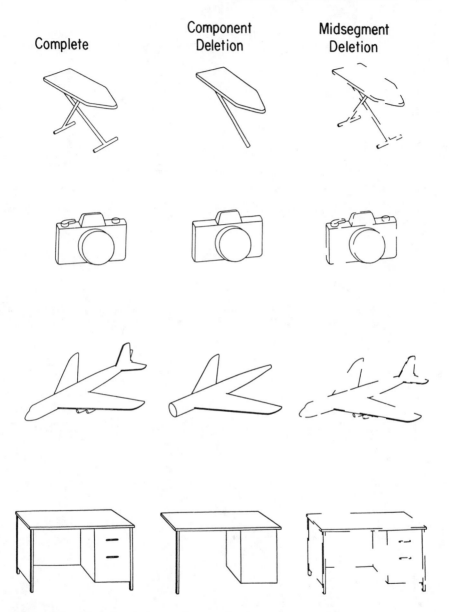

Figure 17.   Sample objects with equivalent proportion of contours removed
either at midsegment or as whole components. From Biederman (1987a), Fig.
22, p. 139.

$$a_{nj\,*}(t) = a_{nj\,*}\left[1 - \sum_{i=1}^{n}\left(\prod_{\substack{l=1 \\ l \neq i}}^{n}\frac{k_l}{k_l-k_i}\right)\bar{c}^{k_it}\right], \quad k_i=1$$

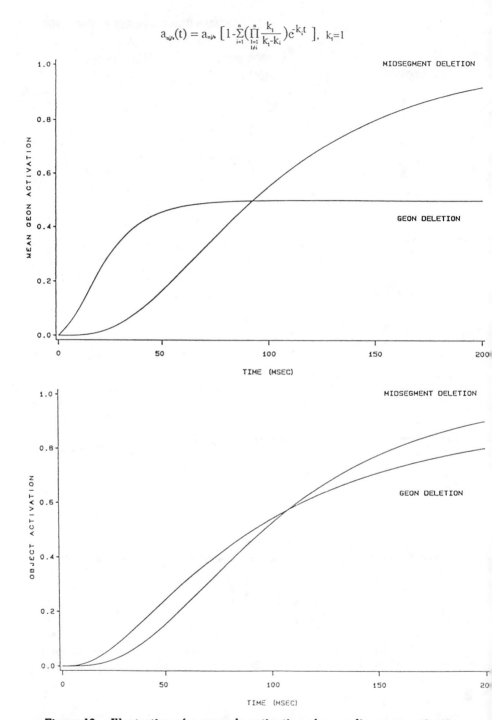

Figure 18. Illustration of a cascade activation of an earlier geon activation stage (top) and a subsequent object representation stage (bottom) for the kind of contour removal illustrated in the second and third columns in Figure

410

initial rapid activation of the components from the partial objects which, however, asymptotes below the activation level of the mid-segment deleted objects. The reason for this is that the missing components have activation levels of zero. Once the filling-in is completed for the objects with midsegment deletion, the complete complement of an object's components are available, providing a better match to the object's representation than is possible with a partial object that had only a few of its components. The net effect is to produce a crossover interaction over exposure duration which produces a similar effect on the next stage, activation of the representation of the object.

This prediction was supported from the results of an experiment [Biederman, Beiring, Ju, Blickle, Gagnon, & Hilton (1987)] which studied the naming speed and accuracy of six- and nine-component objects undergoing these two types of contour deletion. The results, error rates and RTs, are shown in Figures 19 and 20. At brief exposure durations (e.g., 65 msec.) performance with partial objects was better than objects with the same amount of contour removed in midsegment both for errors (Figure 19) and RTs (Figure 20). At longer exposure durations (200 msec.), the RTs reversed, with the objects undergoing midsegment deletion now identified faster than the partial objects.

Orientation Variability

Objects can be more readily identified from some orientations compared to other orientations (Palmer, Rosch, & Chase, 1981). According to the RBC hypothesis, difficult views will be those in which the components extracted from the image are not the components (and their relations) in the representation of the object. Often such mismatches will arise from an "accident" of viewpoint where an image property is not correlated with the property in the 3-D world. For example, when the viewpoint in the image is along the axis of the major components of the object, the resultant foreshortening converts one or some of the components into surface components, such as disks and rectangles which are not included in the componential description of the object (Biederman, 1987a). In addition, surfaces

---

17. The activation of the geons (components) results in activation of the object representations. Geon (component) deletion results in a lower asymptote at the geon activation stage because missing components never get activated prior to object activation. Shown here would be a case where only half an object's geons (weighted by size) are ever activated.

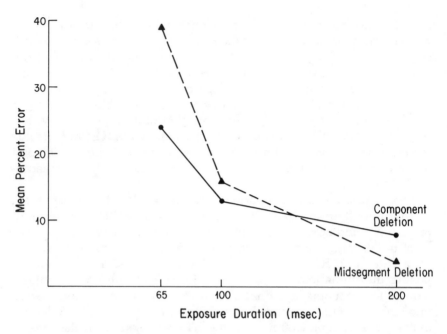

Figure 19.   **Mean percent errors of object naming as a function of the nature of contour removal (deletion of midsegments or components) and exposure duration. The interaction between exposure duration and the ordering of the two conditions of contour removal are as predicted from the object activation stage in Figure 17. From Biederman (1987a), Fig. 23, p. 140.**

may occlude otherwise diagnostic components. Consequently, the components extracted from the image will not readily match the mental representation of the object and identification will be more difficult compared to an orientation which does convey the components (Biederman, 1987a).

   A second condition under which viewpoint affects identifiability of a specific object arises when the orientation is simply unfamiliar, as when a sofa is viewed from below or when the top-bottom relations among the components are perturbed as when a normally upright object is inverted. Jolicoeur (1985) recently reported that naming RTs were lengthened as a function of an object's rotation away from its normally upright position. He concluded that mental rotation was required for the identification of such objects, as the effect of X-Y rotation on RTs was similar for naming and mental rotation. It may be that mental rotation—or a more general imaginal transformation capacity stressing working memory—is required only under the (relatively rare) conditions where the relations among the components

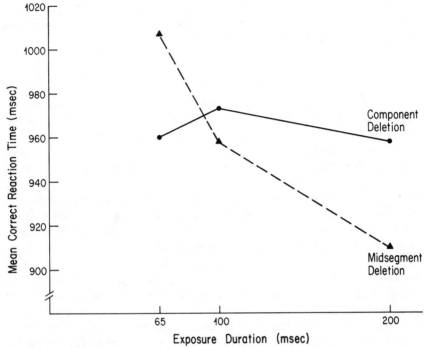

Figure 20.   Mean correct reaction time (msec.) in object naming as a function of the nature of contour removal (deletion at midsegments or components) and exposure duration. The interaction between exposure duration and the ordering of the two conditions of contour removal are as predicted from the object activation stage in Figure 17. From Biederman (1987a), Fig. 24, p. 141.

have to be rearranged. Thus, we might expect to find the equivalent of mental paper folding if the parts of an object were rearranged and the subject's task was to determine if a given object could be made out of the displayed components. RBC would hold that the lengthening of naming RTs in Jolicoeur's (1985) experiment is better interpreted as an effect that arises not from the employment of orientation dependent features but from the perturbation of the TOP-OF relations among the components.

Palmer, Rosch, & Chase (1981) conducted an extensive study of the perceptibility of various objects when presented at a number of different orientations. Generally, a three-quarters front view was most effective for recognition. Their subjects showed a clear preference for such views. Palmer et al. termed this effective and preferred orientation of the object its *canonical orientation*. The canonical orientation would be, from the perspective of RBC, a special case of the orientation that would maximize the match of the components in the image to the representation of the object.

Transfer Between Different Viewpoints

When an object is seen at one viewpoint or orientation it can often be recognized as the same object when subsequently seen at some other orientation in depth, even though there can be extensive differences in the retinal projections of the two views. The componential recovery principle would hold that transfer between two viewpoints would be a function of the componential similarity between the views, as long as the relations among the components were not altered. This could be experimentally tested through priming studies with the degree of priming predicted to be a function of the similarity (viz., common minus distinctive components) of the two views. If two different views of an object contained the same components RBC would predict that, aside from effects attributable to variations in aspect ratio, there should be as much priming as when the object was presented at an identical view. An alternative possibility to componential recovery is that a presented object would be mentally rotated (Shepard & Metzler, 1971) to correspond to the original representation. But mental rotation rates appear to be too slow and effortful to account for the ease and speed in which transfer occurs between different orientations in depth that preserve the upright of an object.

There may be a restriction on whether a similarity function for priming effects will be observed. Although unfamiliar objects (or nonsense objects) should reveal a componential similarity effect, the recognition of a familiar object, whatever its orientation, may be too rapid to allow an appreciable experimental priming effect. Such objects may have a representation for each orientation that provided a different componential description. Bartram's (1974) results support this expectation that priming effects might not be found across different views of familiar objects. Bartram performed a series of studies in which subjects named 20 pictures of objects over eight blocks of trials. (In another experiment [Bartram, 1976], essentially the same results were found with a Same-Different name matching task in which pairs of pictures were presented). In the *Identical condition,* the pictures were identical across the trial blocks. In the *Different View* condition, the same objects were depicted from one block to the next but in different orientations. In the *Different Exemplar* condition, different exemplars, for example, different instances of a chair, were presented, all of which required the same response. Bartram found that the naming RTs for the Identical and Different View conditions were almost equivalent and both were shorter than control conditions, described below, for concept and response priming effects. Bartram theorized that observers automatically compute and access all possible 3-D viewpoints when viewing a given object. Alternatively, it is possible that there was high componential similarity across the different views and the experiment was insufficiently sensitive to detect slight differences from one viewpoint to another. However, in four experiments with colored slides, we (Biederman & Lloyd, 1985) failed to obtain any effect of variation in viewing angle and have thus replicated Bartram's basic effect (or lack of an effect). At this point, our inclination is to agree with Bartram's interpretation, with somewhat different language, but restrict its scope to familiar objects. It should be noted that both Bartram's and our results are inconsistent with a model that assigned heavy weight to the aspect ratio of the image of the object or postulated an underlying mental rotation function.

Different Exemplars Within An Object Class

Just as we might be able to gauge the transfer between two different views of the same object from a componential-based similarity metric, we might be able to predict transfer between different exemplars of a common object, such as two different instances of a lamp or chair.

As noted in the previous section, Bartram (1974) also included a *Different Exemplar* condition, in which different objects with the same name, for example, different model cars, were presented over the various blocks of trials. Under the assumption that different exemplars would be less likely to have common components, RBC would predict that this condition would be slower than the Identical and Different View conditions but faster than a *Different Object* control condition with a new set of objects that required different names for every trial block. This was confirmed by Bartram.

For both different views of the same object, as well as different exemplars (subordinates) within a basic level category, RBC predicts that transfer would be based on the overlap in the components between the two views. The strong prediction would be that the same similarity function that predicted transfer between different orientations of the same object would also predict the transfer between different exemplars with the same name.

The Perceptual Basis of Basic Level Categories

Consideration of the similarity relations among different exemplars with the same name raises the issue as to whether objects are most readily identified at a basic, as opposed to a subordinate or superordinate, level of description. The componential representations described here are representations of *specific*, subordinate objects, although their identification was often measured with a basic level name. Much of the research suggesting that objects are recognized at a basic level have used stimuli, often natural, in which the subordinate level exemplars had componential descriptions that were highly similar to those for a basic level prototype for that class of objects. Only small componential differences, or color or texture, distinguished the subordinate level objects. Thus distinguishing Asian elephants from African Elephants or Buicks from Oldsmobiles requires fine discriminations for their verification. The structural descriptions for the largest components would be identical. It is not at all surprising that with these cases basic level identification would be most rapid. On the other hand, many human-made categories, such as lamps, or some natural categories, such as dogs (which have been bred by humans), have members that have componential descriptions that differ considerably from one exemplar to another, as with a pole lamp vs a ginger jar table lamp, for example. The same is true of objects that differ from their basic level prototype as, for example, penguins or sport cars. With such instances, which unconfound the similarity between basic level and subordinate level objects, perceptual access should be at the subordinate (or instance) level, a result supported by a recent report by Jolicoeur, Gluck, & Kosslyn (1984).

For some categories, such as chairs, one can conceive of an extraordinarily large number of instances. Do we have a priori structural descriptions for all these cases? Obviously not. Although we can recognize many visual configurations as chairs, it is likely that only those for which there exists a close structural description will recognition be rapid. The same caveat that was raised about the Marr & Nishihara's (1978) demonstrations of pipe-cleaner animals and Kolers' chairs earlier must be voiced here. With casual viewing, particularly when supported by a scene context or when embedded in an array of like objects, it is often possible to identify unusual or modified instances without much *subjective* difficulty. But when presented as an isolated object without benefit of such contextual support, recognition of unfamiliar exemplars requires markedly longer exposure durations than those required for familiar instances of a category. The case of nonrigid objects is discussed in Biederman (1985).

DIFFICULT AND EXPERT IDENTIFICATIONS

Primal access is concerned with the initial activation of a representation of an object from that object's image. Often this can be readily achieved from a 100 msec. exposure with naming RTs well under one sec. We have previously considered cases where recognition will be difficult under the componential recovery principle, either because the object was viewed at an "accidental" orientation or because contour had been deleted so that the geons could not be determined.

There are cases, however, where identification is initially difficult not because of an accident of viewpoint or degradation, but because the individual has not been trained to perform the identification. But if all people employ the nonaccidental relations and have either learned or been born with a set of geons, what role could training play in learning an identification? The answer to this question lies in appreciating the conditions under which training will be required. As noted previously (Footnote 7), I believe that people (both children and adults) are prepared to employ different linguistic terms to refer to objects that have different structural descriptions for their largest geons. An important subset of tasks that will require training are those where the student is required to discriminate among exemplars that have many small components but have the same structural descriptions for their largest components.

Consider Figure 21 which shows profile views of seven tanks, four from NATO countries and three from the Warsaw Pact countries, adapted from training materials developed by Kottas and Bessemer (1980). No doubt we have all seen TV footage of NATO maneuvers and May Day Parades, but I suspect that few of us have spontaneously developed a perceptual model for the two classes of tanks. NATO infantry personnel are required to determine friend from foe, and training is required. As suggested by the arrows of the figure, the training largely consists of pointing out small *nonaccidental* differences in the contours between the two classes. A rule can be: if the rear of the turret is completely curved (without notches) then it is a Warsaw Pact tank. From this one-sentence instructional session the reader can now perfectly distinguish the two classes of tanks.

Note that the infantry are not instructed on the metric details of degree of curvature or length of various features. The difficulty in

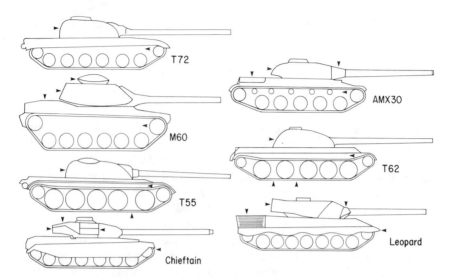

Figure 21.   Four NATO and three Warsaw Pact (Soviet) tanks. The classification can be achieved though a simple nonaccidental contrast: If the rear of the turret is completely curved, then it is a Warsaw Pact tank. From Biederman & Shiffrar (1987), Fig. 4, p. 644.

spontaneously forming a class discrimination is that, with a complex object, the student does not know where to look or what to look for. The instance with tanks is just an example. I believe that whenever possible, competent instructional materials for identifying members within the same basic level category, as with bird guide books, explicitly tell the reader where to look and what to look for. Often, as with the case of birds, the details are with color or shading arrangements. But even here, whenever possible, the guide notes a nonaccidental contrastive difference, rather than a metric property.

Margaret Shiffrar and I recently studied the perceptual learning problems involved in the sexing of day old chicks (Biederman & Shiffrar, 1987). This had been considered an extraordinarily difficult task, requiring months of instruction (E. J. Gibson, 1969). With approximately 15 sec. of instructions indicating where to look and a nonaccidental property to look for (concavity vs. convexity), we were able to train naive subjects to classify the pictures as accurately as expert sexers. Perhaps more important, the figures that the naive subjects tended to misclassify were the same as the experts.

There are, no doubt, cases where perceptual learning does require making fine metric discriminations. But I suspect these are employed only when a contrastive nonaccidental property is not available.

SCENE PERCEPTION

RBC is a theory about the perception of isolated objects. But just as words typically occur in some sentential context, it is often the case that objects occur not in isolation but in a scene-like arrangement with other objects. What role does RBC play in scene perception?

The mystery about the perception of scenes is that the exposure duration required to have an accurate perception of an integrated real-world scene is not much longer then what is typically required to perceive an individual object (Biederman, 1987b). The recognition of a visual array as a scene rather than as a list of objects requires not only the identification of the various entities but also a semantic specification of the interactions among the object and an overall semantic specific of the arrangement. I propose that often quick understanding of a scene is mediated by the perception of *geon clusters*. Such clusters are formed from the largest volume of each of two or three objects in a familiar arrangement. The cluster acts very much as a large object.

In line with this result is the general conclusion that the perception of a scene is not, in general, derived from an initial identification of the individual objects comprising that scene (Biederman, 1987a). That is, in general we do *not* first identify a stove, refrigerator, and coffee cup, in specified physical relations and *then* come to a conclusion that we are looking at a kitchen.

Space does not allow a lengthy review of the experimental basis for this conclusion (see Biederman, 1987b) but direct tests in our laboratory of an object-then-scene route have all been negative. As an index of scene perception we have measured the *violation effect:* The lowered perceptibility of an object when it is placed in an incongruous semantic relation to its setting. For example, a fire hydrant could be placed in a kitchen (where it would be improbable), or placed in the middle of a street (where it would be probable in the scene but in an incongruous position) or moved from the foreground to the background (or vice versa) so that it would appear to be too large or too small (e.g., Biederman, 1981; Biederman, Mezzanotte, & Rabinowitz, 1982). Objects undergoing these violations are less perceptible than when they are in a normal relation to their context.

Degrading the largest and most diagnostic object in a scene does not reduce the violation effect (Klatsky, 1982). Moreover, if the objects are not in a scene like arrangement there is no interference from having an improbable object, say a fire hydrant, in a nonscene arrangement of kitchen objects.

Some demonstrations and experiments by Robert Mezzanotte (1981) provide a possible explanation for these results. Mezzanotte

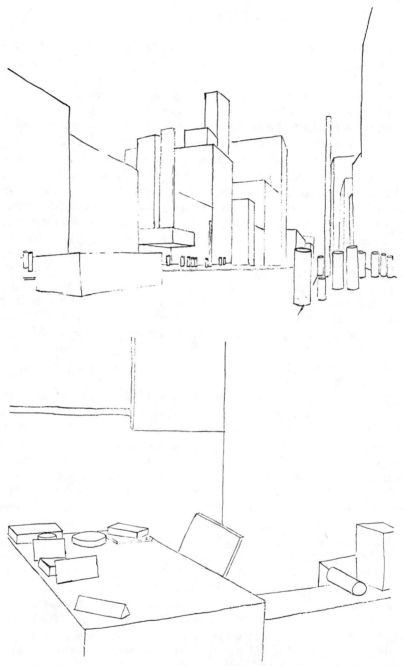

Figure 22.   Office and street scenes drawn by Robert Mezzanotte. None of the objects in these scenes is identifiable, in isolation, as anything other than a geon.

showed that a readily interpretable scene could be constructed from arrangements of single geons that just preserved the overall aspect ratio of the object, such as those shown in Figures 22a and 22b. In these kinds of scenes, none of the entities, when shown in isolation, could be identified as anything other than a simple volumetric body, for example, a brick (Mezzanotte, 1981).

Most important, Mezzanotte found that violation effects for intact objects were readily obtained from such settings, even when he used only a single geon type, the brick. For example, the desk lamp is more difficult to identify when it is undergoing apparent Probability and Size violations, such as those shown in Figure 23, then when it is in a normal relation to its context, as in Figure 24.

Mezzanotte's results are inconsistent with an object-then-scene route in that there are no entities in the scene that are readily identifiable as objects. However, we have recently formulated an explanation, not only of Mezzanotte's results, but all of the phenomena of fast access to a scene representation (Biederman, 1987).

The central point is that people have representations for *geon clusters*. A geon cluster is an arrangement of geons from different objects that preserve the relative size and aspect ratio and relations of the largest visible geon of each object. In such cases, the indi-

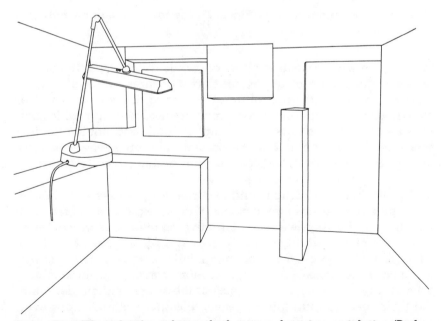

Figure 23.   Example of an object, the lamp, undergoing a violation (Probability and Size) in a scene where none of the entities, when shown individually is identifiable as a kitchen object.

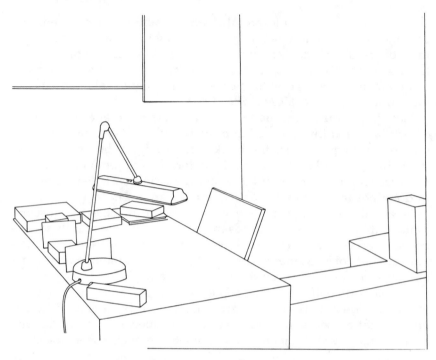

Figure 24. The control scene for Figure 23. The object appears normal given its context.

vidual geon will be insufficient to allow identification of the object. However, just as an arrangement of two or three geons almost always allows identification of an object, an arrangement of two or more geons from different objects may produce a recognizable combination. Figure 25 shows two examples. If this account is true, fast scene perception should only be possible in scenes where such familiar object clusters are present. This account awaits empirical test.

One originally puzzling result, in particular, now seems plausible given this hypothesis of geon clusters. In our experimental research violations of Position were as disruptive on object identification as violations of Probability (e.g., Biederman, Mezzanotte, & Rabinowitz, 1982; Biederman, Teitelbaum, and Mezzanotte, 1983). This was unexpected because Probability violations can be registered merely from an inventory list of the objects in the scene, without determining their specific spatial interactions. Position violations, however, *in addition* to requiring a semantic (inventory) determination of the object and its neighbors, require that the specific spatial relations be interpreted. Given that the magnitude of the violation effect for the

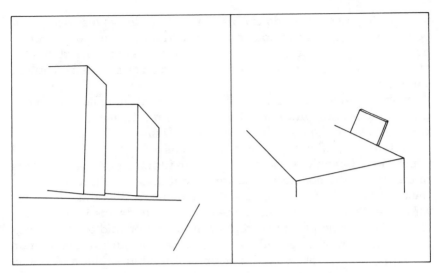

Figure 25.   Examples of Geon clusters for the scenes in figure 22.

different classes of relations indexes the relative accessibility of the relations, Position violations should have showed smaller violation effects in that more information (viz., the relations) would have been required for registration of a Position violation. But if geon clusters determine fast access to a representation of a scene, then it is unlikely for such clusters to exist for combinations of objects undergoing either Probability or Position violations.

Because the clusters are representing the largest visible geon for each object, their identification will be completed more quickly than processes that are dependent upon the determination of the smaller geons of an object. It is in this manner that geon clusters might control primal access for a scene.

CONCLUSION

To return to the analogy with speech perception, the characterization of object perception provided by RBC bears a close resemblance to some current views as to how speech is perceived. In both cases, the ease by which we are able to code tens of thousands of words or objects is solved by mapping that input onto a modest number of primitives—55 phonemes or 36 geons—and then employing a representational system that can code and access free combinations of these primitives. In both cases, the specific set of primitives is derived from dichotomous (or trichotomous) contrasts of a small number (less than 10) of independent characteristics of the input. The ease by which we are able to perceive words or objects may thus derive less from a capacity for judging continuous physical variation than it does from a perceptual system designed to represent the free combination of a modest number of categorized primitives based on simple perceptual contrasts.

In object perception, the primitive components may have their origins in the fundamental principles by which inferences about a 3-D world can be made from the edges in a 2-D image. These principles constitute a significant portion of the corpus of Gestalt organizational constraints. Given that the primitives are fitting simple parsed parts of an object, the constraints toward regularization characterize not the complete object but the object's components. RBC thus provides, for the first time, an account of the heretofore undecided relation between these principles of perceptual organization and pattern recognition. The primitives postulated by RBC may form an integral portion of explanations for skilled perceptual identification and scene perception.

REFERENCES

Attneave, F. (1954). Some informational aspects of visual perception. *Psychological Review, 61,* 183–193.

Attneave, F. (1982). Pragnanz and soap bubble systems. In J. Beck (Ed.), *Organization and representation in visual perception.* Hillsdale, NJ: Erlbaum.

Bartlett, F. C. (1932). *Remembering: A study in experimental and social psychology.* New York: Cambridge University Press.

Bartram, D. (1974). The role of visual and semantic codes in object naming. *Cognitive Psychology, 6,* 325–356.

Bartram, D. (1976). Levels of coding in picture-picture comparison tasks. *Memory & Cognition, 4,* 593–602.

Biederman, I. (1981). On the semantics of a glance at a scene. In M. Kubovy, & J. R. Pomerantz (Eds.), *Perceptual organization.* Hillsdale, NJ: Erlbaum.

Biederman, I. (1987a). Recognition-by-components: A theory of human image interpretation. *Psychological Review, 94,* 115–148.

Biederman, I. (1987b). Matching image edges to object memory. In *Proceedings of the First International Conference on Computer Vision,* (pp. 384–392). IEEE Computer Society, London, England, June, 1987.

Biederman, I., Beiring, E., Ju, G., Blickle, T., Gagnon, D., & Hilton, H. J. (1987). A comparison of the perception of partial vs. degraded objects. Unpublished manuscript, State University of New York at Buffalo.

Biederman, I., & Blickle, T. (1987). The perception of objects with deleted contours. Unpublished manuscript. State University of New York at Buffalo.

Biederman, I., & Hilton, H. J. (1987a). Subjectively modest variations in image processing and within class exemplar variation produce large effects on human classification latencies. Unpublished manuscript. State University of New York at Buffalo.

Biederman, I., & Hilton, H. J. (1987b). Classifying objects that require specification of surface attributes. Unpublished manuscript. State University of New York at Buffalo.

Biederman, I., & Ju, G. (1988). Surface vs. edge-based determinants of visual recognition. *Cognitive Psychology, 20,* 38–64.

Biederman, I., Ju, G., & Clapper, J. (1985). The perception of partial objects. Unpublished manuscript, State University of New York at Buffalo.

Biederman, I., & Lloyd, M. (1985). Experimental studies of transfer across different object views and exemplars. Unpublished manuscript, State University of New York at Buffalo.

Biederman, I., Mezzanotte, R. J., & Rabinowitz, J. C. (1982). Scene perception: Detecting and judging objects undergoing relational violations. *Cognitive Psychology, 14,* 143–177.

Biederman, I., & Shiffrar, M. S. (1987). Chicken sexing: An expert systems and experimental analysis of a difficult perceptual learning task. *Journal of Experimental Psychology: Human Learning, Memory, and Cognition, 13,* 640–645.

Biederman, I., Teitelbaum, R. C., Mezzanotte, R. J. (1983). Scene perception: A failure to find a benefit of expectancy or familiarity. *Journal of Experimental Psychology: Human Learning And Memory, 9,* 411–429.

Binford, T. O. (1971). Visual perception by computer. *IEEE Systems Science and Cybernetics Conference,* Miami, December.

Binford, T. O. (1981). Inferring surfaces from images. *Artificial Intelligence, 17,* 205–244.

Bower, G. H., & Glass, A. L. (1976). Structural units and the reintegrative power of picture fragments. *Journal of Experimental Psychology: Human Learning and Memory, 2,* 456–466.

Brady, M. (1983). Criteria for the representations of shape. In J. Beck, B. Hope, & A. Rosenfeld (Eds.), *Human and machine vision.* New York: Academic Press.

Brady, M., & Asada, H. (1984). Smoothed local symmetries and their implementation. *International Journal of Robotics Research, 3,* 3.

Brooks, R. A. (1981). Symbolic reasoning among 3-D models and 2-D images. *Artificial Intelligence, 17,* 205–244.

Carey, S. (1978). The child as word learner. In M. Halle, J. Bresnan, & G. A. Miller (Eds.), *Linguistic theory and psychological reality.* Cambridge, MA: MIT Press.

Chakravarty, I. (1979). A generalized line and junction labeling scheme with applications to scene analysis. *IEEE Transactions. PAMI,* April, 202–205.

Connell, J. H. (1985). Learning shape descriptions: Generating and generalizing models of visual objects. Unpublished masters thesis, Department of Electrical Engineering and Computer Science, MIT.

Egeth, H., & Pachella, R. (1969). Multidimensional stimulus identification. *Perception & Psychophysics, 5,* 341–346.

Garner, W. R. (1962). *Uncertainty and structure as psychological concepts.* New York: Wiley. ▸

Garner, W. R. (1974). *The processing of information and structure.* New York: Wiley.

Gibson, E. J. (1969). *Principles of perceptual learning and development.* New York: Appleton-Century-Crofts.

Gibson, J. J. (1966). *The senses considered as perceptual systems.* Boston: Houghton-Mifflin.

Guzman, A. (1971). Analysis of curved line drawings using context and global information. *Machine Intelligence 6.* Edinburgh: Edinburgh University Press.

Hoffman, D. D., & Richards, W. (1985). Parts of recognition. *Cognition, 18,* 65–96.

Humphreys, G. W. (1983). Reference frames and shape perception. *Cognitive Psychology, 15,* 151–196.

Intraub, H. (1981). Identification and processing of briefly glimpsed visual scenes. In D. F. Fisher, R. A. Monty, & J. W. Senders (Eds.), *Eye movements: Cognition and visual perception.* Hillsdale, NJ: Erlbaum.

Ittleson, W. H. (1952). *The Ames demonstrations in perception.* New York: Hafner.

Jolicoeur, P. (1985). The time to name disoriented natural objects. *Memory & cognition. 13,* 289–303.

Jolicoeur, P., Gluck, M. A., & Kosslyn, S. M. (1984). Picture and names: Making the connection. *Cognitive Psychology, 16,* 243–275.

Kanizsa, G. (1979). *Organization in vision: Essays on Gestalt perception.* New York: Praeger.

King, M., Tangney, J., Meyer, G. E., & Biederman, I. (1976). Shape constancy and a perceptual bias towards symmetry. *Perception & Psychophysics, 19,* 129–136.

Klatsky, G. J. (1982). Getting to the top: On the access routes to scene schemata. Unpublished doctoral dissertation, Department of Psychology, State University of New York at Buffalo.

Knoll, T. F., & Jain, R. C. (1986). Recognizing partially visible objects using feature indexed hypotheses. *IEEE Journal of Robotics and Automation, RA-2,* 3–13.

Kolers, P. A. (1970). The role of shape and geometry in picture recognition. In B. S. Lipkin & A. Rosenfeld (Eds.), *Picture processing and psychopictorics.* New York: Academic Press.

Kottas, B. L., & Bessemer, D. W. (1980). Comparison of potential critical feature sets for simulator-based target identification training. Final Report U. S. Army Research Institute for the Behavioral and Social Sciences, Fort Knox Field Unit.

Leyton, M. (In press). Symmetry-curvature duality. *Computer Vision, graphics, and image processing.*

Lowe, D. (1984). Perceptual organization and visual recognition. Unpublished doctoral dissertation, Department of Computer Science, Stanford University.

Malik, J. (1987). Interpreting line drawings of curved objects. *International Journal of Computer Vision, 1,* 73–103.

Marr, D. (1977). Analysis of occluding contour. *Proceedings of the Royal Society of London B., 197,* 441–475.

Marr, D. (1982). *Vision.* Freeman: San Francisco.

Marr, D., & Nishihara, H. K. (1978). Representation and recognition of three dimensional shapes. *Proceedings of the Royal Society of London B., 200,* 269–294.

Marslen-Wilson, W. (1980). Optimal efficiency in human speech processing. Unpublished manuscript, Max Planck Institut fur Psycholinguistik, Nijmegen, The Netherlands.

McClelland, J. L. (1979). On the time-relations of mental processes: An examination of systems of processes in cascade. *Psychological Review, 86,* 287–330.

McClelland, J. L., & Rumelhart, D. E. (1981). An interactive activation model of context effects in letter perception, Part I: An account of basic findings. *Psychological Review, 88,* 375–407.

Mezzanotte, R. J. (1981). Accessing visual schemata: Mechanisms invoking world knowledge in the identification of objects in scenes. Unpublished doctoral dissertation, Department of Psychology, State University of New York at Buffalo.

Miller, G. A. (1956). The magical number seven, plus or minus two: Some limits on our capacity for processing information. *Psychological Review, 63,* 81–97.

Neisser, U. (1963). Decision time without reaction time: Experiments in visual scanning. *American Journal of Psychology, 76,* 376–385.

Palmer, S., Rosch, E., & Chase, P. (1981). Canonical perspective and the perception of objects. In J. Long & A. Baddeley (Eds.), *Attention & performance IX.* Hillsdale, NJ: Erlbaum.

Penrose, L. S., & Penrose, R. (1958). Impossible objects: A special type of illusion. *British Journal of Psychology, 49,* 31–33.

Pentland, A. P. (1986). Perceptual organization and the representation of natural form. *Artificial Intelligence, 28,* 293–331.

Pisoni, D. B., Nusbaum, H. C., & Greene, B. G. (1985). Perception of synthetic speech generated by rule, *Proceedings IEEE, 73,* 1665–1676.

Pomerantz, J. R. (1978). Pattern and speed of encoding. *Memory & Cognition, 5,* 235–241.

Rayner, K. (1975). The perception span and peripheral cues in reading. *Cognitive Psychology, 7,* 65–81.

Rock, I. (1983). *The Logic of perception.* Cambridge, MA: MIT Press.

Rosch, E., Mervis, C. B., Gray, W., Johnson, D., & Boyes-Braem. (1976). Basic objects in natural categories. *Cognitive Psychology, 8,* 382–439.

Rumelhart, D. E., McClelland, J. L., and the PDP Research Group (1986). *Parallel distributed processing: Explorations in the microstructure of cognition. Vol. 1: Foundations.* Cambridge, MA: MIT.

Shepard, R. N., & Metzler, J. (1971). Mental rotation of three dimensional objects. *Science, 171,* 701–703.

Snodgrass, J. G., & Vanderwart, M. (1980). A standardized set of 260 pictures: Norms for name agreement, image agreement, familiarity, and visual complexity. *Journal of Experimental Psychology: Human Learning and Memory, 6,* 174–215.

Treisman, A. (In press). Feature analysis in early vision: Evidence from search asymmetries. *Psychological Review.*

Tversky, A. (1977). Features of similarity. *Psychological Review, 84,* 327–352.

Ullman, S. (1984). Visual routines. *Cognition, 18,* 97–159.

Waltz, D. (1975). Generating semantic descriptions from drawings of scenes with shadows. In P. A. Winston (Ed.), *The psychology of computer vision* (pp. 19–91). New York: McGraw-Hill.

Winston, P. A. (1975). Learning structural descriptions from examples. In P. H. Winston, (Ed.), *The psychology of computer vision* (pp. 157–209). New York: McGraw-Hill.

Witkin, A. P., & Tenenbaum, J. M. (1983). On the role of structure in vision. In J. Beck, B. Hope, & A. Rosenfeld (Eds.), *Human and machine vision* (pp. 481–543). New York: Academic Press.

Woodworth, R. S. (1938). *Experimental psychology.* New York: Holt.

15

Visual Object Identification: Some Effects of Image Foreshortening and Monocular Depth Cues*

G. Keith Humphrey

Department of Psychology
University of Western Ontario

Pierre Jolicoeur

Department of Psychology
University of Waterloo

One of the main problems in understanding visual object recognition stems from the fact that an object can give rise to an infinite number of two-dimensional projections when seen from different viewpoints. A central concern, then, is the discovery of information and processes that can be used to derive a stable description of an object despite variations in the two-dimensional image.

Objects can be described in terms of their component parts and in terms of their overall shape. Some early theories of object recognition proposed that the distinctive features of objects can be used to derive a stable object description (Selfridge & Neisser, 1960). According to such views, the features of objects will be mapped onto some long term memory representation of an object and recognition will occur if sufficient features have been processed to index the representation. Recent work that has extended this tradition has sought to describe a vocabulary of three-dimensional primitives from which a rich body of object representations can be constructed (e.g. Biederman, 1985; Marr & Nishihara, 1978). Much of the work has been motivated by an attempt to decompose the problem of representing the overall shape of an object into the shape of its components. For

*The research presented in this chapter was supported by an operating grant (A1643) from the Natural Sciences and Engineering Research Council of Canada to G. Keith Humphrey. Thanks to Bonnie Williams for running the experiments.

example, in Biederman's (1985; this volume) "recognition-by-components" proposal, the problem of viewpoint variance for objects is solved by choosing a system of components which give rise to invariant two-dimensional information despite changes in viewpoint. Thus, according to Biederman, one should not search for invariants at the level of objects, but rather in the component parts of objects. These invariantly perceived parts can then be used to index long-term object representations.

A theory of object recognition has to specify not only a set of primitive components, but also a way of describing the arrangements of these components in objects. Marr and Nishihara (1978) developed a scheme that can lead to the description of an object that is independent of viewpoint and that considers the arrangement of components comprising the object. They argue that the best way to overcome the diversity of two-dimensional projections that arise from variations in viewpoint is to derive a representation that is based on a frame of reference aligned with the salient properties of the object. Thus, changes in the orientation of the object with respect to the viewer will not lead to changes in the object's representation, because the frame of reference is based on the object and it will also change as the viewpoint changes. Marr and Nishihara refer to this type of frame as object-centered. Once this object-centered frame has been assigned, parts of the object can be encoded in relation to this frame. A key concept for our purposes is the axis aligned with the principal direction of this frame, which we will call the intrinsic axis.

According to Marr and Nishihara (1978), different information could be used to select an intrinsic axis. For elongated objects, the axis of maximum elongation will correspond with the intrinsic axis. For objects that are not significantly elongated, other intrinsic properties could be used, such as bilateral symmetry. Thus the object-centered intrinsic axis of a spoon would be aligned with the axis of elongation, whereas that for a face would be aligned with the axis of symmetry. In Marr and Nishihara's scheme, this process of axis selection can be applied to an image as a whole as well as to various component parts of an object.

In summary, the major thrust of these models is to construct a representational system that is essentially invariant over three-dimensional rotations of an object with respect to a viewer. Although the above models should be successful in solving the rotational invariance problem, for a large number of situations some views of an object would still lead to difficult recognition. In particular, views in which the projection of the intrinsic axis of an object is foreshortened in the image would be expected to result in more difficult recognition. In fact, object recognition should be increasingly more difficult as the degree of foreshortening is increased. Two factors, not necessarily mutually exclusive, could account for the expected difficulty due to foreshortening of the intrinsic axis. First, the processes that determine the direction of maximum elongation would be expected to be affected by foreshortening. Second, views in which the intrinsic axis is foreshortened may also result in the occlusion of parts, components and/or features.

Although effects of object rotation in the image plane have been extensively studied (e.g., Jolicoeur, 1985; Maki, 1986; Rock, 1973), little research has been done to investigate the influence of foreshortening of the axis of elongation on object recognition. We briefly review the available evidence concerning the effects of foreshortening. Palmer, Rosch and Chase (1981) have shown that subjects are slower at naming objects photographed from an unconventional viewpoint than they are for those photographed from a typical view. The typical views corresponded to a perspective in which the salient features and major axes were clearly represented. Some of the unconventional photographs used by Palmer et al. (1981) foreshortened a major axis in the object, but also obscured some of the components, so the influence of axis foreshortening per se is difficult to assess. Palmer et al. argued that the typical views maximize the amount of salient information about an object that is available for encoding it. Their results could be interpreted to suggest that the more parts of an object that are visible, the more quickly the object is identified. In this sense their results are consistent with those of Biederman (1985; this volume) as described below.

The ability to recognize an object from a novel perspective has recently been linked to damage of the right hemisphere in man. Damage to the right hemisphere has been associated with problems in a number of visual capacities such as recognition of incomplete line drawings, appreciation of spatial relations between and within objects, perception of the relationship of parts to wholes, and a variety of other deficits (for a discussion see Bradshaw & Nettleson, 1983). In the present context, research by Warrington and Taylor (1973, 1978) is relevant. They found that patients with right parietal damage were selectively impaired in the identification of "unusual" or "unconventional view" photographs of objects compared with "prototypical" or "conventional view" photographs of the same objects (see Figure 1). Also, the right hemisphere damaged patients were impaired on a same or different matching task in which one conventional view photograph was paired with an unconventional view photograph of the same object, or with a different object. The patient's task was to judge whether the two photographs depicted

A   B

Figure 1.   Examples of (A) a lateral view of an object, and (B) a foreshortened view of the same object.

the same object or not. According to Warrington and Taylor, the brain-damaged people were much less able than control subjects or left hemisphere damaged patients to tolerate any deviation from the "prototypical" representation of an object. Many of the "unconventional view" photographs used by Warrington and Taylor (1973, 1978) were foreshortened views of objects. For example, Figure 1 shows photographs of a "conventional" and "unconventional" views of an object, a shoe, similar to some of those presented to their brain-damaged patients.

Marr was intrigued by these observations on right hemisphere damaged patients and the distinction he and Nishihara developed between viewer-centered and object-centered coordinate frames seems to have been in part a result of considering the implications of Warrington and Taylor's research (see Marr, 1982, p. 35). Marr's interpretation was that the right hemisphere damaged patients fail to impose the correct object-centered reference frame on the foreshortened views and thus do not derive the correct object-centered representations. In viewing real objects, additional information in the scene, such as can be obtained from stereopsis, perspective cues, texture gradients, and/or shading could help to specify the natural axis. However, this additional information was lacking in the photographs and, according to Marr, led the patients to the wrong choice for the intrinsic axis.

More recent work has extended the observations of Warrington and Taylor. Humphreys and Riddoch (1984; see also Ratcliff & Newcombe, 1982) have studied a group of four patients with right hemisphere damage. They found that these four patients were selectively impaired in perceiving object shape in foreshortened view photographs of objects as measured by performance on a matching task and in naming the objects. Humphreys and Riddoch (1984) also found that these patients had great difficulty in a task requiring them to judge the orientation of objects whose principal axis was foreshortened in photographs. They suggested that the patients had a strong tendency to impose a two-dimensional frame of reference on the foreshortened views. To test this notion, Humphreys and Riddoch (1984) attempted to manipulate the assignment of the reference frame by including some depth cues in the photographs. They found that the addition of cues giving strong linear perspective to the backgrounds on which these objects were photographed facilitated performance in these patients. They argued that the presence of these cues to depth helped the patients impose a three-dimensional reference frame on the foreshortened-view photographs. Humphreys and Riddoch (1984) suggest that, in the absence of these depth cues, the patients rely on the reference frame provided by the edges of the photograph to determine the principal axis orientation. Recognition errors arise because the descriptions of the foreshortened object relative to an axis aligned with the edge of the picture frame does not match the description relative to the intrinsic axis.

Recent work by Warrington and James (1986) has questioned whether the results from right hemisphere damaged patients and normal subjects on identifying foreshortened views can be best explained in terms of a failure to derive the intrinsic axis of the objects represented. They suggest, based on evidence obtained with a new technique, that the results are more consistent with a distinctive-features model of object recognition and that foreshortening impedes recognition because it obscures features. Also, it should be noted that the performance of a fifth patient (with bilateral damage) tested by Humphreys and Riddoch (1984) was more affected by manipulations that occluded features than the foreshortening manipulations that impaired performance in the other four patients.

Experiments with normal subjects have also shown that occluding or removing components of objects interferes with recognition. Biederman (1985; this volume) has conducted a number of experiments with line drawings of objects showing that removal of information crucial to the formation of component parts severely slows or makes recognition impossible. Specifically, if contours were deleted at regions of concavity, the perception of component parts of objects was hindered, making recognition of the objects virtually impossible. Biederman has also demonstrated that removal of components, leaving the rest of the contour outline of an object intact, slows the recognition process. Such results are consistent with the notion that recognition depends on features or components being mapped into some long-term representation of an object and that recognition will occur if sufficient components have been processed to index the representation.

EXPERIMENTS

The research reported in the present chapter was carried out for two principle reasons. First, we wished to discover whether foreshortening effects could be obtained with views in which the relevant parts or components of an object were never occluded. Should such effects be found, this would suggest that all effects of foreshortening cannot be attributed to the occlusion of parts or features. Second, the effects of cues to depth extraneous to the object were also investigated. If the depth cues affect the perception of foreshortened views of objects, it would seem difficult to explain the effect based on the notion that the only effect of foreshortening was to reduce the number of available features, parts or components, given that the number of parts was constant across the conditions.

The experiments all employed the same task, set of objects, and the same backgrounds used to control extraneous depth information. These aspects of the experiments are described first, following which we describe the individual experiments.

Stimuli

Line drawings of 32 objects with an obvious elongated major axis were used in the study (for examples, see Figure 2). Both "profile" views and foreshortened views of the objects were produced. The objects were videorecorded from slightly above the surface on which they rested. For the profile views, the objects were placed at a 45 degree angle to a completely lateral view in which the axis of elongation would be perpendicular to the videocamera. For the foreshortened views, the objects were placed at an 80 degree angle to the lateral view. The videorecords were traced from the screen of a videomonitor and were photographed as slides. All of the objects were readily nameable in both views when shown to subjects in nonspeeded conditions. An additional four drawings produced from the same perspectives served as warm up stimuli in the experiments.

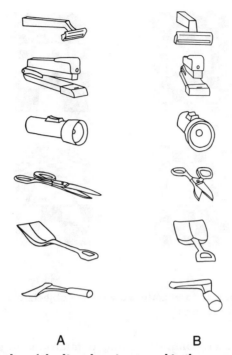

A   B

Figure 2.   Examples of the line drawings used in the experiments described in the chapter. (A) shows "profile" views of objects, and (B) shows foreshortened depictions of the same objects.

Depth Backgrounds

Two checkerboard patterns were also constructed and used to create two conditions of background depth information. One of the patterns was a checkerboard showing perspective convergence, perspective compression, and changing size of checks from bottom to top. This checkerboard appeared to be receding in depth and we will refer to this pattern as the depth background. The other checkerboard pattern contained no convergence, no compression or variation in check size, and yielded no impression of depth. This pattern will be referred to as the flat background. Figure 3 shows examples of each of the backgrounds. The pictures were presented using slide projectors and a rear-projection screen. The objects subtended, on average, 1.5 degrees of visual angle. The backgrounds subtended 6.6 degrees in the horizontal and 8.3 degrees in the vertical dimensions.

General Procedure

The task was to name each of the objects as quickly as possible with the first appropriate name that came to mind. Naming latency was

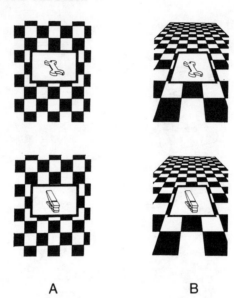

A   B

Figure 3.   Examples of foreshortened views of objects placed on the (A) flat and (B) depth backgrounds.

recorded by a microcomputer using a microphone and voice-activated relay. Timing began with the opening of a tachistoscopic shutter and ended with the triggering of the voice-activated relay. The picture remained in view until the subject responded or 10 seconds elapsed. Errors were recorded by the experimenter and trials for which the subjects did not make a response in 10 sec. or named the object incorrectly were treated as errors. Subjects were given four practice trials before timing began. All subjects who participated were naive with respect to the purpose of the experiments and participated only once. Subjects were not given any information about the stimuli other than that they were line drawings of common objects.

Subjects

Twelve new subjects with normal or corrected-to-normal vision participated in each experiment except for Experiment 1 in which there were 24 subjects.

EXPERIMENT 1

The purpose of Experiment 1 was to establish that the foreshortened views were more difficult to identify than the profile views. Separate groups of 12 subjects were presented with profile views and with foreshortened views and their naming latencies were measured. The objects were presented without the backgrounds. Statistical analyses revealed that the naming latencies were reliably longer for the foreshortened perspective than for the profile views ($p <.01$) The mean naming latency was 1460 msec. for the foreshortened views and 1167 msec. for the profile perspective. Furthermore, the error rate for the profile views (5%) was less than that for the foreshortened views (12%; $p <.01$). These results suggest that foreshortening effects are not due solely to component, part, or feature occlusion.

Having established that the foreshortened perspective views were more difficult to identify, we felt that they would be the most likely to be affected by the background patterns. In Experiments 2 to 5, we investigated the effects of depth backgrounds on the identification of the foreshortened views (see Figure 3). For each subject, half of the trials were presented with the depth checkerboard pattern and half with the flat checkerboard pattern. These trials were intermixed haphazardly throughout the session for each subject.

EXPERIMENT 2

In the first experiment investigating the potential effects of depth backgrounds, the background and object were presented simultaneously. As in Experiment 1, the subjects saw the stimuli presented on a rear-projection screen using binocular vision. The analysis did not reveal a statistically reliable difference between the backgrounds ($p = .52$). The mean naming latency for objects on the flat background was 1679 msec and was 1608 msec for the depth background. The error rate was 12% for the flat background and was 15% for the depth background, which were not statistically different. Parenthetically, we wish to note that Biederman (1981a), using an object identification task quite different from ours, especially because it did not involve foreshortened views, found that pictorial depth information did not improve performance relative to conditions without such cues. Biederman found that depth information actually hindered performance, as measured by the proportion of errors, relative to conditions without depth. His backgrounds and objects were presented simultaneously as well.

EXPERIMENT 3

In the previous experiment, we questioned each of the subjects immediately after the experimental trials to see if they had noticed that there were different backgrounds. A number of subjects (5) said that they had not noticed whether the backgrounds differed or not. This observation suggested that the subjects may not be processing much of the background very deeply. If so, it may not be surprising to find that the backgrounds had little effect on the speed of object identification.

In the present experiment, the backgrounds were presented three to four seconds before the objects appeared in the central region of the backgrounds (see Figure 3) and stayed visible until the subject responded. We thought that such a manipulation would allow the subjects to attend to the background and then to use this information in the object-naming task.

Again, there was no reliable difference between the naming latencies for objects presented on the depth backgrounds and those presented on the flat backgrounds ($p = .35$). Mean naming latency was 1429 msec. for objects presented on the depth backgrounds and 1491 msec. for the objects presented on the flat backgrounds. The error rates were not reliably different either. They were 12% for the depth background and 17% for the flat background. The subjects had clearly noticed the difference in the backgrounds as revealed by questioning after the experiment, but this did not translate into a differential effect of the backgrounds on naming latency.

EXPERIMENT 4

Both of the experiments with the backgrounds described above were done with binocular viewing and a full view of the projection screen and its edges. Perhaps the information from binocular vision, movement parallax, and the edges of the screen were reducing any differential effect that the backgrounds could exert on naming latencies. The present experiment eliminated these sources of depth information by having the subjects view the pictures through a small aperture with one eye at the end of a rectangular tube embedded in the wall of a box. Because only one eye was used, there could not be any binocular cues to the flatness of the depth background. Motion parallax was eliminated by the small aperture and the edges of the screen were occluded by the rectangular tube. The other eye could not see the projection screen because of the wall of the viewing box. Under these conditions the depth background gave a strong three-dimensional appearance. Similar arrangements have been used in "trompe l'oeil" demonstrations (Pirenne, 1975). Such demonstrations are constructed using photographs taken with pinhole cameras that yield pictures in precise linear perspective, with the center of the pinhole being the center of perspective. Viewing such photographs through a small aperture yields a strong three-dimensional impression (Pirenne, 1975).

Subjects were seated in front of the viewing apparatus and asked to look through the aperture with whichever eye they preferred. As in the previously described experiment, the backgrounds were presented three to four seconds before the objects appeared.

In contrast with the previous two experiments, objects appearing on depth backgrounds were named more rapidly (1392 msec) than those seen on the flat background (1533 msec; $p < .05$). The error rate was 8% for the objects on the depth background and 10% for the objects presented on the flat background. These error proportions were not significantly different.

Because the naming latency difference in the present experiment was the first statistically reliable result showing an effect of the background, we felt it was necessary to replicate this finding. Another group of 12 subjects was run under the same conditions. The mean naming latency for the objects presented against a depth background was 1341 msec. and 1551 msec. for those seen against the flat background. This difference was statistically reliable ($p < .01$). The error rate was 9% for the objects seen on the depth background and 10% for those seen on the flat background.

EXPERIMENT 5

In a final experiment we used the apparatus restricting the view to one eye, but the objects and backgrounds were presented simultaneously. As in the Experiment 2, which also had simultaneous presentation of objects and backgrounds, there was no reliable difference in the naming latency as a function of background ($p = .19$). The mean naming latency was 1264 msec. for the objects presented on a depth background and 1379 msec for those presented on the flat background. The error rate was 10% for those on the depth background and 8% for those on the flat background.

COMPARISON OF EXPERIMENTS 2 TO 5

Although the experiments described above were run separatedly and as such subject assignment was not random across experiments, we thought it would be instructive to compare the experiments despite problems that could arise in nonrandom assignment. Only the set of conditions used in Experiment 4 produced a reliable differential effect on naming latency as a function of background, however, the results for each of the experiments were in the appropriate direction—shorter latency to name objects seen on depth backgrounds. An overall analysis including all four experiments showed an effect of background ($p < .01$), but did not yield an interaction of the different experimental conditions with the backgrounds. This result could be seen to weaken the claim that the conditions of Experiment 4 were unique in yielding a differential effect. Nevertheless, the result strengthens the claim that the backgrounds influence object identification latency.

Besides the overall effect of background, there was an effect of viewing conditions. The aperture viewing conditions led to more rapid naming than the free viewing conditions ($p < .05$). This result may have occurred because the subjects' attention was more focused in the aperture conditions. Finally, there was a interaction ($p < .05$) of the viewing conditions with the background SOA (stimulus onset asynchrony; background and objects presented simultaneously, versus background presented before the objects). This interaction apparently occurred because the simultaneous presentation conditions of Experiment 2 produced much longer naming latencies than in the other three experiments.

An overall analysis of the error data was also performed. There was a significant effect of the viewing conditions ($p < .05$) with the aperture viewing condition having a lower error rate than the free viewing conditions. There was also an interaction of the background-object SOA with the background conditions ($p < .05$). Simultaneous presentation of background and objects tended to lead to a higher error rate with the depth backgrounds than with the flat backgrounds (see also Biederman, 1981a). In contrast, if the backgrounds were presented before the objects, there tended to be fewer errors with the depth backgrounds than with the flat backgrounds.

The most important result of these comparative analyses for present purposes was the overall effect of the background manipulation. In general, depth backgrounds led to shorter naming latency than flat backgrounds.

DISCUSSION

The foreshorted views of objects were more difficult to identify, as indicated by naming latency and error, than the profile views. These results are consistent with the findings of Palmer et al. (1981) and with the research on right hemisphere damaged patients (Humphreys & Riddoch, 1984; Ratcliff & Newcombe, 1982; Warrington &

Taylor, 1973; 1978). Given that we were especially careful not to occlude object parts or components in the foreshortened views, the results could be interpreted as suggesting that the foreshortened views were more difficult to identify primarily because of the obscuring of the intrinsic axis. Thus, our experiments go some way towards showing that foreshortening effects may not be entirely due to effects at the level of components or features. However, an alternative interpretation is that foreshortening impaired the perception of each of the foreshortened components without necessarily involving difficulty with the assignment of a frame of reference. This latter interpretation is certainly a viable alternative, but at some point recognition must involve more than the registration of the components or features of an object. Recognition also involves assigning an overall orientation based on the intrinsic axis of the object and coding the components or features of an object in relation to that axis. So although our demonstration is far from definitive, we believe that at least part of the effect of foreshortening results from a difficulty is assigning the intrinsic axis.

The results obtained with the backgrounds are consistent with the idea that some of the effects of foreshortening occur because of a difficulty in assigning the intrinsic axis. Because the object views were identical under the two background conditions the results are difficult for a solely component or feature based theory of object recognition to handle. The results of these experiments are consistent with those of Humphreys and Riddoch (1984) who found that the presence of monocular depth cues helped the patients with right hemisphere lesions in tasks involving foreshortened views of objects. They suggested that these cues facilitated the assignment of a three-dimensional reference frame on the foreshortened views. In the present experiments, the differential effect of the two backgrounds on object identification could also be interpreted as influencing the frame of reference and thus the assignment of the intrinsic axis used by the subjects in identifying the line drawings.

Any complete theory of object recognition will have to consider both the components making up an object and the arrangement of these components to form objects. Many objects can be represented in terms of their overall shape and in terms of a combination of components and as such there may be a variety of "routes" to visual object recognition (Humphreys & Riddoch, 1984). In the everyday world we normally encounter objects in a rich semantic context which can have marked effects on recognition (Biederman, 1981b; Palmer, 1975). We also encounter objects in a rich visual context in which information specifying depth is central. Our experiments demonstrate that at least some types of depth information which are part of our visual ecology do have an influence on object identification.

REFERENCES

Biederman, I. (1981a). Do background depth gradients facilitate object identification? *Perception, 10*, 573–578.

Biederman, I. (1981b). On the semantics of a glance at a scene. In M. Kubovy & J. R. Pomerantz (Eds.), *Perceptual Organization*. Hillsdale, NJ: Lawrence Erlbaum Associates.

Biederman, I. (1985). Human image understanding: Recent research and a theory. *Computer Vision, Graphics, and Image Processing, 32*, 29–73.

Bradshaw, J. L., & Nettleson, N. C. (1983). *Human cerebral asymmetry*. Englewood Cliffs, NJ: Prentice-Hall.

Humphreys, G. W., & Riddoch, M. J. (1984). Routes to object constancy: Implications from neurological impairments of object constancy. *The Quarterly Journal of Experimental Psychology, 36A*, 385–415.

Jolicoeur, P. (1985). The time to name disoriented natural objects. *Memory and Cognition, 13*, 289–303.

Maki, R. H. (1986). Naming and locating the tops of rotated pictures. *Canadian Journal of Psychology, 40*, 368–387.

Marr, D. (1982). *Vision*. San Francisco: W. H. Freeman and Company.

Marr, D., & Nishihara, H. K. (1978). Representation and recognition of the spatial organization of three-dimensional shapes. *Proceedings of the Royal Society of London B, 200*, 269–294.

Palmer, S. E. (1975). The effects of contextual scenes on the identification of objects. *Memory and Cognition, 3*, 519–526.

Palmer, S., Rosch, E., & Chase, P. (1981). Canonical perspective and the perception of objects. In J. Long & A. Baddeley (Eds.), *Attention and performance IX*. Hillsdale, NJ: Lawrence Erlbaum Associates.

Pirenne, M. H. (1975). Vision and art. In E. C. Carterette & M. P. Friedman (Eds.), *Handbook of perception, Vol. 5. Seeing*. New York: Academic Press.

Ratcliff, G., & Newcombe, F. (1982). Object recognition: Some deductions from the clinical evidence. In A. W. Ellis (Ed.), *Normality and pathology in cognitive function*. New York: Academic Press.

Rock, I. (1973). *Orientation and form*. New York: Academic Press.

Selfridge, O. G., & Neisser, U. (1960). Pattern recognition by machine. *Scientific American, 203*, 60–68.

Warrington, E. K., & James, M. (1986). Visual object recognition in patients with right hemisphere lesions: Axes or features? *Perception, 15*, 355–366.

Warrington, E. K., & Taylor, A. M. (1973). The contribution of the right parietal lobe to object recognition. *Cortex, 9*, 152–164.

Warrington, E. K., & Taylor, A. M. (1978). Two categorical stages of object recognition. *Perception, 7*, 695–705.

16
Stable Representation of Shape*

Robert J. Woodham

Laboratory for Computational Vision
Department of Computer Science
University of British Columbia

INTRODUCTION

Computational vision is the study of intelligent systems that produce descriptions of a world from images of that world. The purpose is to determine those aspects of the world that are required to carry out some task. For most tasks, shape is a necessary component of any description produced. Thus, shape representation is a central concern to designers of computer vision systems.[1]

In a general purpose vision system, the mapping from signal input to final shape description is too complex to be treated as a function in a single representation. Shape analysis requires many levels of intermediate representation. Identifying those levels and estab-

---

* This report describes research done at the Laboratory for Computational Vision of the University of British Columbia. Support for the Laboratory's research is provided, in part, by the UBC Interdisciplinary Graduate Program in Remote Sensing, by the Natural Sciences and Engineering Research Council of Canada (NSERC) under grants SMI-51, A0383 and E0008, and by the Canadian Institute for Advanced Research. The work on shape representation also was supported, in part, by NSERC grant A3390.

Technical details on convex polyhedra and the mixed volume derive from the Ph.D. thesis of J. J. Little. The author also thanks J. J. Little and A. K. Mackworth for many discussions on aspects of shape representation.

The author is a Fellow of the Canadian Institute for Advanced Research.

[1] To avoid confusion, the term *representation* is used to identify a formalism, or language, for encoding a general class of shapes. The term *description* is restricted to mean a specific expression in the formalism that identifies an instance of a particular shape, or class of shapes, in the representation.

lishing the constraints that operate both within and between levels is the fundamental challenge of computational vision research. Each level of representation must consider both the processes that derive the representation and the processes that compute with the representation. At the level of the signal, one deals with descriptions that can be derived directly from the image. This leads initially to representations for the 2D shape of image patterns. Interpreting image properties as scene properties leads to representations for the visible surfaces in the scene. Finally, recognition of distinct objects and their spatial arrangement requires representations for 3D shape that are independent of viewpoint. Computational vision thus distinguishes three levels of representation: 2D image, visible surfaces, and 3D objects (Marr, 1982; Barrow & Tenenbaum, 1981a; Brady, 1982; Binford, 1982; Horn, 1986). The principal shape representations considered at each of these levels are discussed in Woodham (1987).

This chapter identifies stability as one of several design criteria that a shape representation should satisfy. The definition of stability adopted here is the standard one from numerical analysis. That is, a computation is stable if small changes in the input produce correspondingly small changes in the output. This definition is well-routed in mathematics but it is difficult to make precise when dealing with symbolic descriptions. One small step in this direction is the use of the mixed volume as a measure of the similarity between two convex polyhedra. Technical details are found in the Ph.D. thesis of Little (1985c) and related publications (Little, 1983; 1985a, 1985b). Here, we interpret Little's results, compared to alternatives, in terms of stability.

The next section outlines a general approach to vision research and identifies one particular research strategy to follow. The following section describes design criteria for shape representation. The next section introduces orientation-based representations for 3D shape and describes a prototype vision system suitable for automatic bin-picking. The section following that provides the necessary background for convex polyhedra. The next section discusses similarity measures for convex polyhedra. Concluding remarks are in the final section.

AN APPROACH TO VISION RESEARCH

A complete theory of human vision must account for the relationship between the natural world and human visual perception. It is here taken as a given that the natural world consists of 3D objects and that perceptions ultimately can be represented as symbolic descriptions. In this view, the 2D image acts as an intermediate representation that mediates between the 3D world and visual perception. Figure 1 illustrates. To understand the relationship between the 3D world and perception, there are four components to consider, as suggested by the four labeled arrows in Figure 1.

Arrow 1 characterizes the mapping from the 3D world to the 2D image. Given a spatial arrangement of objects made of a particular set of materials and illuminated in a particular way, the laws of physical optics determine the image. Geometric equations determine where each point on a visible surface will appear in the image and corresponding radiometric equations determine its brightness and color. This is properly the domain of computer graphics, al-

3D WORLD   2D IMAGE   PERCEPTION

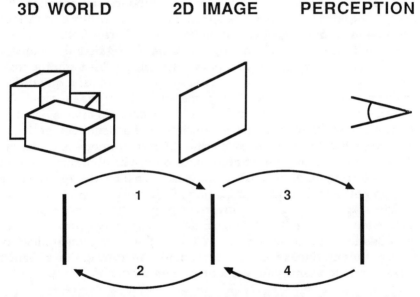

Figure 1. Visual perception of the natural world. The natural world consists of 3D objects made out of different materials illuminated in different ways. An image is a spatially varying brightness pattern. Visual perception consists of symbolic descriptions of the natural world computed from the image. There are four mappings to describe. Arrow 1 is the mapping from the 3D world to the 2D image. Arrow 2 is the inverse mapping from the 2D image to the 3D world. Arrow 3 is the mapping from the 2D image to perception. Arrow 4 is the inverse mapping from a perception to the 2D image. Since the mappings 1 and 3 are, in general, many-to-one, the inverse mappings 2 and 4 determine equivalence classes, respectively, of worlds that produce identical images and of images that produce identical perceptions.

though it is relevant to point out that computer graphics is primarily concerned with producing "realistic" images, to convey information, as opposed to images that necessarily depict physical reality.

Arrow 2 characterizes the inverse mapping from 2D image to 3D world. All so-called "shape from" methods are formulated as problems at this level, including: shape from binocular stereo (Baker & Binford, 1981; Grimson, 1981); shape from shading (Horn, 1977; Ikeuchi & Horn, 1981; Woodham, 1981, 1984; Horn & Brooks, 1986) and photometric stereo (Woodham, 1980); shape from contour (Witkin, 1981; Barrow & Tenenbaum, 1981b; Kanade, 1981; Binford, 1981; Brady & Yuille, 1984); shape from motion (Hildreth, 1984) and optical flow (Horn & Schunck, 1981); and shape from texture (Witkin, 1981; Kender, 1982).Since the mapping from the 3D world to the 2D image (arrow 1) is many-to-one, the inverse mapping (arrow 2) is underconstrained. That is, there are many 3D worlds that produce the identical 2D image. In most situations, the inverse problem is ill-posed in that there is no unique, stable solution (Poggio & Torre, 1984).

Arrow 3 characterizes the mapping from the 2D image to perception. This is the domain of perceptual psychology. The mapping from the 2D image to perception also is many-to-one in that many different 2D images produce identical perceptions. Familiar examples include color and texture metamers and edge contrast effects, such as the Cornsweet illusion. Many constraints on the visual perception of 2D images derive from the assumption that a 2D image is an image of a 3D world. Thus, one cannot deal only with the perception of images (arrow 3). Necessarily, one must consider the relationship between 3D worlds and their 2D images (arrows 1 and 2).

Arrow 4 characterizes the inverse mapping from a perception to the 2D image. Again, this inverse mapping is underconstrained. But, the question can be posed as, "What is the equivalence class of images that produces a given perception?" Answering this question corresponds to identifying perceptual metamers.

The Research Strategy

"Shape from" methods define the computational task in terms of arrow 2 of Figure 1. That is, the task is to determine the 3D shape of objects from their 2D projection onto images. Although each method differs considerably in precise detail, all share a common characterization as computational tasks. Each embodies the following steps:

- *Identify the visual task.* This involves picking a task domain and a class of locally computable image features for the domain that provides (i.e., make plural) cues to 3D shape.
- *Derive mathematical equations that describe how the world determines the image.* The equations are based on the laws of optics and, in general, consider both geometry and radiometry (arrow 1 of Figure 1). The equations determine the mapping from scene to image. Shape analysis, however, requires a solution to the inverse problem. That is, one must determine the mapping from image to scene (arrow 2 of Figure 1).
- *Demonstrate that the inverse problem is underconstrained.* It is usually straightforward to demonstrate that the problem is locally underconstrained. In general, the problem is also globally underconstrained although this can be more difficult to demonstrate.
- *Identify additional constraints that lead to a unique stable solution to the inverse problem.* Image features determine equivalence classes of possible scene features. Conceptually, a unique stable solution is obtained when a suitable metric is applied to the equivalence classes to select a single preferred solution. The metric is often expressed as a performance index designed to achieve smooth, regular, or minimal energy solutions. Identifying a suitable performance index is not a trivial matter. There are many possible measures to consider for a given visual task. Some degree of mathematical rigor is generally required to demonstrate that a particular choice does, in fact, lead to a unique stable solution. Finally, even when the existence of the desired solution is established, it is still necessary to develop an algorithm to compute the solution.
- *Show that the solution thus obtained agrees with human perception.* Whatever the metric, the correct physical solution cannot be obtained in all cases. Human perception does not always correspond to the correct physical solution either. One level of agreement with human perception is to demonstrate that the computed solution agrees with human perception for the chosen visual task. At a second level, one also compares known algorithms for computing the solution to plausible mechanisms for biological implementation.

SHAPE REPRESENTATION

Several authors propose design criteria that general-purpose shape representations should satisfy (Binford, 1982; Woodham, 1987; Marr &

Nishihara, 1978; Brady, 1983; Mokhtarian & Mackworth, 1986). No single representation proposed to date satisfies all of the criteria. Nevertheless, the criteria provide a useful framework to discuss representations that have been proposed. The design criteria are summarized below:

- *The representation of shape must be computable using only local support.* The ability to derive the representation from the input data is the minimal requirement. Local support further stipulates that the representation can be computed locally. This is required to deal with occlusion and to perform detailed inspection. It is also of practical importance since processes that derive the representation can then be implemented efficiently.
- *The representation of shape must be stable.* That is, small changes in the input should cause only small changes in the result. Images are subject to noise. Thus, stability is an important criterion for processes that derive initial descriptions from an image. Stability is also an important criterion for subsequent levels of representation because, without stability, it is difficult to define an effective measure of similarity to compare descriptions.
- *The representation of shape must be rich in the sense of information preserving.* Images are two-dimensional, while objects are three-dimensional. Image projection loses information. An image defines an equivalence class, usually infinite, of worlds that project to the identical image. A representation is rich if it does not arbitrarily restrict or extend this equivalence class. Rich representations are needed to describe a large class of objects, including objects that may never have been seen before.
- *The representation must describe shape at multiple scales.* Representations at multiple scales are useful for several reasons. First, representations at multiple scales suppress detail until it is required. Descriptions at a coarse scale relate to overall shape. Detail emerging at finer scales includes features that are more local. A pinhole in a metal casting is not significant when the task is to identify the part. But, it is critical when the task is to inspect the part for defects. Second, objects must be representable at different levels of detail. This can be accomplished using a hierarchical representation of shape that also takes into account the difference in object appearance owing to scale. For example, a forest is made up of individual trees. A forest can be represented hierarchically as a particular spatial arrangement and species composition of individual trees. At a coarser scale, the forest must still be represented as a forest, even when the individual trees are

no longer discernible. Third, in the presence of noise, there is an inherent trade-off between the detectability of an image feature and its precise localization in space. By working at multiple scales, it is possible to optimize this trade-off dynamically, as required. Fourth, a coarse to fine analysis can introduce significant computational speed-up in methods for shape analysis requiring search or convergence. Fifth, to be useful, a representation should be storage efficient. Representations at multiple scales are needed to be both storage efficient and rich.

- *The representation must define an object-based semantics for shape description and segmentation.* In general, comparison of 2D shape descriptions fails because there is no stable similarity measure to use. Large changes in shape description follow from minor changes in either the spatial configuration of the objects in the world, the viewpoint, or the illumination. Shape analysis requires representations in which 3D shape is explicit so that spatial relationships between surfaces can be computed easily. This is necessary to segment complex shapes into simpler components, to predict how objects will appear, and to deal with occlusion.

- *The representation of shape must correspond to human performance on the task.* Earlier, it was noted that an image defines an equivalence class of worlds that project to the identical image. Similarly, a representation defines equivalence classes of images that produce identical descriptions in the representation. Human perception also defines equivalence classes of worlds and images that produce identical perceptions. A representation of shape corresponds to human performance on some task if two conditions are satisfied. First, images that produce distinct descriptions in the representation are perceived as distinct in the task. Second, images that produce identical descriptions in the representation are perceived as identical in the task. A correspondence to human performance is difficult to achieve, in part because much remains to be understood about human perception. Nevertheless, developing this correspondence is a major motivating factor for current work in computational vision.

ORIENTATION-BASED REPRESENTATION OF SHAPE

A *depth map* is one way to represent the shape of a visible surface. A depth map determines distance to the surface along parallel rays on a dense, regularly spaced grid. A range finder is a sensor that produces surfaces descriptions of this form. Several "shape from" methods, including shape from binocular stereo and shape from motion, also produce surface descriptions in the form of a depth map. The depth map representation has certain deficiencies when the task is to determine 3D object identity, position, and attitude. For one thing, depth maps do not transform in a simple way when the object rotates or, equivalently, when the viewpoint changes. Thus, it is difficult to compare a sensed depth map directly to stored object models.

Alternatively, one can represent the shape of a visible surface by specifying surface orientation on a dense, regularly spaced grid. This representation has been called a *needle diagram* by Horn (1986). Photometric stereo (Woodham, 1980) produces surface descriptions of this form. Other "shape from" methods, including shape from shading and shape from texture gradient, also produce surface descriptions in the form of a needle diagram. A depth map description can be transformed into a needle diagram by numerical differentiation. The needle diagram itself still depends on both the position and attitude of the object in view. It is possible, however, to transform a needle diagram in a simple way to compute an *orientation histogram*. The orientation histogram corresponds to a hemisphere of the object's Extended Gaussian Image (EGI), as we shall see. Object identity and attitude is readily determined by comparing a sensed orientation histogram to object models stored using the EGI representation.

Consider the continuous case illustrated in Figure 2. Let U be the unit sphere, termed the *Gaussian sphere*. Each point on the Gaussian sphere identifies the unit vector formed by joining the origin to that point. Let E be a portion of a surface S bounded by a closed curve. The image of E under the *Gauss map*

$$G(p) = \omega, \; p \in S, \; \omega \text{ unit normal at } p$$

is $G(E)$, the *Gaussian image of E*.

The Gauss map allows us to define three related concepts. First, the *Gaussian curvature* at p, denoted by $K(p)$, is

$$K(p) = \lim_{|E| \to 0} \frac{|G(E)|}{|E|}$$

where E is a compact region on S enclosing p. Second, the *area function* at ω, denoted by $A(\omega)$, is

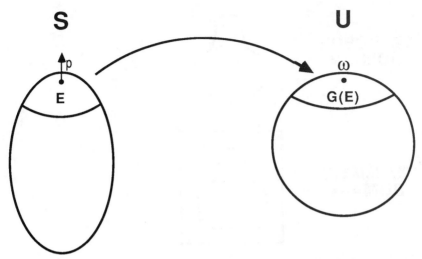

Figure 2.   The Gauss map. Let p be a point on a surface S. The unit surface normal vector at p determines a point ω on the Gaussian sphere U. $G(E)$ is the Gaussian image of E under the Gauss map $G(p) = \omega$.

$$A(\omega) = \lim_{|R| \to 0} \frac{|G^{-1}(R)|}{|R|}$$

where R is a compact region on U enclosing ω. Third, a 3D object's *Extended Gaussian image* (EGI) is simply its area function defined on U. The EGI of an object records the variation of surface area with surface orientation, using U as the reference system. If S has strictly positive curvature, $p \in G^{-1}(\omega)$ is unique. That is to say, the EGI uniquely represents convex objects (up to translation) and is thus information preserving for this class of objects.

In the case of polyhedra, the set of surface orientations is finite. Let $\{\omega_i\}$ be the set of unit (outward) surface normals of the planar faces of a given convex polyhedron. Then its EGI can be represented as $\{A(\omega_i)\}$ where $A(\omega_i)$ is the area of the face with orientation ω_i. One can imagine translating each ω_i to a common point of application and scaling it so that its length is $A(\omega_i)$. This produces another representation, equivalent to the EGI, that has been called a *spike model* (Horn, 1986).

The EGI is invariant to translation and can be normalized to be invariant also to scale. Rotations are easy to deal with since an object and its EGI rotate together. The EGI is easy to derive from other representations of three-dimensional objects. (See Horn, 1984 for a primer on extended Gaussian images.)

3D SCENE DOMAIN

2D IMAGE DOMAIN

VISIBLE SURFACE

3D OBJECT DOMAIN

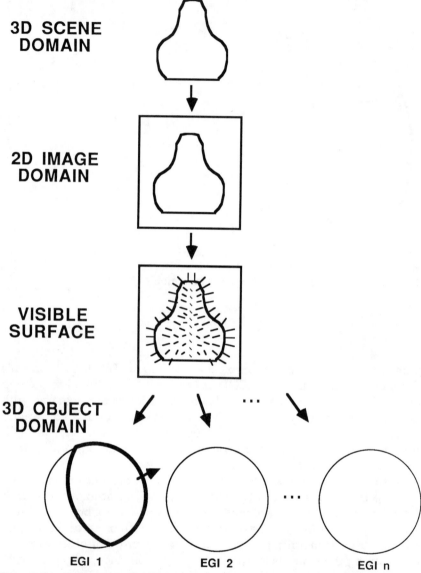

EGI 1   EGI 2   EGI n

Figure 3.   A 3D object, made of a given material and illuminated in a given way, determines the 2D image. Using photometric stereo, or other techniques, a description of the visible surface is computed from the image. Here, the description is of surface orientation at each visible point, called a needle diagram. Each of several 3D object's is represented by its Extended Gaussian Image (EGI). The needle diagram determines a hemisphere of the object's EGI. Matching the visible hemisphere to known EGI's determines both object identity and object attitude. This, for example, allows a robot to pick mixed parts out of a bin.

Figure 3 illustrates a prototype vision system for automatically picking parts out of bin. Versions of this system have been built and reported in the literature (Horn, 1984; Horn & Ikeuchi 1984; Ikeuchi, Nishihara, Horn, Sobalvarro, & Nagata, 1986). A 3D object, made of a given material and illuminated in a given way, determines the 2D image. Using photometric stereo, or other techniques, a description of the visible surface is computed from the image in the form of a needle diagram. The needle diagram is converted into a discrete orientation histogram, using a standard tesselation of the Gaussian sphere U. The EGI's of known objects are stored internally in the form of discrete orientation histograms. The measured orientation histogram determines a hemisphere of the object's EGI. Object recognition is achieved by matching the visible hemisphere to the correct stored EGI. The three degrees of freedom in object attitude also are determined by the position and orientation of the visible hemisphere at the correct match. This allows a robot to pick mixed parts out of a bin.

Figure 3 provides the overview. Unfortunately, the matching computation can be ill-conditioned. Recently, the mixed-volume has been used as the similarity measure. The result is more robust than direct EGI matching and can support efficient multiresolution attitude determination.

CONVEX POLYHEDRA

The discussion of convex polyhedra given here follows Little 1985. (See Lyusternik, 1963; Grunbaum, 1967) for a more comprehensive treatment.)

Basic Definitions

A *plane* can be represented as the set

$$\{x \mid <\omega,x> = c\}$$

where $<\cdot,\cdot>$ denotes vector inner product, ω is a unit normal vector to the plane and c is a scalar constant. A plane also defines the *half space* given by

$$\{x \mid <\omega,x> \le c\}$$

The intersection of a finite number of half spaces forms a *convex polyhedron* denoted by

$$\bigcap_{i=0}^{n} \{x \mid <\omega_i, x> \leq c_i\} \tag{1}$$

A bounded polyhedron is termed a *polytope*. A 2D polytope is called a *polygon*. A 3D polytope is called a *polyhedron*.

The *support function*, $H(\omega)$, of a convex polytope is defined as the perpendicular distance to the closest tangent plane, with normal vector ω, that touches but does not pass into the interior of the polytope. Distance is measured from an arbitrary origin chosen inside the polytope. Figure 4 illustrates the geometric construction of the

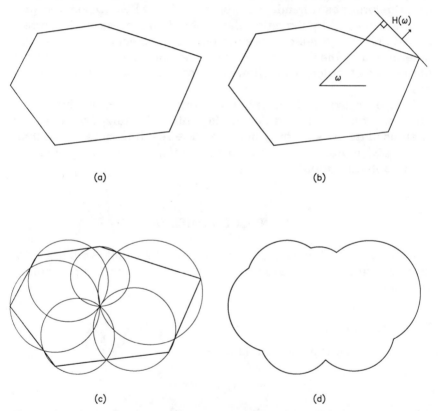

(a)   (b)

(c)   (d)

Figure 4. Geometric construction illustrating the definition of a support function, $H(\omega)$, shown here for the 2D case. A sample polygon is shown in (a). Orientation, ω, is defined with respect to an origin, interior to the polygon, and a reference direction, as shown in (b). For any ω, consider the tangent line having normal vector in the direction ω. $H(\omega)$ is the distance of closest approach of the tangent line to the origin. The points determined by $H(\omega)$ all lie on a set of circles, as shown in (c). Each circle passes through the origin and one vertex of the polygon. The resulting support figure is shown in (d).

support function, $H(\omega)$, for a sample polygon. The polygon is shown in Figure 4(a). The orientation, ω, is defined with respect to an origin, interior to the polygon, and a reference direction, as shown in Figure 4(b). For any ω, consider the tangent line having normal vector in the direction ω. $H(\omega)$ is the distance of closest approach of the tangent line to the origin. The points of closest approach all lie on a set of circles, as shown in Figure 4(c). Each circle passes through the origin and one vertex of the polygon. The resulting support figure is drawn in Figure 4(d). (For polyhedra, the support function, $H(\omega)$, determines a piece-wise spherical support figure.) Even though the set of orientations, $\{\omega_i\}$, corresponding to faces is discrete, the support function, $H(\omega)$, is a defined for all ω.

The volume of a convex polytope can be expressed in terms of its area function, $A(\omega_i)$, and support function, $H(\omega_i)$. For the 2D case, the area of the polygon is $\frac{1}{2} <H(\omega_i),A(\omega_i)>$. For the 3D case, the volume of the polyhedron is $\frac{1}{3} <H(\omega_i),A(\omega_i)>$. In all cases, the measure is independent of choice of origin in the definition of $H(\omega)$.

Let $\{\omega_i\}$ be a fixed set of orientation vectors. Then, a convex polytope can be uniquely represented in terms of $\{\omega_i\}$ either by specifying the corresponding set $\{c_i\}$ or the corresponding set of areas $\{A(\omega_i)\}$. If $\{c_i\}$ is given, the polytope is determined by equation (1). At first, it might seem that some information is lost if only $\{A(\omega_i)\}$ is given since the position of the surface normals is not preserved. Said another way, the adjacency information of the faces is not explicit. Minkowski first showed in 1897 that the EGI (i.e., $\{A(\omega_i)\}$) uniquely determines (up to translation) a convex polyhedron. Reconstructing the convex polyhedron from its EGI was first demonstrated using an iterative algorithm developed by Little (1983).

Now, two polyhedra P and Q are *homothetic* if

$$P = \{x \mid x = \lambda \cdot y + t, y \in Q, \lambda \in R^1, t \in R^3\}$$

That is, two polyhedra are homothetic if they differ only by a translation and by a scaling. (Two polyhedra are not homothetic if they differ by a rotation.) If two polytopes are homothetic, their EGI's can be made equal by a scaling.

The *convex sum* (or *mixture*) of two polyhedra P and Q is

$$\lambda \cdot P + (1 - \lambda) \cdot Q = \{\lambda \cdot x + (1 - \lambda) \cdot y \mid x \in P, y \in Q, 0 \le \lambda \le 1\}$$

Let $R = \lambda \cdot P + (1 - \lambda) \cdot Q$ where $0 \le \lambda \le 1$. The support function, $H_R(\omega)$, of the convex mixture, R is the convex sum of the support functions $H_P(\omega)$ and $H_Q(\omega)$. That is,

$$H_R(\omega) = \lambda \cdot H_P(\omega) + (1 - \lambda) \cdot H_Q(\omega)$$

For convex polygons, a similar relationship holds for the EGI, $A_R(\omega)$, of the convex mixture, R, in terms of the EGI's $A_P(\omega)$ and $A_Q(\omega)$. That is,

$$A_R(\omega) = \lambda \cdot A_P(\omega) + (1 - \lambda) \cdot A_Q(\omega)$$

Unfortunately, for convex polyhedra this relationship no longer holds. In general, there is no expression for $A_R(\omega)$ in terms of $A_P(\omega)$ and $A_Q(\omega)$ alone.

The Mixed-Volume

The volume, $V(R)$, of the convex mixture, R, also is more difficult to express. For the 3D case, one obtains

$$V(R) = \lambda^3 V(P) + 3\lambda^2(1 - \lambda)V_3(P,Q) + 3\lambda(1 - \lambda)^2 V_3(Q,P) + (1 - \lambda)^3 V(Q) \quad (2)$$

where

$$V_3(S,T) = \frac{1}{3} <H_S(\omega_i), A_T(\omega_i)>$$

is called the *mixed volume* of S and T. $V_3(S,T)$ is the inner product of the support function of S with the area function of T. For the 3D case, the mixed volume is not, in general, symmetric (i.e., $V_3(S,T) \neq V_3(T,S)$). For the 2D case, the mixed volume is symmetric.

As above, let $R = \lambda \cdot P + (1 - \lambda) \cdot Q$ be the convex sum of two convex polyhedra P and Q. Then, by the *Brunn-Minkowski Theorem*,

$$V(R)^{1/3} \geq \lambda \, V(P)^{1/3} + (1 - \lambda)V(Q)^{1/3} \quad (3)$$

with equality if and only if P and Q are homothetic. Combining equations (2) and (3), we obtain

$$V_3(P,Q)^3 \geq V(Q)^2 \, V(P)$$

with equality holding if and only if P and Q are homothetic. The mixed volume captures the relationship between the shapes of two polytopes. When the mixed volume is minimal, the polytopes are homothetic. Mixing the two does not cause a shape change, only a scaling. Little's iterative algorithm (Little, 1983) minimizes the mixed volume to reconstruct object shape, determined by $H(\omega_i)$, from the given EGI, $A(\omega_i)$.

SIMILARITY MEASURES FOR CONVEX POLYHEDRA

When the task is object recognition, the similarity measure should be invariant to translation and scaling, since these typically vary with viewpoint rather than with object identity. Rotation also is a viewpoint dependent measure. When the task requires the determination of object attitude, the similarity measure must be sensitive to rotation. The mixed volume captures exactly these properties.

Initially we were mislead into thinking the adjacency structure of a polytope was important. (Adjacency structure determines which faces share an edge, which edges meet at a vertex and which vertices lie on a face.) Interestingly, Little's reconstruction method, based on the mixed volume, does not deal explicitly with adjacency structure. Adjacency structure is determined as a consequence of the minimization and can change many times as the algorithm iterates. It is also important to note that adjacency structure is not a stable property of a polytope. Small changes in the relative positions of faces can produce large changes in the adjacency structure.

Determining Attitude by Comparing Area Functions

Determining the attitude of a known object is equivalent to finding the rotation, ϕ, that brings the known area function into correspondence with the sensed area function. Let $\{A(\omega_i)\}$ be the sensed area function of the visible surface of an object. Let $\{A_\phi(\omega_i)\}$ be the area function of the prototype object with rotated attitude ϕ. At each sampled attitude ϕ, the measure of similarity is given by

$$\sum_i (A_\phi(\omega_i) - A(\omega_i))^2$$

Determining the best match corresponds to finding the attitude, ϕ, that minimizes this measure. (Alternatively, one can find the attitude, ϕ that maximizes $<A_\phi(\omega_i), A(\omega_i)>$.)

A direct comparison of area functions is the method used by Horn and Ikeuchi (1984). But, there are difficulties. As the resolution is increased, effectiveness of area matching decreases. The tesselation of the Gaussian sphere, U, becomes finer resulting in more empty cells in the orientation histogram of the known object. Thus, even when the attitude difference between the sensed and the known area functions is small, the match may be poor. In fact, the match may be poor even at the correct attitude. Said another way, the similarity measure used is not stable.

Determining Attitude with Mixed Volumes

Let $\{A(\omega_i)\}$ be the sensed area function of the visible surface of an object. Let $\{H_\phi(\omega_i)\}$ be the support function of the known object with rotated attitude ϕ. Rotating an object preserves volume. Thus, the mixed volume

$$<H_\phi(\omega_i), A(\omega_i)>$$

is minimized at the attitude, ϕ, that brings the known object into correspondence with the sensed object.

For polytopes, the area function, $A(\omega)$, is non-zero for only finitely many values of ω. When the attitude of the sensed object is slightly different from that of the known object, the correlation of area functions can be zero. (It is possible to consider smoothing the area functions directly, to improve correlation, as Brou has observed (1984)). The support function, $H(\omega)$, is a continuous function of ω and the mixed volume achieves this smoothing in a more rigorous way.

Little's experiments (1985) support this approach. The effects of the magnitude of the difference in attitude between object and prototype are significantly smaller for the mixed volume method. This suggests it is possible to trade-off resolution on the Gaussian sphere, U, with the number of test attitudes, ϕ, to achieve an efficient and accurate coarse-to-fine determination of object attitude. The justification for the mixed-volume method depends on the area function and support function being defined over the entire Gaussian sphere. But, it appears to work well in practice when only the visible hemisphere of the sensed object is available.

CONCLUDING REMARKS

The prototype bin-picking system demonstrates that robust, practical machine vision systems can be designed and built. The approach is based on a careful analysis of the physics of imaging and on the view of machine vision as an inverse problem. This resulted in the concepts of photometric stereo, the Extended Gaussian image and mixed volumes.

The research strategy exemplified by this work is applicable to a broader class of vision tasks. The discussion of the research strategy and of design criteria for shape representations is intended to suggest that this is so. We have only just begun, however, and much remains to be learned.

Stability, in particular, is relevant to a number of the design criteria discussed above. It is not yet clear how the computations of early vision, which are primarily numeric in form, can interface with knowledge representations, which are primarily symbolic in form.

Currently, we do not know how to define stability for symbolic representations. The basic mechanism to compare two symbolic expressions is to test for equality. More general matching of expressions is possible when syntactic transformations are allowed to reduce the comparison to an equality test. For example, the substitution of terms in one expression for variables in the other to make the expressions identical is called *unification*. Algorithms exist to find the most general (i.e., simplest) unifier of any finite set of unifiable expressions, or to report failure if the set cannot be unified. Finally, when transformation includes the ability to do deductive inference, two symbolic expressions can be considered equivalent if each implies the other. Stability, as it is currently defined, requires the ability to quantify similarities and differences. If a significant portion of a computational vision system is to be symbolic, rather than numeric, it will be necessary to extend the notion of stability to cover the symbolic domains too.

REFERENCES

Baker, H. H., & Binford, T. O. (1981). Depth from edge and intensity based stereo. *Proceedings of the 7th International Joint Conference on Artificial Intelligence*, pp. 631–636, Vancouver, B.C.

Barrow, H. G., & Tenenbaum, J. M. (1981a). Computational vision. *Proceedings of the IEEE*, 69, 572–595.

Barrow, H. G., & Tenenbaum, J. M. (1981b). Interpreting line drawings as three-dimensional surfaces. *Artificial Intelligence*, 17, 75–116.

Binford, T. O. (1981). Inferring surfaces from images. *Artificial Intelligence*, 17, 205–244.

Binford, T. O. (1982). Survey of model-based image analysis systems. *International Journal of Robotics Research*, 1(1), 18–64.

Brady, M., & Yuille, A. (1984). An extremum principle for shape from contour. *IEEE Transactions on Pattern Analysis and Machine Intelligence*, 6, 288–301.

Brady, M. (1982). Computational approaches to image understanding. *ACM Computing Surveys*, 14, 3–72.

Brady, M. (1983). Criteria for representations of shape. In: J. Beck, B. Hope & A. Rosenfeld (Eds.), *Human and machine vision* (pp. 39–84). New York: Academic Press.

Brou, P. (1984). Using the Gaussian image to find the orientation of objects. *International Journal of Robotics Research, 3*(4), 89–125.

Grimson, W. E. L. (1981). *From images to surfaces.* Cambridge, MA: MIT Press.

Grunbaum, B. (1967). *Convex polytopes.* New York: Wiley.

Hildreth, E. C. (1984). Computations underlying the measurement of visual motion. *Artificial Intelligence, 23,* 309–354.

Horn, B. K. P. (1977). Understanding image intensities. *Artificial Intelligence, 8,* 201–231.

Horn, B. K. P. (1984). Extended Gaussian images. *Proceedings of the IEEE, 72,* 1671–1686.

Horn, B. K. P. (1986). *Robot vision.* Cambridge, MA: MIT Press/McGraw-Hill.

Horn, B. K. P., & Brooks, M. J. (1986). The variational approach to shape from shading. *Computer Vision Graphics and Image Processing, 33,* 174–208.

Horn, B. K. P., & Ikeuchi, K. (1984). The mechanical manipulation of randomly oriented parts. *Scientific American, 251,* 100–111.

Horn, B. K. P., & Schunck, B. G. (1981). Determining optical flow. *Artificial Intelligence, 17,* 185–203.

Ikeuchi, K., & Horn, B. K. P. (1981). Numerical shape from shading and occluding boundaries. *Artificial Intelligence, 17,* 141–184.

Ikeuchi, K., Horn, B. K. P., Nagata, S., Callahan, T., & Feingold, O. (1984). Picking up an object from a pile of objects. In M. Brady & R. Paul (Eds.), *Robotics research: The first international symposium* (pp. 139–162). Cambridge, MA: MIT Press.

Ikeuchi, K., Nishihara, H. K., Horn, B. K. P., Sobalvarro, P., & Nagata, S. (1986). Determining grasp configurations using photometric stereo and the PRISM binocular stereo system. *International Journal of Robotics Research, 5*(1), 46–65.

Kanade, T. (1981). Recovery of the three-dimensional shape of an object from a single view. *Artificial Intelligence, 17,* 409–460.

Kender, J. R. (1982). A computational paradigm for deriving local surface orientation from local texture properties. *Proceedings of the IEEE Workshop on Computer Vision: Representation and Control,* pp. 143–152, Rindge, NH.

Little, J. J. (1983). An iterative method for reconstructing convex polyhedra from extended Gaussian images. *Proceedings of the 3rd National Conference on Artificial Intelligence,* pp. 247–250, Washington, D.C.

Little, J. J. (1985a). Extended Gaussian images, mixed volumes, and shape reconstruction. *Proceedings of the ACM Symposium on Computational Geometry,* pp. 15–23, Baltimore, MD.

Little, J. J. (1985b). Determining object attitude from extended Gaussian images. *Proceedings of the 9th International Joint Conference on Artificial Intelligence,* pp. 960–963, Los Angeles, CA.

Little, J. J. (1985c). Recovering shape and determining attitude from extended Gaussian images. TR-85-2, UBC Dept. of Computer Science, Vancouver, BC.

Lyusternik, L. A. (1963). Convex figures and polydedra. New York: Dover Publications.

Marr, D. (1982). Vision: A computational investigation into the human representation and processing of visual information. San Francisco: W. H. Freeman.

Marr, D., & Nishihara, H. K. (1978). Representation and recognition of the spatial organization of three dimensional structure. Proceedings of the Royal Society of London B, 200, 269–294.

Mokhtarian, F., & Mackworth, A. (1986). Scale-based description and recognition of planar curves and two-dimensional shapes. IEEE Transactions on Pattern Analysis and Machine Intelligence, 8, 34–43.

Poggio, T., & Torre, V. (1984). Ill-posed problems and regularization analysis in early vision. AI-Memo-773, MIT AI Laboratory, Cambridge, MA.

Witkin, A. P. (1981). Recovering surface shape and orientation from texture. Artificial Intelligence, 17, 17–45.

Woodham, R. J. (1980). Photometric method for determining surface orientation from multiple images. Optical Engineering, 19, 139–144.

Woodham, R. J. (1981). Analyzing images of curved surfaces. Artificial Intelligence, 17, 117–140.

Woodham, R. J. (1984). Photometric method for determining shape from shading. In S. Ullman & W. Richards (Eds.), Image understanding 1984 (pp. 97–125). Norwood, NJ: Ablex.

Woodham, R. J. (1987). Shape analysis. In Encyclopedia of Artificial Intelligence, S. Shapiro (Ed.), pp. 1039–1048, New York, NY: John Wiley & Sons.

17
Adequacy Criteria for Visual Knowledge Representation

Alan K. Mackworth*

Department of Computer Science
University of British Columbia

INTRODUCTION

The proper study of artificial intelligence is the design of computational systems that represent, use, and acquire knowledge to perceive, reason, communicate, and act. Under that definition, knowledge representation is the heart of artificial intelligence. Past and future success in building systems for vision, problem solving, planning, and language depends critically on progress in knowledge representation. Workers in the field have been prolific in proposing and exploiting a variety of knowledge representation schemes such as grammars, semantic nets, programs, logics, schemas, rules, constraints, and neural nets. However, as we explore in the world of knowledge representation we need navigational tools: the analogs of chart, compass, log, and sextant. In this paper a framework for evaluating knowledge representation schemes is presented.

A BRIEF HISTORY

Chomsky's (1965) early work on syntactic structures was explicitly motivated by adequacy criteria. In developing the theory of transformational grammar the devices of anomaly, paraphrase, and ambiguity were constantly exploited as experimental probes. By regard-

---

* Fellow, Canadian Institute for Advanced Research

ing a particular grammatical framework as a representation of grammatical knowledge Chomsky was asking, "Is this representation sufficiently powerful to capture these phenomena?"

In fact, he was able to go beyond sufficiency to necessity. For a structurally ambiguous sentence *any* adequate grammatical framework must necessarily provide multiple syntactic interpretations. For a structurally anomalous sentence *any* adequate grammar must fail to provide an interpretation. Two sentences that are mutual paraphrases must share common structure in their syntactic interpretations. The imaginative use of these probes allowed the delimitation of the necessary boundary conditions of any adequate grammar. In particular, the many varieties of anomaly were particularly fruitful.

Chomsky proposed the problematic competence/performance distinction but concentrated the adequacy arguments on the competence of the "ideal" speaker/hearer's generative (rule-based) mental representations. This approach directly inspired Clowes (1971) and Huffman (1971) to specify representations for simple blocks world scenes that would similarly reject ill-formed scenes and discover multiple interpretations of ambiguous pictures. Well-chosen examples highlight distinctions, representations, and chains of reasoning that any adequate system must make, possess, and follow to be able to discriminate well-formed from ill-formed scenes, for example, as illustrated in Figure 1.

McCarthy and Hayes (1969) further distinguished three kinds of adequacy for mental representations of the world: "metaphysical," "epistemological," and "heuristic." The focus of their paper is on epistemological adequacy: "A representation is called epistemologically adequate for a person or machine if it can be used practically to express the facts that one has about the world." On this basis they argued for various logical systems as declarative representation languages.

In the early 1970s, other researchers were building and advocating systems that exploited procedural representations of knowledge; the ensuing declarative/procedural controversy featured a loud debate between two entrenched camps. Despite attempts to find a synthesis of the two positions (Winograd, 1975) the best perspective on the controversy is afforded by realizing that the two camps were implicitly relying on criteria that emphasized different forms of adequacy of knowledge representation, thus ensuring the incoherence of the debate.

In computer science the distinction between the specification level and the implementation level of description of a computational task is common. It is also common to identify these levels, roughly,

Figure 1.   What's wrong with this scene?

with "what" and "how." The implementation level is often recursively described as a series of levels such that each level is implemented or realized on top of the virtual machine provided by the level immediately below it. This recursion terminates at the physical machine level—the real hardware—that provides primitive computational operations.

Marr (1982) discussed three levels:

1.   Computational theory
2.   Representation and algorithm
3.   Hardware implementation

He emphasized the one-many relationships from level 1 to level 2 and from level 2 to level 3. Marr said that level 1 describes the "what" and "why" of the computational task, level 2 specifies "how" at an abstract level, and level 3 specifies "how" at the concrete level of an artificial or biological machine. One can criticize this approach (surely representations are required at level 1 and 3 as well as at level 2) but the crucial point that Marr makes is that we cannot say that we understand how a machine is carrying out an information processing task until we have adequate descriptions at all levels.

ADEQUACY CRITERIA

These attempts to clarify the notion of adequacy often interact in oblique ways: each author emphasizes a different idiosyncratic aspect. We have not yet arrived at a coherent theory with commonly accepted set of terms and criteria. This paper is an attempt to establish a framework for that theory.

The claim is that adequacy criteria can be categorized as *descriptive adequacy* and *procedural adequacy* criteria (Mackworth, 1977) and that both must necessarily be satisfied by an adequate knowledge representation scheme.

DESCRIPTIVE ADEQUACY

Descriptive adequacy criteria are concerned with the extent to which the mental representation adequately describes or represents situations in the world. Eleven descriptive adequacy criteria are outlined here.

D1. *Capacity.* If the computational task requires that an unbounded number of possible situations in the world be distinguished then a descriptively adequate finite representation scheme (Hayes, 1974) must necessarily embody a generative and recursive set of rules (in some language) that can generate an unbounded number of configurations.

D2. *Primitives.* The set of rules must generate descriptions of the legitimate primitive objects in the world and their possible attributes and relationships.

D3. *Composition.* Composition rules generate descriptions of structured objects, their components and their attributes and relationships.

D4. *Specialization.* Specialization rules generate possible refinements of object classes.

D5. *Subworlds.* If the world of interest consists of two or more distinct subworlds then the representation scheme must maintain that distinction. For example, for visual knowledge representation, the two-dimensional image domain and the three-dimensional scene domain are mutually exclusive. A descriptively adequate visual knowledge representation can only avoid elementary category errors, such as confusing an edge with the line that depicts it, by categorizing or typing objects as belonging to one domain or the other. (Any real visual knowledge representation will make several finer distinctions.) If the distinction between image and scene is maintained then there must be two sets of rules describing the primitives in both domains and the composition rules needed to form composite objects.

D6. *Depiction.* A visual knowledge representation must also carry information about the *depiction relation:* how objects in the scene domain appear in the image domain. (The use of the term "relation of representation" for the depiction relation is misleading, as we shall see.)

D7. *Equivalence Classes.* For three-dimensional worlds the depiction relation is a many-to-one mapping function from the scene domain to the image domain, confounding the subdomains of lighting and surface reflectance, orientation, shape and position, with viewpoint and other attributes of the imaging situation (Mackworth, 1983b).

Even for a simple image like that in Figure 2 there are an infinite number of scenes that could serve as legal interpretations. The representation, to serve descriptive adequacy, must provide a finite representation of all and only those scenes which, technically, constitute an equivalence class. The configuration in the representation scheme is a description of that equivalence class. Much of the history of computational vision research can be seen as a continuing investigation into this aspect of descriptive adequacy.

D8. *Detail.* An adequate visual representation provides descriptions at a variety of physical scales, both in the image and in the scene.

D9. *Stability.* In a stable representation scheme a minor change in the world causes, at most, a minor change in the representa-

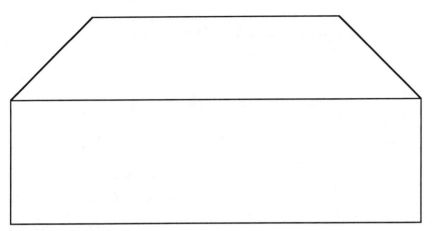

Figure 2.   A simple image.

       tion. The detail and stability criteria are decoupled in the sense that it is possible to satisfy one and not the other.

D10.   *Invariance.* It is often desirable for the representation to be essentially invariant under transformations of the world just as the deep structure of a sentence is essentially invariant under the passive transformation. The paraphrase technique is useful for evaluating invariance.

D11.   *Correct.* It should not go without saying that the representation should be correct. The *relation of representation* is the relation between a situation and the configuration describing it. By insisting that the relation of representation be a total function from situations onto configurations we ensure, for example, that the representation of any unambiguous situation is a *canonical* configuration not allowing or requiring any arbitrary choices. Under *correctness* we include anomaly and ambiguity. An anomalous situation should have no coherent configuration while an ambiguous situation should have two or more. Anomaly and ambiguity are useful for evaluating correctness and, further, may provide diagnostic advice on how to enhance the adequacy of the representation.

    This framework allows us to look retrospectively at all computational vision research and judge the extent to which descriptive adequacy is achieved.

    As a familiar illustration, consider the Huffman-Clowes labeling scheme shown in Figure 3. It satisfies, to a greater or lesser extent, criteria D1 (Capacity), D2 (Primitives), D3 (Composition), D5 (Sub-

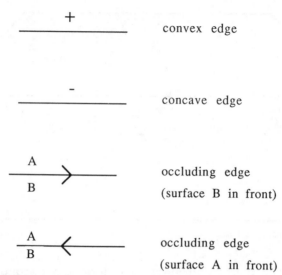

Figure 3. The Huffman-Clowes labels for an edge.

worlds), D6 (Depiction), D7 (Equivalence Classes), and D11 (Correctness). For an image of a single line, it generates the four interpretations of Figure 3, which is an extensional representation of an equivalence class of scenes. Each of the four interpretations corresponds to an intensional representation in that the degree of convexity/concavity of the edge is not specified. A description of the image in Figure 2, in terms of that scheme, is shown in Figure 4: it stands for an infinite set of legal scenes that produce that image. Unfortunately, the labeling of Figure 5 is also allowed by that

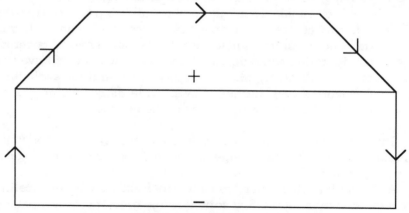

Figure 4. A correct interpretation of Figure 2.

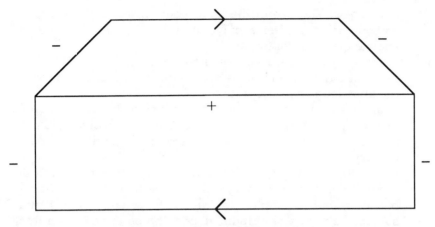

Figure 5.   An anomalous interpretation of Figure 2.

scheme but does not correspond to any legal scene. In order to eliminate such spurious interpretations we were forced to provide richer
descriptions of this simple visual world that include information on
possible surface orientations and positions (Mackworth, 1977).

The design of Mapsee2, a system that interprets hand drawn
sketch maps, was explicitly motivated by adequacy criteria (Havens
& Mackworth, 1983). Criteria D1–D6 and D11 were, on the whole,
satisfied for this domain through the use of schema based knowledge representation. Essentially in that system a partially instanti-
ated schema instance corresponds to the evolving description of an
equivalence class.

Recent work in shape description (Mackworth & Mokhtarian, 1984;
Mokhtarian & Mackworth, 1986) provides a representation (curvature
scale space) that satisfies D1–D3 and D8 (Detail), D9 (Stability), D10
(Invariance), and D11 (Correctness).

This framework allows us to look retrospectively at research into
the use of knowledge representation in vision systems but more importantly, it encourages the use of explicit adequacy criteria as a
guiding research strategy.

PROCEDURAL ADEQUACY

Procedural adequacy criteria assess aspects of the use and acquisition of the knowledge embodied in a representation scheme. Five
procedural adequacy criteria are outlined here.

P1. *Soundness.* The use of the knowledge is sound if the system embodying it produces only interpretations allowed by the knowledge representation.

P2. *Completeness.* The use of the knowledge is complete if the system embodying it produces all the interpretations allowed by the knowledge representation.

Soundness and completeness are adopted, by analogy, from theorem proving where an inference strategy is sound and complete if it proves only and all the theorems logically implied by the axioms.

P3. *Flexibility.* Another procedural adequacy consideration is the flexibility of use of the available information. An ideal image-based system would regard all of its potential information sources as inputs or outputs, depending on the availability of information (Mackworth, 1983a). The system would allow control flow from image to scene (image analysis) or from scene to image (image synthesis), bidirectionally or multidirectionally if more than two domains or information sources were available. Or, indeed, information sources may start as partially specified by a symbolic description and the system would refine that description as it used the other information sources and the constraints embedded in the domain knowledge representation. Under this view, the dichotomous classification of potential information sources/sinks as inputs or outputs becomes obsolete. Until we achieve the ideal, we also discuss under procedural adequacy the control facilities provided in the knowledge representation language and the ease of reprogramming the system from, say, analysis to synthesis (Stanton, 1972).

P4. *Acquisition.* A key aspect of procedural adequacy is the knowledge acquisition process. Is the knowledge representation suitable for knowledge acquisition? Does acquisition occur through evolution, learning, teaching, or explicit programming? Are there appropriate, efficient, and flexible algorithms for knowledge acquisition?

P5. *Efficiency.* Under efficiency criteria we can include standard computational complexity arguments. Using such arguments we should first establish a lower bound on the time (and/or space) complexity of the task or problem itself, that is, a lower bound on the resources that *any* correct algorithm must consume. This is the inherent complexity of the task. Then any proposed algorithm may be analyzed for its worst or average case performance. The worst case performance of an *optimal* algorithm will equal the complexity of the problem.

A vision program that interpreted an image by exhaustive analysis-by-synthesis enumerating the members of the infinite set of all possible images until one matched the input would not satisfy the efficiency criterion of procedural adequacy. A variety of ways of controlling that search and reducing the resources consumed are available such as the use of image features to index into the knowledge representation.

However, many of the interesting problems in AI are NP-complete—that is the problems themselves apparently inherently require exponential resources (as a function of the size of the input). For example, the Huffman-Clowes labeling problem itself has recently been shown to be such a problem (Kirousis & Papadimitriou, 1985).

Also under efficiency we should discuss the use of parallelism. Many new models and technologies of computation challenge the standard von Neumann model of computation. Many of these display coarse or fine-grained parallelism often on a massive scale; these affect radically the complexity measures and criteria used.

THE INTERACTION BETWEEN DESCRIPTIVE AND PROCEDURAL ADEQUACY

Apart from the claim that a good meta-framework for knowledge representation allows us to design and build better knowledge representations, what are the other practical implications of the point of view advocated here? The most interesting implications arise from the interactions between descriptive and procedural adequacy.

Given two knowledge representation schemes equal from the perspective of descriptive adequacy we should, of course, choose the one that more nearly meets the criteria of procedural adequacy. Or, more constructively, design a system to more nearly meet those criteria.

The twin sets of criteria are, of course, often apparently in conflict. The approach more usually taken in AI has been to favor procedural efficiency if not full procedural adequacy (which includes soundness, completeness, acquisition, and flexibility of use of knowledge) by hand coding special purpose procedures into application programs. The current trend is away from that approach as the full dimensions of the adequacy issue become apparent.

However, there are fundamental theoretical obstacles to achieving simultaneously full descriptive and procedural adequacy that are as important to theories of knowledge representation as comparable *laws of impotence*, such as Heisenberg's Uncertainty Principle and Einstein's Laws of Relativity, are to physics.

For example, Levesque and Brachman (1985) consider an aspect of descriptive adequacy ("expressiveness") and an aspect of procedural adequacy ("tractability") to be involved in a fundamental tradeoff. Full first order logic is only semi-decidable. Certain subsets of first order logic are decidable but, putatively, only in exponential time while more restricted subsets are decidable in polynomial time.

Horn clause form is a restriction of first order logic that requires each sentence to have one of the two forms:

$$P \leftarrow Q \wedge R \wedge \ldots$$

or W

where P,Q,R,W are predicate symbols (which may take terms as arguments). Prolog programs are sets of sentences that take this form. This limitation on expressive power (sentences such as $P \vee Q$ are not allowed) combined with assumptions such as the unique names assumption (Reiter, 1978), and negation as failure (Clark, 1978) brings some advantages in procedural adequacy. In particular, although the logic is still undecidable, the resolution strategy used is relatively efficient and the synthesis/analysis flexibility of the system is useful; the same knowledge can be used in several "directions", although the user may optimize the knowledge base for use in a certain procedural direction.

Given that many interesting tasks (such as the labeling problem mentioned earlier) seem to be inherently exponential the strategy of weakening descriptive adequacy to achieve efficient (polynomial) algorithms is not available. A complementary strategy is to search for efficient *approximation* algorithms for such tasks. An approximation algorithm might, for example, provide a necessary but not always sufficient test for, say, the acceptability of a picture. For the labeling problem, the Waltz (1972) arc consistency "filtering" algorithm is such a test which runs in linear time (Mackworth & Freuder, 1985).

Generalizations of this approach known as network consistency algorithms can be used for many NP-complete tasks and there is a general network consistency approach that provides a spectrum of levels of consistency (applying tighter and tighter tests of consistency) at increasing computational cost. For many (but, provably, not for all) problems the cheaper, lower levels of consistency are sufficient. For example, if the constraint graph is a tree then simple arc consistency (running in linear time) is both a sufficient and a necessary test for the existence of a solution (Mackworth and Freuder, 1985).

Mapsee3 explicitly constructs a constraint graph among schema instances. It uses a hierarchical arc consistency algorithm, exploiting the tree-structured specialization hierarchy, to search efficiently for an interpretation of the sketch map (Mackworth, Mulder, & Havens, 1985).

Parallelism may offer attractive solutions to some problems but others are not amenable to that approach. In particular, the relaxed form of constraint satisfaction known as arc consistency is apparently inherently sequential. Kasif (1986) has shown that arc consistency is log-space complete for P, the class of problems solvable on a single Turing machine in polynomial time. The implication of this is that it is unlikely that arc consistency can be solved in-polylogarithmic time with a polynomial number of processors.

CONCLUSION

In summary we have developed explicit criteria for evaluating the descriptive and procedural adequacy of visual knowledge representations. These criteria may be used to evaluate representation schemes and to design better ones but, as pointed out, there are theoretical obstacles to satisfying fully all the criteria simultaneously.

REFERENCES

Chomsky, N. (1965). *Aspects of the theory of syntax*. Cambridge, MA: MIT Press.

Clark, K. L. (1979). Negation as Failure. In H. Gallaire & J. Minker (Eds.), (pp. 293–322) *Logic and data bases*. New York: Plenum.

Clowes, M. B. (1972). On seeing things. *Artificial Intelligence*, 2, 79–112.

Havens, W. S., & Mackworth, A. K. (1983). Representing knowledge of the visual world. *IEEE Computer*, 16, 90–96.

Hayes, P. J. (1974). Some problems and non-problems in representation theory. AISB Summer Conference, University of Sussex, pp. 63–79.

Huffman, D. A. (1971). Impossible objects as nonsense sentences. *Machine Intelligence*, 6, (p. 295–323) Meltzer & D. Michie (Eds.), American Elsevier, NY.

Kasif, S. (1986). On the parallel complexity of some constraint satisfaction problems. Proceedings of the Fifth National Conference on Artificial Intelligence, Philadelphia, pp. 349–353.

Kirousis, L. M., & Papadimitriou, C. H. (1985). The complexity of recognizing polyhedral scenes. *26th Annual Symposium on Foundations of Computer Science*, IEEE Computer Society, October 21–23, pp. 175–185.

Levesque, H. J., & Brachman, R. J. (1985). A fundamental tradeoff in knowledge representation and reasoning (revised version). In R. J. Brachman & H. J. Levesque (Eds.), *Readings in knowledge representation*, (pp. 41–70). Morgan Kaufmann Publishers, Inc., Los Altos, CA.

McCarthy, J., & Hayes, P. (1969). Some philosophical problems from the standpoint of artificial intelligence. *Machine Intelligence 4*, (pp. 463–502). B. Meltzer and D. Michie (eds.), American Elsevier, NY.

Mackworth, A. K. (1977). How to see a simple world: An exegesis of some computer programs for scene analysis. *Machine Intelligence, 8*, E. W. Elcock and D. Michie (eds.), pp. 510–540. New York: John Wiley.

Mackworth, A. K. (1983a). Recovering the meaning of diagrams and sketches. *Proceedings Graphics Interface '83*, Edmonton, Alberta, Canada, pp. 313–317.

Mackworth, A. K. (1983b). Constraints, descriptions and domain mappings in computational vision. In O. J. Braddick & A. C. Sleigh (Eds.), *Physical and biological processing of images* (pp. 33–40). Berlin: Springer-Verlag.

Mackworth, A. K., & Freuder, E. C. (1985). The complexity of some polynomial network consistency algorithms for constraint satisfaction problems. *Artificial Intelligence, 25*(1), 65–74.

Mackworth, A. K., & Mokhtarian, F. (1984). Scale-based descriptions of planar curves. *Proceedings of the Canadian Society for Computational Studies of Intelligence*, London, Ont., 114–119.

Mackworth, A. K., Mulder, J., & Havens, W. S. (1985). Hierarchical arc consistency: Exploiting structured domains in constraint satisfaction problems. *Computational Intelligence, 1*(3), 118–126.

Marr, D. (1982). *Vision San Francisco: W. H. Freeman.

Mokhtarian, F., & Mackworth, A. K. (1986). Scale-based description and recognition of planar curves and two-dimensional shapes. *IEEE Transactions on Pattern Analysis and Machine Intelligence, 8*, (1) 34–43.

Reiter, R. (1978). On closed world data bases. In H. Gallaire & J. Minker (Eds.), *Logic and data bases* (pp. 55–76). New York: Plenum.

Stanton, R. B. (1972). The interpretation of graphics and graphic languages. In F. Nake & A. Rosenfeld (Eds.), *Graphic languages* (pp. 144–159). Amsterdam: North-Holland Publishing Co.

Waltz, D. L. (1972). Understanding line drawings of scenes with shadows. In P. H. Winson (Ed.), *The psychology of computer vision* (pp. 19–91). New York: McGraw-Hill.

Winograd, T. (1975). Frame representations and the declarative/procedural controversy. In D. G. Bobrow & A. Collins (Eds.), *Representation and understanding: Studies in cognitive science* (pp. 185–210). New York: Academic Press.

Author Index

Subject Index